Biomethane through Resource Circularity

The Circular Economy in Sustainable Solid and Liquid Waste Management

Series Editor: Dr. Sadhan Kumar Ghosh,

Professor, Mechanical Engineering, & Chief Coordinator, Centre for Sustainable Development and Resource Efficiency Management, Jadavpur University, Kolkata, India

Biomethane through Resource Circularity: Research, Technology, and Practices

Edited by Sadhan Kumar Ghosh, Michael Nelles, H.N. Chanakya, and Debendra Chandra Baruah

The Circular Economy in Construction Industry

Sadhan Kumar Ghosh, Sannidhya Kumar Ghosh, Benu Gopal Mohapatra, and Ronald Mersky

Effective Waste Management and Circular Economy: Legislative Framework and Strategies

Sadhan Kumar Ghosh, Sasmita Samanta, Harish Hirani, and Carlos RV Silva Filho

For more information about this series, please visit:
https://www.routledge.com/The-Circular-Economy-in-Sustainable-Solid-and-Liquid-Waste-Management/book-series/CESSLWM

Biomethane through Resource Circularity

Research, Technology, and Practices

Edited by

Sadhan Kumar Ghosh

Professor, Mechanical Engineering & Chief Coordinator, Centre for Sustainable Development and Resource Efficiency Management, Jadavpur University, Kolkata, India

Michael Nelles

Professor, Director, Department Waste and Resource Management, Rostock University & Scientific General Manager, DBFZ — the German Centre for Biomass Research Leipzig, Germany

H. N. Chanakya

Chief Research Scientist, Centre for Sustainable Technologies (CST), IISc., Bangalore, India

Debendra Chandra Baruah

Professor, Department of Energy, Tezpur University, Tezpur, Assam, India

CRC Press
Taylor & Francis Group
Boca Raton London New York

CRC Press is an imprint of the
Taylor & Francis Group, an **informa** business

First edition published 2022
by CRC Press
6000 Broken Sound Parkway NW, Suite 300, Boca Raton, FL 33487-2742

and by CRC Press
4 Park Square, Milton Park, Abingdon, Oxon, OX14 4RN

CRC Press is an imprint of Taylor & Francis Group, LLC

Library of Congress Cataloging-in-Publication Data
A catalog record has been requested for this book

ISBN: 978-1-032-06900-5 (hbk)
ISBN: 978-1-032-06904-3 (pbk)
ISBN: 978-1-003-20443-5 (ebk)

DOI: 10.1201/9781003204435

Typeset in Times
by KnowledgeWorks Global Ltd.

Dedication

Dedicated to the millions of doctors and nurses, medical service providers and first responders, and engineers and scientists involved in vaccine development, administration, storage, and distribution. To the teachers and psychology practitioners who are supporting students in this most stressful situation. To the waste handlers, the workers handling the dead in crematories and cemeteries, the NGOs, and those helping to combat the pandemic situation.

Dedicated to the millions who have been affected and are struggling to combat the pandemic due to this unprecedented outbreak of COVID-19.

Wishing for the eradication of COVID-19 from around the world.

Contents

Introduction

Rapid population growth in developing countries has kept per capita energy consumption low compared to that of highly industrialised developed countries. In industrialised countries, people use four to five times more energy than the global average and nine times more than the average for the developing countries. Therefore, energy is one of the most important factors to global prosperity. The dependence on fossil fuels as the primary energy source has led to global climate change, environmental degradation, and human health problems. Renewable energy technologies are clean sources of energy that have a much lower environmental impact than conventional energy technologies.

Humanity has been following a linear model of production and consumption since the industrial revolution. Researchers report that global material use has tripled over the past four decades, with annual global extraction of materials growing from 22 billion tonnes in 1970 to 70 billion tonnes in 2010. The latest report from the International Resource Panel, "Global Material Flows and Resource Productivity" provides a comprehensive, scientific overview of this important issue. It shows a great disparity of material consumption per capita between developing and developed countries. This has tremendous implications for achieving the United Nations Sustainable Development Goals in the next 10 years. Global material use has been accelerating. Material extraction per capita increased from 7 to 10 tonnes from 1970 to 2010, indicating improvements in the material standard of living in many parts of the world. Raw materials have been transformed into goods that are afterward sold, used, and turned into waste that is often unconsciously discarded and unmanaged. On the other hand, the circular economy is an industrial model that is regenerative by intention. One of the goals of the circular economy is to have a positive effect on the planet's ecosystems and to stop the excessive exploitation of natural resources. Energy extraction from renewable sources gives a tremendous push to the resource-consumption and circular-economy initiatives.

The global annual generation of biomass waste is in the order of 140 Gt, which needs to be utilised using environmentally benign and economically viable methods. Major parts of this waste generated from land-based activities, including agriculture, are a burgeoning problem since their disposal, utilisation, and management practices have yet to be made sustainable and efficient. Particularly in developing countries, most biomass residues are left in the field to decompose or are burned in the open, resulting in significant environmental impacts. Biomass-derived renewable energy (bioenergy) is one of the most common potential forms of renewable energy. Biomethane, methane produced from anaerobic decomposition of organic materials obtained from the waste stream of biomass, is one of the most promising options of bioenergy, with multifaceted benefits including alternatives to fossil fuel, true contribution to circularity, and ensurance of total cleanliness. Biomethane technology also offers solutions concerning poor energy access and benefits local communities by generating gaseous fuel and organic manure. Anaerobic technology-based biomethane generation is also promising for treatment of industrial wastewater.

There are several challenges that hinder the successful implementation of biomass-based renewable energy, including biomethane production, despite the immense potential for technology scale-up. This inspires more research, discussion, and technology development for successful conversion of biomass waste to biomethane. Undoubtedly, these require discussions and fine-tuning of the relevant factors along with effective supply chain network design and effective business models for a sustainable future.

Preface

Biomethane through Resource Circularity: Research, Technology, and Practices aims to focus the renewable energy conversion pathway of bioresources through anaerobic digestion technology promoting circularity of resources. Some contemporary research outcomes concerning technology and resources of biomethane generation for a sustainable future are highlighted comprehensively in this book. The book covers three specific aspects divided into three sections: (i) energy, society, and circularity nexus of biomethane technology, presented through five chapters, (ii) research and development of biomethane technology in the present day context, presented through seven chapters, and (iii) feedstock and application of biomethane technology, presented through eight chapters.

Human engagement, both at the domestic front and in industrial operations, are inevitably accompanied by waste generation. Solid waste generation is unavoidable, even in compliance with the goal or concept of cleaner production, which requires that a higher percentage of raw materials be converted into products. But the recycling option of cleaner production can be considered an appropriate means of combating the menace of solid waste. This usually involves the collection of the waste and its reuse in the same or a different part of production (on-site recovery and reuse) or collection and treating wastes so that they can be sold to consumers or other companies. In line with this, biogas technology employs the use of anaerobic digestion of wastes to produce methane-rich gas known as biomethane. This has been an emerging technology that has become a major focus of interest in waste management throughout the world. It is an identified veritable option in the integrated waste management of municipal solid waste involved in waste-to-energy transformation.

Sustainability expectations on all renewable energy options are huge. The biomethane route has another interesting dimension of circularity, with a provision of recirculation of nutrients into the soil in addition to the production of clean fuel. The waste-to-energy concept is truly implemented by this proven option of bioenergy, and thus a potential contributor of cleanliness and social welfare.

Continuous effort for perfection of extending the horizon of know-how is achieved by extensive research and development covering biomethane technology, feedstock, reactor design, and applications. Microbial decomposition of biomasses is a highly complex biochemical phenomenon. Diversity of feedstock characteristics coupled with seasonal variation of thermal environment enhances the complexity of reaction kinetics. Academic analysis of research and development outcomes concerning the series of contextual issues of biomethane production from a wide range of bioresources with varying characteristics has the potential to reduce the gap and offer intended benefits.

A wide range of biomass materials is potential feedstock for biomethane generation. Understanding the relevant factors on the supply chain of such raw materials with a view to ensure societal benefits, addressing the burning issues of energy access, total cleanliness, and security of soil and crop nutrients, has immense national and global importance. The series of these chapters throws some light on the contemporary issues concerning research, technology, and applications related to waste biomass to biomethane conversion. This book covers a wide spectrum of possible energy generations from a variety of non-conventional biomasses which will be popular among readers and give valuable input to the target audience.

This book presents the results of research, technology, practice, and case studies adopted for resource circularity in biomethane, which will be helpful for researchers, policymakers, and practitioners to implement low-cost technologies.

9 December 2021

Prof. Sadhan Kumar Ghosh

Prof. Michael Nelles

Dr. H. N. Chanakya

Prof. Debendra Chandra Baruah
Editors

Acknowledgments

The editors acknowledge the support of the chairman and the organising committee of 8th IconSWM 2018, 9th IconSWM-CE 2019, and 10th IconSWM-CE 2020, which was attended by 44 countries, and the governing council of the International Society of Waste Management, Air and Water for allowing the editors and contributing authors to contribute the articles for this book.

The editors acknowledge the support and encouragement of the following organisations with deep gratitude.

Members of the Secretariat of the International Society of Waste Management, Air and Water
UNCRD, Japan and IPLA, Global Secretariat at ISWMAW
Representatives who joined from UNEP, UNDP, World Bank in the IconSWM-CEs
Researchers of the INDIA H20 Horizon 2020 research project supported by DBT, Government of India and European Union
Ocean Plastic Turned into an Opportunity in Circular Economy (OPTOCE) Project funded by SINTEF, Norway
Indo-Hungary Industrial Research Project at Mechanical Engineering Department, Jadavpur University supported by DST, Government of India and Hungarian Government
Members of the Centre for Sustainable Development and Resource Efficiency Management, Jadavpur University
Quality Management Consultants
CRC Press/Taylor & Francis Group editors involved in this publication

Those who supported and encouraged the editors of the book are: Mrs. Pranati Ghosh, Mrs. Stephanie Nelles, Mrs. Madhuri Gore, and Mrs. Pompa Baruah.

The editors express their gratitude to the individuals whose support and encouragement are thankfully acknowledged.

Editors

Dr. Sadhan Kumar Ghosh has been a professor of mechanical engineering since 1998 and chief coordinator for the Centre for Sustainable Development and Resource Efficiency Management at Jadavpur University, India. He also served as Dean of the Faculty of Engineering and Technology, as well as Head of Mechanical Engineering. He was the director of CBWE, Ministry of Labour and Employment, Govt. of India and Larsen & Toubro Ltd. He is a renowned personality in the fields of waste management, circular economy, green manufacturing, supply, chain management, sustainable development, co-processing of hazardous and MSW in cement kilns, plastics waste and e-waste management and recycling, and management system standards – ISO and TQM, and has had three patents approved. Prof. Ghosh is the founder chairman of the IconSWM, president of the International Society of Waste Management, Air and Water (ISWMAW), and the chairman of the Consortium of Researchers in International Collaboration. He has received several awards in India and abroad, including the distinguished visiting fellowship by the Royal Academy of Engineering, UK to work on "Energy Recovery from MSW." He has written 9 books, 40 edited volumes, and more than 230 national and international articles and book chapters. He is an associate editor of Elsevier's *Journal of Waste Management* and the *International Journal of Materials Cycle and Waste Management*, and editor-in-chief of IconSWM-ISWMAW Secretariat. His significant contributions have been placing Jadavpur University on the world map of research on waste management. He is a consultant and international expert of the United Nations Centre for Regional Development/Department of Economic and Social Affairs; Asian Productivity Organization, Japan; China Productivity Council; South Asia Cooperative Environment Programme, Sri Lanka; Institute for Global Environmental Strategies, Japan; and similar organisations. His international research funding includes European Union Horizon 2020, Erasmus plus, UK-India Education and Research Initiative, Royal Society–DST, Global Challenges Research Fund UK, Royal Academy of Engineering, and the Georgia Government. He is the leader of the collaborative international research project on "Global Status of Implementation of Circular Economy (2018–2022)" by ISWMAW, involving experts from 44 countries. He was the convener of ISO TC 61 WG2 and member in the Indian mirror committee of ISO TC 207 and ISO TC 275. He is an expert committee member of government initiatives and was the state-level advisory committee member of Plastics Waste (Management & Handling) Rules, 2011; expert committee member for the preparation of standards for RDF for utilisation set-up by the Ministry of Housing and Urban Affairs, government of India; and Chair Elect of 11th IconSWM 2021, held in Dec. 2021. He is available at: sadhankghosh9@gmail.com and www.sadhankghosh.com.

Dr. Michael Nelles has been a full professor of Waste and Resource Management in the Faculty of Agricultural and Environmental Sciences at the University of Rostock, Germany since 2006. Prof. Nelles has also been the scientific director of the German Biomass Research Center (DBFZ) in Leipzig since 2012, is an environmental engineer, and studied technical environmental protection at the Technical University of Berlin. From 1994–1999 he was the vice director of the Department of Waste Management at the Montanuniversität Leoben in Austria. From 2000–2006 he was a professor of Environmental Engineering at the University of Applied Science in Göttingen (Germany). His research activity is based on fundamental and applied aspects of waste management with focus on technological, environmental, and economic aspects of mechanical, biological, and thermal treatment systems of waste and biomass in different recycling and recovery routes. He is a member of national and international advisory boards of organisations in the field of waste management and biomass utilisation, and is also a board member of different national and international conferences and journals. He is the author of over 400 articles and chapters in books and journals. His international activities focus on the Asian region, in particular China as he was a guest professor in Beijing, Hefei, Shanghai, and Shenyang, and won the National Friendship Award won in 2011.

Dr. H. N. Chanakya, Chief Scientist, Centre for Sustainable Technologies, Indian Institute of Science, Bangalore, India has over 50 publications and is an expert in the area of biological waste management. He extensively worked on highly rated biomethanation of liquid and solid wastes and developed several highly rated biomethanation technologies for the treatment of organic solid wastes, leaf litter, agro-residues, agro-processing effluents, and more. In addition, he has evolved processes for reuse of sewage to restore water bodies and recycling of domestic grey-water. Toward making biomethanation processes a lot more sustainable, he has developed several value-added by-products such as fibre, pest repellents, stored-grain disinfectants, mushrooms, lignin removal strategies, etc. from biomethanation residue and digester liquid. He has served as Technical Advisor to various regulatory and implementing bodies in Karnataka such as KSPCB, Bangalore City Corporation (BBMP), Bangalore Water Supply and Sewerage Board (BWSSB), and more, as well as at the national level to DST, DBT, MNRE, etc. He is the nominated member of the Karnataka State Planning Commission.

Dr. Debendra Chandra Baruah is a professor in the Energy Department at Tezpur University, Assam, India. He has contributed significantly to the fields of farm mechanization, energy conservation, rural energy planning and management, biomass residue utilization, waste management, and clean energy generation, and published more than 100 articles, and guided 20 PhD scholars. Fourteen national and international collaborative projects have been successfully handled by him. Some notable research outputs of Prof. Baruah have been: development of spatial energy models for crop harvesting, a generation of useful know-how related to energy management in agriculture, energy demand forecasting for mechanized rice farming, modeling of tea-drying behavior, development of low-cost solar air heaters, development of spatial models for assessment of biomass/bio-waste using GIS-remote sensing targeted to achieve a rural circular economy, and development of multi-purpose biomass cookstoves with provision of waste heat recovery and identification of non-edible biodiesel feedstock. He has visited the US, UK, Bangladesh, Thailand, South Africa, Singapore, Belgium, and Greece on academic and research purposes. Prof. Baruah has served as HoD, Dean of Students, Welfare, Director of Internal QA Cell, and Coordinator of IPR Cell at Tezpur University.

Contributors

Ogouvidé Akpaki
University of Kara
Kara, Togo

N. Anand
Birla Institute of Technology and Science Pilani
Hyderabad Campus
Telangana, India

Aries O. Ativo
Central Bicol State University of Agriculture
Philippines

Gnon Baba
University of Lomé
Lomé, Togo

K. S. Badarinarayan
M S Engineering College
Bengaluru, India

Ravi Bajaj
Indian Institute of Technology, Kharagpur
West Bengal, India

Srinivasan Balachandran
Durgapur Women's College, Durgapur
West Bengal, India

Alomoy Banerjee
Birla Institute of Technology and Science Pilani
Hyderabad Campus
Telangana, India

Debendra Chandra Baruah
Tezpur University, Tezpur
Assam, India

Dipal Baruah
Girijananda Chowdhury Institute of Management and Technology
Tezpur, Assam, India

Aman Basu
Visva-Bharati, Santiniketan
West Bengal, India

Swati Bhatia
Birla Institute of Technology and Science Pilani
Hyderabad Campus
Telangana, India

Indranil Bhui
Visva-Bharati, Santiniketan
West Bengal, India

Varadha V.P. Bhuvana
Centre for Rural Energy
Gandhigram Rural Institute
Deemed to be University
Gandhigram, India

Anita Biswas
Visva-Bharati, Santiniketan
West Bengal, India

H. N. Chanakya
Indian Institute of Science
Bangalore, India

Shibani Chaudhury
Visva-Bharati, Santiniketan
West Bengal, India

Ravi Kumar D.
Birla Institute of Technology and Science Pilani
Hyderabad Campus
Telangana, India

B. I. Devanand
Centre for Rural Energy
Gandhigram Rural Institute
Deemed to be University
Gandhigram, India

Brajesh Kumar Dubey
Indian Institute of Technology, Kharagpur
West Bengal, India

Ravi K. Dwivedi
Maulana Azad National Institute of Technology, Bhopal
Madhya Pradesh, India

P. Sankar Ganesh
Birla Institute of Technology and Science Pilani
Hyderabad Campus
Telangana, India

Kapil Garg
Indian Institute of Technology, Kharagpur
West Bengal, India

Sadhan Kumar Ghosh
Jadavpur University
West Bengal, India

Romann Glowacki
German Biomass Research Centre
Leipzig, Germany

Ingo Hartmann
German Biomass Research Centre
Leipzig, Germany

Moonmoon Hiloidhari
Independent Researcher, Powai
Mumbai, India

Ashritha J.
Indian Institute of Science
Bangalore, India

Sandhya Jayakumar
Bruhat Bengaluru Mahanagara Palike
Bengaluru, India

S. Kanyapushpanjali
Central Leather Research Institute
Chennai, India

V. Kirubakaran
Centre for Rural Energy
Gandhigram Rural Institute
Deemed to be University
Gandhigram, India

Hemanth Kumar K. J.
Vidya Vardhaka College of Engineering
Mysore, India

Nitale M'Balikine Krou
University of Kara
Kara, Togo

S. M. Elavaar Kuzhali
Central Leather Research Institute, Chennai, India

Ankita Laha
Visva-Bharati, Santiniketan
West Bengal, India

Volker Lenz
German Biomass Research Centre
Leipzig, Germany

Jan Liebetrau
German Biomass Research Centre
Leipzig, Germany

Gbemisola I. Megbope
Covenant University
Ota, Nigeria

Bharat Modhera
Maulana Azad National Institute of Technology
Bhopal, Madhya Pradesh, India

V. Mozhiarasi
Central Leather Research Institute
Chennai, India

Franziska Müller-Langer
German Biomass Research Centre
Leipzig, Germany

Satya Narra
University of Rostock
German Biomass Research Centre
Leipzig, Germany

Michael Nelles
DBFZ Leipzig and University of Rostock
German Biomass Research Centre
Leipzig, Germany

Oluwatosin J. Ogundare
Covenant University
Ota, Nigeria

Chukwuebuka N. Ojukwu
Covenant University
Ota, Nigeria

David O. Olukanni
Covenant University
Ota, Nigeria

Aastha Paliwal
Indian Institute of Science
Bangalore, India

Sagarika Panigrahi
Indian Institute of Technology, Kharagpur
West Bengal, India

Sumangala Patil
M S Engineering College
Bengaluru, India

Lynlei L. Pintor
Ecosystems Research and Development Bureau
Laguna, Philippines

Chinmay Pradhan
Utkal University,
Bhubaneswar, Odisha, India

Jyothilakshmi R.
M. S. Ramaiah Institute of Technology,
Bengaluru, India

S. Raghuraman
Birla Institute of Technology and Science
Pilani
Hyderabad Campus
Telangana, India

P. M. Benish Rose
Central Leather Research Institute
Chennai, India

Srinjoy Roy
Birla Institute of Technology and Science Pilani
Hyderabad Campus
Telangana, India

Dasappa S.
Birla Institute of Technology and Science Pilani
Hyderabad Campus
Telangana, India

Debaprasad Sarkar
Barrackpore Rastraguru Surendranath
College
Barrackpore, India

Krishanu Sarkar
Visva-Bharati, Santiniketan
West Bengal, India

Sutripta Sarkar
Barrackpore Rastraguru Surendranath
College
Barrackpore, India

Preseela Satpathy
Utkal University
Bhubaneswar, Odisha

Nandhini S. Selva
Centre for Rural Energy
Gandhigram Rural Institute
Deemed to be University
Gandhigram, India

Amit Kumar Sharma
Maulana Azad National Institute of
Technology
Bhopal, Madhya Pradesh, India

S. V. Srinivasan
Central Leather Research Institute
Chennai, India

T. Thavasivamanikandan
Centre for Rural Energy
Gandhigram Rural Institute
Deemed to be University
Gandhigram, India

Daniela Thrän
German Biomass Research Centre
Leipzig, German

Frank Uhlenhut
University of Applied Sciences
Emden, Germany

D. Weichgrebe
Leibniz Universität Hannover
Hanover, Germany

Section I

Biomethane: Energy, Society, and Circularity Nexus

1 Changing Focus on Bioenergy through Resource Circulation

A Review for India and Europe

Debendra Chandra Baruah, H. N. Chanakya, Sadhan Kumar Ghosh, and Michael Nelles

1.1 INTRODUCTION

The global bioenergy market size stood at USD 344.90 billion in 2019 and is projected to reach USD 642.71 billion in 2027, exhibiting a compound annual growth rate of 8.0% during the forecast period of 2020–2027. The factors that were fuelling market growth before the COVID-19 pandemic include raising concern about the safety of the environment and increased demands of energy. Global bioenergy market growth witnesses higher growth rate due to the cost-effectiveness. Approximately 87% of energy demand is satisfied by energy produced through consumption of fossil fuels as of 2014, while the International Energy Agency (IEA) predicts that this share will fall to 75%, the total consumption of fossil fuels will continue to rise, adding another 6 Gt of carbon to the atmosphere by 2035. Souza et al. (2015) reported that over the past 2 years, 137 experts from 24 countries and 82 institutions have collaborated to analyse a range of issues related to the sustainability of bioenergy production and use. The resulting assessment was launched at a symposium at the World Bank in September, 2015 (Souza et al., 2015) highlighting key findings, opportunities, and challenges for sustainable energy in developing regions, as well as the role of bioenergy in the 2030- and 2050-time horizons. For the production of significant quantity of bioenergy, utilising the biomass needs a robust policy instrument as well as a tradition of biomass utilisation in the country. Some bioenergy policy-relevant issues may include:

- Need of bioenergy in the country
- Land availability to produce bioenergy without affecting food security, and rather enhancing it
- Biomass availability and its supply chain
- Bioenergy scaling up without inducing unintended land use changes
- Bioenergy efficiency
- Bioenergy impacts for the environment and reduction of greenhouse gas emissions
- Competition between biofuels and low fossil fuel prices
- Financing and commercialisation schemes
- Research and development to achieve maximum benefits from bioenergy.

This study reviews the concept of resource circulation and circular economy in the areas of biomass utilisation converting to bioenergy in the European Union (EU) and India. The chapter presents glimpses of the historical review of bioenergy resources and recirculation, promotion and challenges in the conversion from traditional to modern bioenergy, future of biomethane programmes, focus on biogas technology and the applications and the scopes for modernisation, and a strength, weaknesses, opportunities and threats (SWOT) analysis considering an Indian bioenergy perspective.

DOI: 10.1201/9781003204435-2

1.2 BIOENERGY – RESOURCES AND ITS CIRCULATION

Generation of methane (the simplest hydrocarbon) by anaerobic decomposition of organic matter has been known to the mankind since 17th century. However, its applications and technological advancement are relatively new to the global communities. The following subsections will discuss the historical backdrop of bioenergy resources and circulation in India and the EU.

1.2.1 Historical Review for India

Treatment of sewage and its use as an alternate fuel have been the primary applications during the early part of developmental history. Anaerobic digestion (AD) technology and research and development (R&D) in AD has a long history in India, with the first biogas plant established in Matunga (Mumbai) in 1897 (Sathianathan, 1989; Abbasi et al., 2012). Moreover, AD of agro-residues became a key area of early R&D beginning at the Indian Institute of Science (circa 1909–1920) and leading to the first R&D paper on AD of various wastes and agro-residues (Fowler and Joshi, 1920). The anaerobic-aerobic processing of agro-residues (Bangalroe method of composting) was evolved to achieve a near total recycling of plant nutrients with low losses. This allowed recycling of crop and animal waste from one cropping season to the next, with greater attention to nitrogen returned to cropland (Fowler and Joshi, 1920), which quickly spread to the researchers and implementers in India (Ramachandra Rao, 1996), as well as other parts of the world.

AD has been considered a method or a system to provide many outputs, wherein the primacy was accorded to multiple output(s). This, however, changed over time and the globally important concerns of the period, as depicted in Figure 1.1 (Chanakya, 2009).

In Figure 1.1, the boxes in the bottom row (blue) indicate the forces and problems that are being addressed during the specified period and the boxes in yellow show the types of R&D activities or issues addressed (middle row) and the top row shows the outcomes. This is an update version of the figure presented in Chanakya (2009).

The evolution of the current AD resources, practices, and technologies is analysed as an indirect approach to understand how "sustainability" and "circular economy principles" were brought in over a large time frame extending "a little over a century". Although the techniques of having pure cultures of many of the *Archea* and related obligate anaerobic microorganisms had not been developed until the late 1960s and 1980s, respectively, research on AD had been continuously carried out by addressing the underlying transformations. AD and biogas plants (BGP) have been envisaged

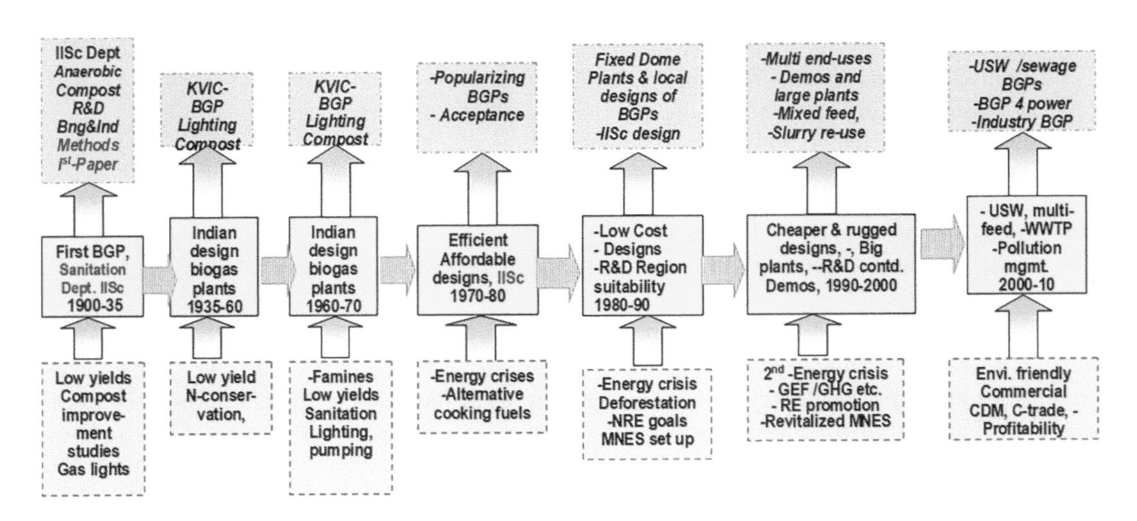

FIGURE 1.1 Various forces that have driven the biogas R&D and dissemination in India.

to solve problems related to N recovery and reuse, combined with sanitation (Fowler and Joshi, 1920; Ramachandra Rao 1996; Acharya, 1935; Acharya, 1961), rural and peri-urban lighting using gas (Sathianathan, 1989), clean cooking fuel (MNES, 1982), rural water supply, non-grid illumination and sustainable development (Rajabapaiah et al., 1993), GHG emission control and avoidance (Johansson et al., 1993), non-farm rural livelihoods (Chanakya, 2009 CNG/SNG (compressed natural gas/substitute natural gas), rural sanitation and even environmental clean-up (Sathianathan, 1989). Most of the early R&D and technology development efforts had been focused on AD of animal wastes, mainly driven by rural sanitation and N-conservation goals. R&D on agro- and biomass residues, including municipal solid waste (MSW), came to prominence after the 1990s.

Scientific and technological efforts of AD during this period had progressed along three paths. They attempted to meet three different goals, namely, (i) AD of animal wastes for rural energy/sanitation, (ii) AD of segregated MSW, and (iii) AD of agro-residues and other herbaceous biomass. Between 1920 and 1960, early R&D and technology development addressed cattle manure, a resource plenty found across rural India. Evolution of the family biogas plant technology and the spread of biogas technology using animal dung as main feedstock in rural India are considered a reasonable success in India (nearly 5 million plants) and China (26.7 million plants, 2007) (MNES, 1998; Chen et al., 2010). A pattern of annual growth of household biogas plants during two periods of references are presented in Figures 1.2 and 1.3 (Kharbanda and Qureshi, 1985; Lohan et al., 2015). Yet, they had a limited reach and did not meet the cooking energy needs of all rural homes. Later, it was estimated that based on availability of cattle manure, manure BGPs was feasible only for up to about 17 million rural homes (Khandelwal, 1990; Ramana and Sinha, 1995). Meeting the cooking needs of all rural families would need BGPs that could use alternative feedstock (Chanakya and Balachandra, 2012; Balachandra and Ravindranath, 2009), which would lead to meet the cooking and lifeline energy needs for sustainable development for all of rural India (Reddy and Reddy, 1994; Jagadish et al., 1998). The science of AD for biomass feedstock, rice straw, sugar cane trash, terrestrial, aquatic weeds, etc. evolved slowly and received emphasis only recently.

AD science for biomass is critical of technology development (Chanakya et al., 2004; Chanakya, 2009). However, much science and technology development has been driven by R&D in wastewater

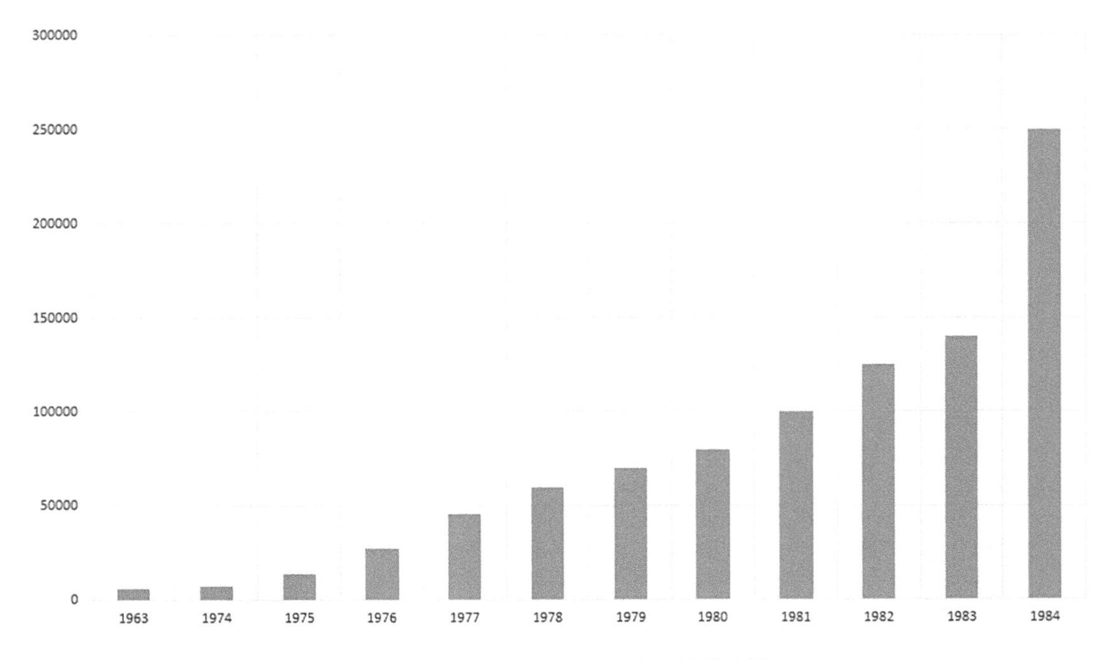

FIGURE 1.2 Growth of household biogas plants in India during 1963–1984.

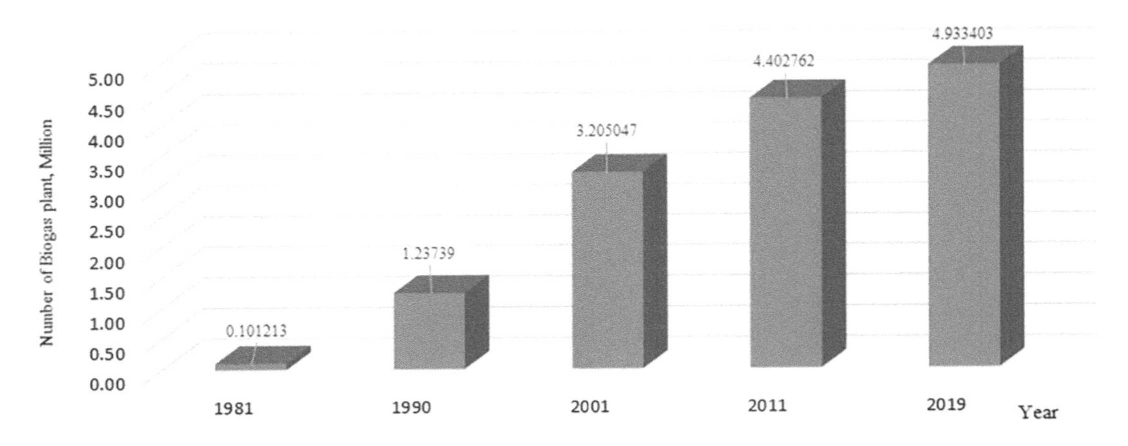

FIGURE 1.3 Growth of biogas installation in India during 1981–2019.

– a system that does not deal with ligno-cellulosic and its attendant complexities. The mounting energy costs of aerobic wastewater treatments led to a large-scale shift towards adopting AD options, where a primary AD step brought down the organic loads by nearly 90% with surplus energy for secondary treatment. Yet, in terms of science and technology for process improvements of AD for agro-residues and other ligno-cellulosic biomass has been slow or poorly relatable. Firstly, there was very little of soluble materials in typical agro-residues and biomass feedstock and, secondly, they could rarely be maintained as a stable slurry to enable digestion, just as in the case of sewage sludge or animal wastes (Rajabapaiah et al., 1981). This has necessitated a major shift in AD towards ligno-cellulosics in the recent past.

1.2.2 Historical Review for the EU

Bioenergy has traditionally been the most important renewable energy carrier in Europe and the EU, and has remained thus to date, notwithstanding the rise of other options, such as wind and solar power. Both production and consumption of bioenergy have been increasing rather steadily over the past two decades. Prior to the industrial revolution, bioenergy constituted the only available energy carrier for heating, as well as transport (in the form of feed for mount, draught, and pack animals) in most areas, not just in Europe, but globally.

During the 19th century (in some parts of Europe, as early as the later 18th century), increasing industrialisation brought about a dramatic rise in the exploitation of fossil fuels, as well as the advent of electricity and motorised mobility. This resulted in plant-based energy sources losing importance (Pfeiffer and Thrän, 2018). However, developments regarding bioenergy production and use had already gained momentum again in the early 20th century. Pfeiffer and Thrän (2018) see this as mainly due to two developments:

Firstly, increasing population, especially urbanisation, created problems of sanitation and water management. The application of the already-known biomass fermentation process to wastewater treatment, with biogas initially no more than a side product, resulted from this need.

As early as the 1890s, such sewage gas was first used to fuel street lamps in the United Kingdom (Bond and Templeton, 2011). In Germany, digesters for wastewater treatment were built during the second and third decade of the 20th century. By the late 1920s, sewage gas was being fed into the first public gas systems and, a decade later, used in several municipal vehicle fleets. During the same decade, Danish wastewater treatment plants started to use the gas produced during treatment to heat the process tanks (Mathieu and Eyl-Mazzega, 2019). However, at that point in time, interest in the technology remained limited to a handful of larger cities, and the overall potential was considered limited. A further field of application tentatively opened in the 1940s, when the first plant

TABLE 1.1
Selected Milestones of Scientific/Technological Developments in Biogas, Biofuels and Gasification between ca. 1900 and 1945

1896s	Sewage gas first used to fuel streetlamps in a UK city (Exeter)
Around 1900	City and synthesis gas mainly produced by coal gasification
1910s–1920s	First special digesters for anaerobic wastewater treatment
1920s	Invention of first technologically mature wood gas generation for mobile use
1925	Invention of Fischer-Tropsch synthesis of diesel-like fuel from coal
1927	First feeding-in of surplus sewage gas into a public gas network by a wastewater treatment plant
1927	World's first hydrogenation plant (Bergius Pier process)
1937	First operation of municipal vehicle fleets with biogas in several German cities
1944	First agricultural plant producing biogas from solid manure, agricultural waste, and residues

Source: Adapted from Pfeiffer and Thrän (2018).

was built, which produced biogas in an agricultural setting, i.e., from solid manure and agricultural residues and wastes.

Secondly, technological developments meant that solid biofuels could be used in quite new areas, and this again was eagerly taken up, especially in times of crises – either because of fuel shortages or out of political considerations (e.g. to achieve energy autarky before and during wartimes). The most prominent of these developments is probably the use of wood gas generators. According to Pfeiffer and Thrän (2018), about 9,000 wood gas vehicles alone were in use in Europe by the end of the 1930s.

Other such developments include ethanol gained from paper processing or wood pyrolysis and vegetable oil used in diesel engines, as well as processes such as the Fischer Tropsch synthesis and the Bergius Pier process for liquid fuels. Both of the latter were originally based on coal and used accordingly during those years, but they are applicable to biomass as well. Since their production was found to be much less economic, interest in these processes remained limited at that point in time.

In a nutshell, many of the technological developments, as well as economic considerations that are familiar from the present, can be said to go back, in essence, nearly a century. An overview of some important technological milestones in these early years is given in Table 1.1.

1.3 PROMOTION AND CHALLENGES

Traditional to modern bioenergy: promotion and barriers (Technological and cultural parameters affecting the spatially varying changes/shifts – in EU context)

As described, the technological foundation for a transition from traditional bioenergy uses, such as wood stoves, to modern applications was lain early on in Europe. However, the widespread uptake of these technologies in Europe was delayed by the end of the war and its shortages. Prices for fossil fuels were low in the 1950s and 1960s, crude oil consumption rose sharply not just in Europe, but globally (Figure 1.4), and energy autarky was not a burning issue any longer, either. The 1940s did see the first agricultural biogas plant invented and actually established. It already used a mixture of solid animal manure, vegetable waste, and sewage sludge, and the digestate was used as an agricultural fertiliser.

Further development was slow: only a double-digit number of further plants was built in Germany during the following two decades – the country that was, and still is, leading in Europe with regard to biogas production and number of biogas plants installed (Eurostat, 2021). The combination of low oil prices and low efficiency of these first biogas plants limited interest of both farmers and politicians.

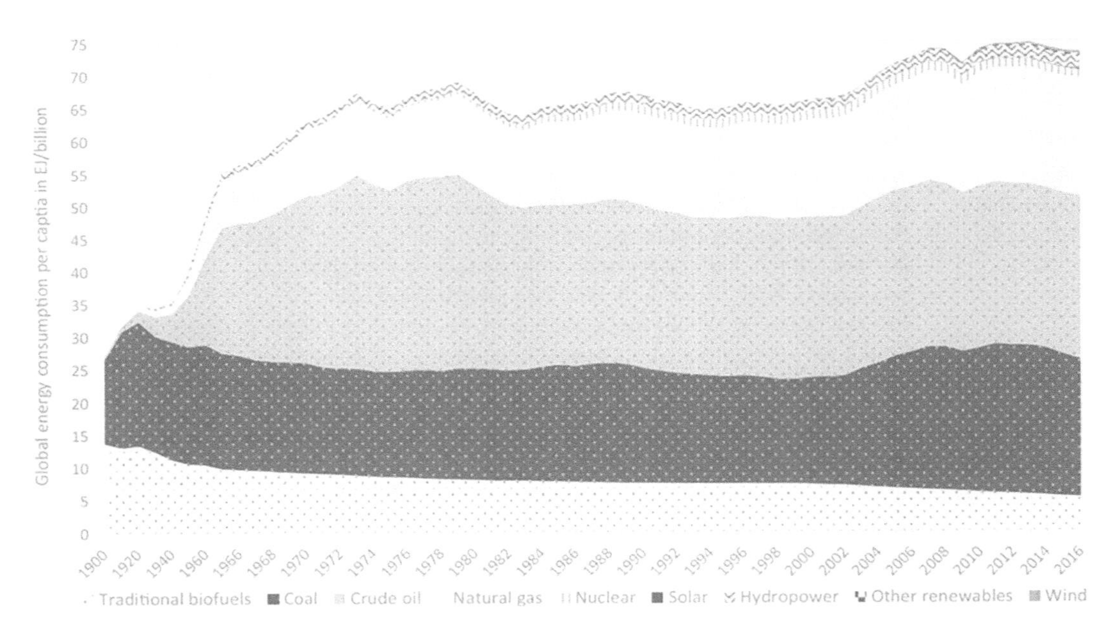

FIGURE 1.4 Development of global energy consumption per capita in EJ (ExaJoule). (Adopted from Pfeiffer and Thrän, 2018.)

With surging oil prices, the energy crisis of the 1970s reignited fears over energy security and, consequently, interest in researching alternatives. However, bioenergy was only one option considered, and even though new plants were built, their number remained relatively low. On the technological side, the shift of focus towards gas output resulted in increasing use of liquid manure, and co-substrates, which increased gas yields. In the 1980s, AD was increasingly used throughout Europe to treat industrial wastewater, as well as municipal waste (European Biogas Association, 2019). In agricultural settings of Germany, the focus was on the digestate as an improved fertiliser, especially in organic farming (Pfeiffer and Thrän, 2018), otherwise, on Combined Heat and Power (CHP) for agricultural use. In Denmark, the main application was CHP production for local towns (Mathieu and Eyl-Mazzega, 2019).

The turning point for bioenergy came in the 1990s, with rising concerns over environmental and health damage caused by the extraction and processing of fossil fuels, as well as concerns over climate change that arose later. Governments in a number of European countries began to promote renewable energies, with different focal points and frameworks such as fee-in tariffs for electricity from renewables. In Germany, such a tariff was introduced in 1991, and the number of biogas plants began to rise sharply from 139 in 1992 to 1,050 in 2000 (Statista, 2020). Italy introduced a generous feed-in tariff, the highest in Europe, for electricity from small biogas plants in 2008, which resulted in an impressive surge in plants from 510 in 2010 to 1,264 only two years later. In Denmark, a similar development happened when a feed-in subsidy was introduced for injecting biogas into the gas grid in 2012. With this, a new market opened, which also allowed for larger plants to be built (Mathieu and Eyl-Mazzega, 2019).

An additional stimulus resulted from the "set-aside incentive scheme for arable land", which the European Commission introduced in 1988 in order to reduce agricultural overproduction (European Commission, 1988). The production of energy crops was allowed on these lands and became a major source of biogas raw materials, such as in Germany and Italy, but not in Denmark, where manure remained the all-important input materials. From circa 2012 or 2013 onwards, governments gradually moved away from promoting energy crops for both biogas and transport biofuels, due to the "food-versus-fuel debate", and focused incentives on residues and wastes.

1.4 IMPLEMENTATION CHALLENGES IN INDIA

Wise bioresource use to meet a majority of the rural and urban energy needs and sustainable development goals (SDGs) in India is therefore a unique opportunity, especially when the futures are built on existing experiences in this area. The major challenges lie in converting this opportunity to a reality through creating a policy environment that drives the system towards employing a sustainably generated biomass energy-centric future. Such a system needs to use rural resources for energy and energy services without exploitation, deprivation, or resource depletion. The next challenge requires us to facilitate a technical system that enables generation of required energy conversion and carrier technologies and deploys them early commercially (Figure 1.4, boxes in rose). Some of these critical ones are:

a. Methane-fed fuel cell power for intermittent use
b. Truck-mounted methane collection and liquefying system powered by biogas itself
c. Creation of CNG/SNG outlets in right proportion in rural and decentralised settings
d. Conversion of farm machinery to run on methane/biogas with and without compression

The third challenge is to create a policy- and technology-spread environment, wherein most farmers in rural areas will convert their hitherto unused and poorly used bio- and crop-residues to methane in modern biomethanation plants, such that it meets their primary need of 1–1.5 m^3 biogas for domestic purposes and the surplus is set for export to urban areas in a manner similar to milk clusters. Surplus biogas is delivered in a methane tanker, which purifies, compresses, and stores the gas after making necessary accounting and payments digitally. Science and technology interventions required to make methane-centric development and sustainability as indicated by boxes in orange and links in bold red lines in Figure 1.5. Boxes in green/grey and links in green dotted lines are emerging or existing situation and resource-use centres. This figure is an updated figure from Chanakya and Balachandra (2012).

A fourth challenge is to create a policy environment that favours the reduction of abusive, wasteful, and poorly productive energy usage in urban areas and instead make even urban areas energy positive – AD of MSW, for instance. Thus, creating a policy environment that favours the adoption of energy efficient solutions in urban areas without affecting people's aspirations would be a challenge that needs to be overcome in the low-C scenario. The next challenge is shifting to compressed biogas (CBG) for a majority of urban mobility needs. The Sustainable Alternatives towards Affordable Transportation (SATAT) initiative promotes CBG and provides long-term support for any investments in that direction. However, a policy environment that facilitates the choice of public transport would be ideal to lower the fossil fuel footprint and substitute it by the CBG.-based transport.

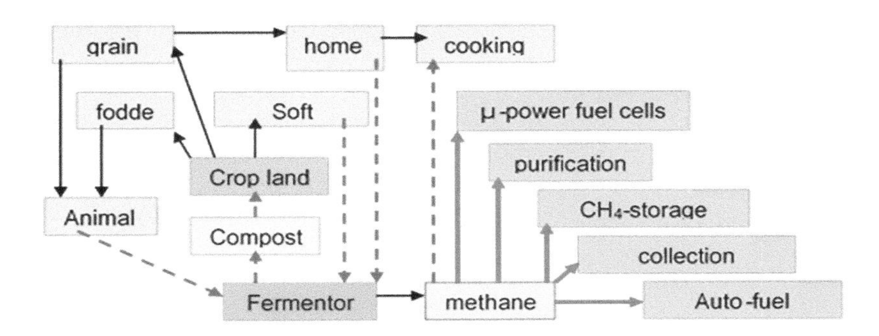

FIGURE 1.5 Existing and proposed flow of soft residues and methane.

1.5 FUTURE OF BIOMETHANE PROGRAMMES

1.5.1 The Future of Biomethane Programmes in India

The emerging fossil-fuel shortages are being addressed by tapping methane from non-edible soft biomass to the mutual benefits of rural-urban India and will make the developing world reach a new high in the sustainability. The many science and technology and "resource management" gaps have now been addressed in a large way by creating aggregators. The poorly utilised or inefficiently converted soft biomass is being incentivised in multiple ways. Today, soft biomass is generally burnt or left to decompose in the fields and this is being reversed. All non-edible or non-fodder crop and weed biomass residues, or soft biomass residues, generated in rural and peri-urban India are generally managed through one of these practices – converted to compost, left un-harvested in the field, or burnt directly in the field to reduce bulk. Estimates suggest it to be around 600–800 million dry tons annually (Jagadish et al., 1998). All these could be firstly fermented anaerobically to derive methane (and energy) and finally consigned to existing or better end uses, reducing environmental damage on one side and also enabling farmers to add to their income sources. The required technology to do so is shaping up in different forms while attending to the fermentation nature of the local biomass resource, be it paddy straw, maize stalks, or the like. Their widespread adoption will lead to realising multiple millennium development goals (MDGs) and SDGs related to empowerment in rural areas, waste management, and sanitation, alleviation of poverty, reducing vulnerability, and clean energy for all.

Current development patterns indicate that rural homes will continue to use commercially procured power only for essential and entertainment needs. All these can be provided by decentralised small-fuel cells. Export of surplus methane generated from agro-residues as a transport fuel will augment rural development. The methane energy carrier could be organised and sustained along similar lines as the successful milk production model of rural India. When such possibilities exist, most non-urban and peri-urban families can and will have larger than required methane fermenters that produce some level of surplus biogas. Producing and selling 3–5 m^3 surplus methane would mean ₹100–150/day of assured daily income to families, especially women. Bigger plants would improve economy, as well as organic matter recycling. Other revenue streams could be added to valorise digester liquid and digester residue, as discussed later on. This also suggests that a similar potential exists in other parts of tropical and subtropical Asia. For countries like India, 40% of primary energy comes from biomass. The same will now meet development goals, poverty alleviation, and soil-C restoration needs, as well as a good part of urban needs.

1.5.2 The Future of Biomethane Programmes in the EU

Biogas consumption increased by a factor of 25 between 1990 and 2017, Bioenergy Europe (2019) in presented in Figure 1.6. For the future, the outlook in Europe is mixed and largely dependent on political frameworks, less on technological developments, as highlighted via three examples in the following.

In Germany, subsidies for electricity feed-in became less generous from 2012 onwards, which slowed down installation of new plants. In 2017, a tender model replaced the fixed feed-in tariffs, which meant that plant operators faced less reliable economic conditions, and, unsurprisingly, target volumes were not reached by far in the first two auctions (Table 1.2).

The number of organic waste-based and small manure-based biogas plants is still growing due to more favourable conditions. However, most of the increase in capacity currently comes from enlargement of existing plants (Mathieu and Eyl-Mazzega, 2019). Between 2009 and 2014, a bonus for upgrading biogas to biomethane resulted in a sharp increase in such units, but this has slowed since the phase-out of the bonus. Around 90% of the biomethane is currently used in CHP plants. Very little is used in the transport sector, even though biomethane has the highest GHG mitigation potential of all available biofuels, and uses cheap substrates (Mathieu and Eyl-Mazzega, 2019). Gaseous fuels are comparatively badly established in Germany (Daniel-Gromke et al., 2018), but

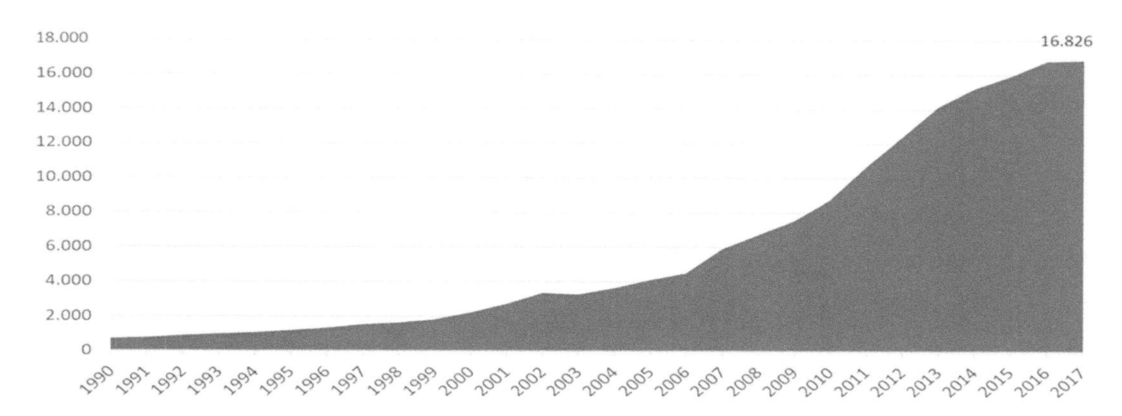

FIGURE 1.6 Gross domestic energy consumption of biogas in the EU28 (in ktoe; Bioenergy Europe, 2019).

there is some potential for its application in heavy-duty road as well as in marine transport, in the form of liquefied biogas. However, CHP will probably remain the most efficient use, particularly if more biogas plants can be fitted with biomethane upgrading technology: by the end of 2018, about 8,980 biogas plants were in operation in Germany, 203 of which were equipped for upgrading. Biomethane has potential for sector coupling, with the gas infrastructure as well as with transportation or for industrial processes.

In Denmark, a tender scheme came into force in 2020. As for Germany, this is expected to slow down further development. However, Mathieu and Eyl-Mazzega (2019) still see possibilities for expansion, as only 10% of domestic manure goes to biogas production. Since the introduction of subsidies for grid injection, economic frameworks have favoured large companies, which resulted in large energy providers, not the farmers, owning the upgrading installations. While this is seen as negative from a social perspective, it does mean investment hurdles are somewhat lower than in other countries. In transport, prospects are possibly even worse than in Germany, as taxes for bio-CNG vehicles have increased and the fuel itself is taxed the same as natural gas, higher than fossil diesel. On the other hand, the integration into the gas grid is much more advanced, with biomethane making up about 10% of grid transportation volume.

In Italy, production stagnated after the generous feed-in tariffs for energy crops were cut. From 2013, the government also shifted its focus from electricity generation to biomethane, with the aim of promoting application in transport, for co-generation, and for grid injection, but a lack of secondary legislation prevented impacts until recently. However, Italy does have the largest natural gas vehicle fleet in Europe – three quarters of the total amount on the continent – and thus by far the most favourable infrastructure for biomethane use in transport. Consequently, the political target for advanced biofuel consumption in 2030 is set to be fulfilled to 75% by biomethane (0.8 Mtoe).

Actual biomethane production, as well as injection into the gas grid, is only just starting to take off in Italy: as of 2019, six biomethane-producing plants were in operation, three of which

TABLE 1.2
Results of German Biomass Auctions in 2017 and 2018

Date of Auction	Volume of Auction (MW_e)	Number of Awarded Plants (n)	Awarded Installed Capacity (MW_e)
01/09/2017	122.4	21 biogas plants + 1 biomethane-CHP	27.5
01/09/2018	225	79	77

Note: Mathieu and Eyl-Mazzega, 2019, with data from Bundesnetzagentur

were connected to the grid, and six more were about to start production. For Italy, Mathieu and Eyl-Mazzega (2019) see the main barriers not as economic, but in the still unfinished regulatory framework and in public resistance to biogas production.

1.6 FOCUS ON BIOGAS TECHNOLOGY, APPLICATIONS, AND SCOPE FOR MODERNISATION IN INDIA

At the national level, India has come out with several programmes, policy initiatives, inter-ministerial action plans, incentive schemes, and physical-financial targets that address several SDG goals as well as international commitments to climate change issues. Some of the major ones are listed in this section. All of these drive technology and applications in the direction of conversion of various wastes and residues through AD to biogas and a plethora of by-products. SDGs then impose several driving factors that will soon ensure that biogas plants are not merely producing biogas (although it is the most important output) but other by-products that also meet the goals of a circular economy leaving very little secondary wastes and instead convert them to various value-added products that strengthen the circular economy nature of the endeavour on the one side and sustainability indicators on the other side.

It may be noted that the predominant biomethanation program that India pushed until the 1990s was the "cow-dung" based biogas plants that paved way for rural sanitation and hygiene, assured life-line energy in rural India, and removal of drudgery and morbidity arising from cooking on low-value agro-residues and woody biomass. From a potential/feasible target of 10–12 million biogas plants, rural India has already built over 5 million family-sized biogas plants – in business terms about 50% market saturation – a no-mean achievement that has occurred through dissemination and incentivisation rather than through legal or other forms of compulsion. Some of the recent programs, policy initiative, and incentivisation efforts of the country include:

- SATAT-assured pick up of CBG with price revision every 3 years (currently at USD 0.7/kg).
- Electricity policy (100% procurement from W2E plants)
- Motor vehicle rules (use of compressed biogas, CBG in motor vehicles)
- Biogas standards (for automotive and piped-network applications)
- Make in India
- Swachh Bharat mission (Urban/Rural, conversion of human/urban wastes to biogas)
- Smart city projects (encouraging conversion of MSW/sewage to biogas)
- Power for all (24 × 7 power for all – decentralised applications).
- GHG voluntary targets (substitution by agro-residue derived CBG)
- Reduction in oil import (decentralised sale of CBG and in rural India)
- Doubling farmers' income at the farm level
- Scale dependent subsidy (region-wise and W2E BGPs under the ministry of new and renewable energy)

Biogas from paddy straw will soon become the largest single agro-residue derived biogas with the installation of over 25 100-tpd-size biogas plants that are being installed at various straw-surplus and straw-rich locations across the country and are at various stages of completion. A kg of paddy straw produces about 400 L biogas (dry basis), which means 2.5 kg straw will be needed per m^3 biogas produced if costed at ~₹10/m^3 raw material costs of biogas. In this way, ~6.25 kg straw/kg CH_4 will be needed giving it ~₹25/kg methane as raw material costs (inclusive of 10% sacrificial energy in the processing). The current selling price is ₹50/kg compressed biogas, which in turn would substitute 1.5 L petrol (~₹150 current market price). Such a high margin and opportunity costs is expected to drive even mid-sized biomass-based biogas plants to be quite economic to run and operate in a decentralised manner, while also enabling local users to shift from fossil fuel to the more sustainable biogas produced locally with local agro-residues.

Unlike the past dominated by animal-waste-based biogas (gobar gas), the current biogas landscape will depend more on various agro-residues, town wastes, or organic fraction of MSW. Today, many 5-tpd plants of various indigenous designs are already built and run at various urban locations in the country. Most of them convert biogas to electricity and supply electricity to the local grid because during their "formative years" in the 2000s, power generation was the most remunerative revenue model. However, in the last decade, with the rapid spread of solar photo-voltaics, the power costs have fallen by over 60%, making power production an uneconomic end-use for biogas produced from MSW or any other source. As a consequence, a greater emphasis has been placed in the conversion of biogas to automotive fuel – CBG – whose opportunity costs and margin-over-material costs are mentioned in the previous paragraph. To make such plants even more sustainable and feed into a circular economy, large numbers of value-added by-products have been developed across the country, as shown in Figure 1.7. All the three outputs of a typical biogas plant can now be valorised into value-added by-products, depending upon the location and stage of economy, such that biogas sales would not be the only profit centre from typical large biogas plants. Figure 1.7 presents multiple technologies to use the three BGP outputs for more circular and sustainable technology. Biogas plants now provide a wider canvas of outputs to ensure near zero waste discharge as well as higher sustainability. Besides, R&D to convert CH_4 to DME, methanol, etc. are also underway to ensure a multiplicity of outputs from a biogas.

In terms of socioeconomic sustainability, simple household biogas plants, which are the mainstay of the large number of biogas plants in the country at present, if adapted to use agro-residues as well and also produce a few of the multiple outputs projected in the previous figure, they can make villages a lot more sustainable and development friendly and reduce the GHG footprint - a position seen as an important milestone defining our development program. Reuse of by-products will feed into the circular economy. As seen from Figure 1.7, it is possible to have biogas plants that are capable of multiple outputs and value streams. In such cases, we could consider one specialised output,

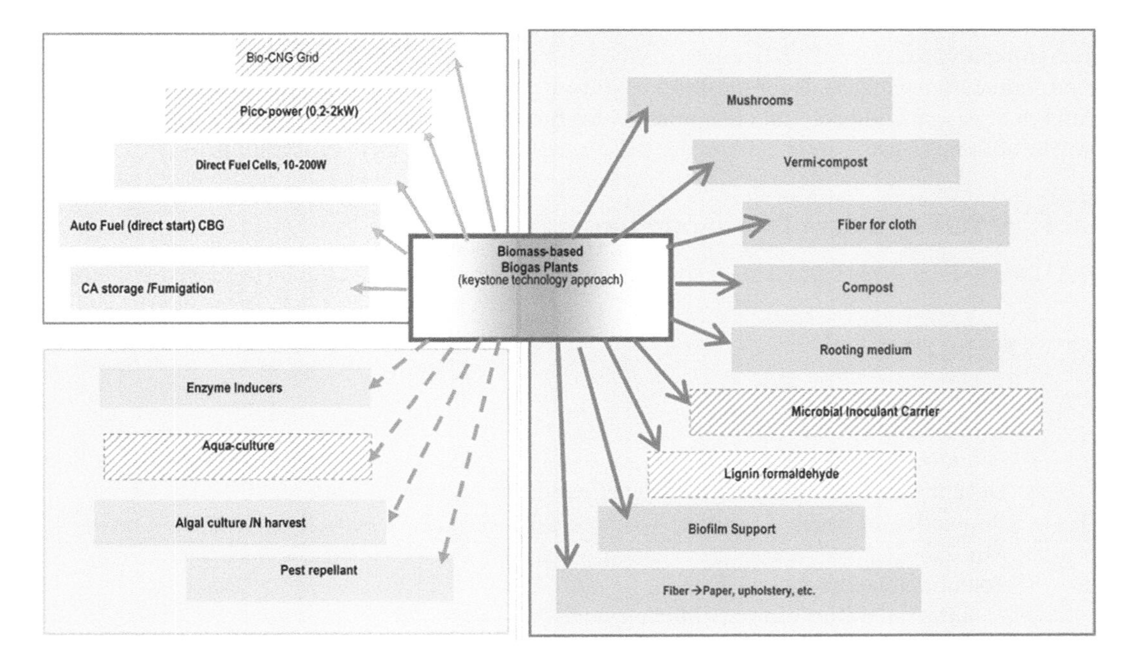

FIGURE 1.7 Multiple technologies to use the three BGP outputs now make the biogas technology a lot more circular and sustainable. (Legends: green **solid arrows** = digested residue; **broken arrow** = digester liquid, grey **solid arrows** = surplus gas. Technology status: shaded = evolved at CST; cross-hatched = other institutions or yet to be done.)

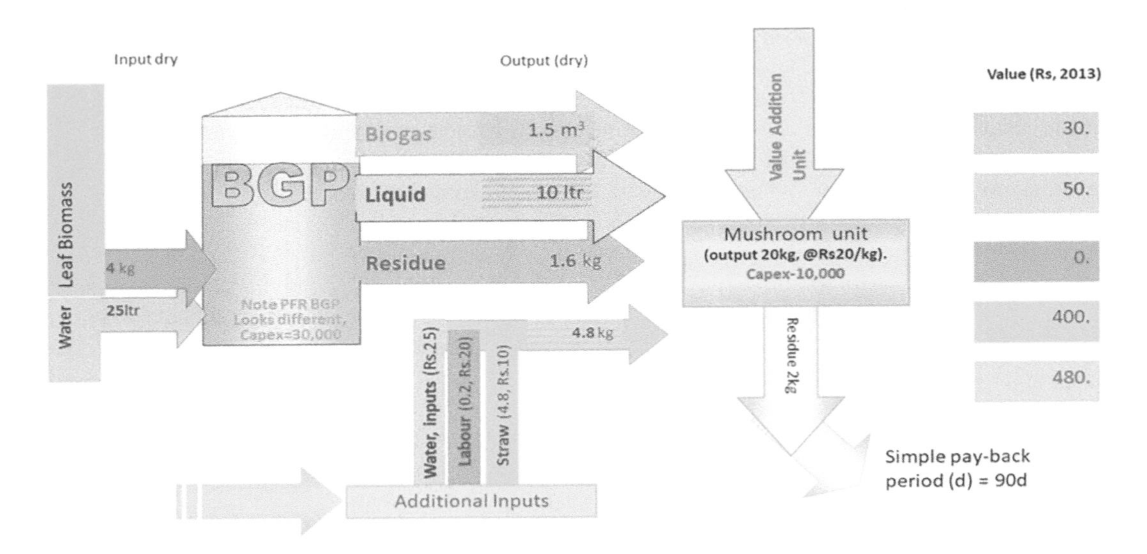

FIGURE 1.8 Analytical framework of sustainability and circular economy potential from the co-production of biogas and mushrooms from typical agro-residues/street sweepings (predominantly leaf litter) is analysed above.

for example, a rural biogas plant that, in addition to producing biogas from rural residues (farm litter), attempts to produce additional outputs. We examine the case of agro-residue or leaf litter biogas plant that in addition to producing biogas also produces mushrooms from the biogas digestate. Figure 1.8 presents an analytical framework of sustainability and circular economy potential from the coproduction of biogas and mushrooms from typical agro-residues and street sweepings, predominantly leaf litter. The digestate from agro-residue, including leaf-litter rich in lignin, are a good source of raw materials for mushroom cultivation. The partially digested feedstock excludes a plethora of fungal and bacterial contaminants for further digestion, leaving behind only white and brown rot fungi to grow, treat, and render these agro-outputs and their valorisation.

1.6.1 SWOT of Biogas Technology: Indian Perspective

SWOT analysis has been carried to identify the scopes for improvement in Indian bioenergy initiatives.

1.7 STRENGTHS

- Multiple benefits
  - Conservation of bio-resources
  - Pollution control
  - Decentralised fuel production, even at remote and isolated locations
  - Production of organic fertiliser with provision of nutrient recycling (N, P, K, Cu, Fe, Mn, Zn)
  - Potential to address/contribute
    - Integrated Rural Energy Management
    - Organic farming
    - Doubling of farmer income
    - Total sanitation mission (Swachh Bharat Gramin Mission) Gobar-Dhan
    - National mission of climate change
    - International commitments (Nationally Determined Contributions/SDGs)

- Proven technologies with almost all scale of options
  - Household AD plant – 2 m^3 to 10 m^3/day biogas production (direct use of farmers)
  - Industrial-scale plant for electricity – kW to MW (indirect benefits of farmers – opportunities to sell surplus farm residues)
  - Industrial scale for bioCNG – several thousand m^3 (indirect benefits of farmers – opportunities to sell surplus farm residues)
- AD technology is matured with uses/experiences/R&Ds and have adequate **indigenous** knowhow covering
  - Chemistry/microbiology of reaction kinetics
  - Suitability of raw materials
  - Reactor design
  - Nutrient analysis of by-product (digestate slurry)
  - Benefits (economy, environment, soil health, GHG mitigation, livelihood improvement)

1.8 WEAKNESSES

- Still remains as "subsidy driven traditional idea" rather than a "commerce driven modern item"
- Majority of the installed systems remain out of order, mostly due to
  - Absence of after-installation care
  - Inadequate realisation of hidden benefits
  - Availability of cheaper alternatives
- Growth of AD is not proportional to the government's effort (promotional scheme) and funding
- Sound policy of monitoring/performance assessment of installed systems with
  - Lack of reliable tools for monitoring and data access
  - Lack of reliable tools for measurement of operating parameters (gas flow, pH, gas species)
- Lack of policy implications
  - On nutrients recycling aspects (standard for biogas digestate as organic fertiliser needed)
  - Potential GHG mitigation

1.9 OPPORTUNITIES

- Convert the "subsidy driven traditional idea" to a "commerce driven modern item of multifaceted benefits"
- Market opportunities with portfolios of
  - Products (biogas, bioCNG, organic fertiliser)
  - Technologies (reactor, accessories viz., sensors, online measuring, data handling and management software, apps)
  - Service provider/agents with smart and reliable solutions for millions of distributed systems
- Saving of financial resources of the government

1.10 THREATS

- Further delay of appropriate indigenous technology solutions could lead to
  - Reliance on foreign products
  - Load on government exchequer without desired result
  - Irreversible false perception on the AD technology

1.11 CONCLUSION

The value of energy recovery is in turning waste into a "resource" that generates usable energy and thereby adding value in the chain of the circular economy. The use of biomass from natural resources and the related production and consumption processes have environmental, economic, and social consequences within countries and beyond national borders. Resource circularity helps in resource conservation. For the bioenergy improvement process dealing with biomass, there will always be challenges where the options are available to use food grains as one of the feedstock, which intervenes into food security. Bioenergy resource use and circulation need to be tapped judiciously, as this may be one of the best options among the environmentally friendly energy recovery technology from biomass.

REFERENCES

Abbasi, T.; Tauseef, S.M.; Abbasi, S.A.: 'A brief history of anaerobic digestion and "biogas"'. SpiBrf. In Envr. Sci. 2, DOI: 10.1007/978-1-4614-1040-9_2, pp. 11–23, (2012).

Acharya, C.N.: 'Preparation of fuel gas and manure by anaerobic fermentation of organic materials'. Indian Coun. Agric. Res. 15, (1961) New-Delhi.

Acharya, C.N.: 'Studies on the anaerobic decomposition of plant materials. IV. The decomposition of plant substances of varying composition'. Biochem. J., pp. 1459–1467, (1935).

Balachandra, P.; Ravindranath, N.H.: 'Sustainable bioenergy for India: Technical, economic and policy analysis'. Energy. 34, pp. 1003–1013, (2009).

Bioenergy Europe: Statistical Report 2019 – Report Biogas. Bioenergy Europe, Brussels, (2019).

Bond, T.; Templeton, M.R.: History and future of domestic biogas plants in the developing world. Energy for Sustainable Development. 15, pp. 347–354, (2011).

Chanakya, H.N.: 'Sustainable Anaerobic Digestion Technologies for Developing Countries'. Intl Workshop in Anaerobic Digestion (Old story for Today and Tomorrow), ENSAM-LBE, Narbonne France, pp. 50–60, (2009).

Chanakya, H.N.; Balachandra, P.: 'Enabling Bioenergy Deployment for LCE: Methane for Cooking in India' (Chapter 10). (Eds. Srinivasan, A.; Ling, F.; Nishioka, S.; Mori, H), Earthscan, London. ISBN: 9781844078615, p. 192, (2012).

Chanakya, H.N.; Rajabapaiah, P; Modak, J.M.: 'Evolving biomass-based biogas plants: The ASTRA experience'. Curr. Sci. 87, No. 7, (2004).

Chen, Y.; Yang, G.; Sweeney, S.; Feng, Y.: 'Household biogas use in rural China: A study of opportunities and constraints'. Ren. Sustain. Energy Rev. 14, pp. 545–549, (2010).

Daniel-Gromke, J.; Rensberg, N.; Denysenko, V.; Stinnre, W.; Schmalfuß, T.; Scheftelowitz, M.; Nelles, M.; Liebetrau, J.: Current developments in production and utilisation of biogas and biomethane in Germany. Chem Ing. Tech. 90, Nos. 1–2, pp. 17–35, (2018).

European Biogas Association: Biogas Basics. https://www.europeanbiogas.eu/category/publications/, accessed 11.03.2021, (2019).

European Commission: Commission Regulation (EEC) No 1272/88 of 29 April 1988 laying down detailed rules for applying the set-aside incentive scheme for arable land. https://op.europa.eu/, Publications Office of the EU, accessed 11.03.2021, (1988).

Eurostat: Supply, transformation and consumption of renewables and wastes. https://ec.europa.eu/eurostat/, accessed 11.03.2021, (2021).

Fowler, G.; Joshi, G.V.: 'Studies in fermentation of cellulose'. J. Indian Inst. Sci. 3, p. 39, (1920).

Jagadish, K.S.; Chanakya, H.N.; Rajabapaiah, P.; Anand, V.: 'Plug flow digesters for biogas generation from leaf biomass'. Biomass Bioenergy. 14, pp. 415–423, (1998).

Johansson, T.B.; Kelly, H.; Reddy, A.K.N.; Williams. R.H.:(Eds.): Renewable Energy—Sources for Fuels and Electricity, Island Press, NY, pp. 1–69, (1993).

Khandelwal, K.C.: 'Biogas technology development and implementation strategies—Indian experience', Proceedings of International Conference on Biogas Technology and Implementation Strategies, Bremen Overseas Research and Development Agency (ed.), Bremen, Germany and Pune, India, Jan 10–15, (1990).

Kharbanda, VP; Qureshi, M.A.: 'International association of energy economics'. Energy J. (6), (1985).

Lohan, S.K. et al.: 'Biogas: A boon for sustainable energy development in India's cold climate'. Ren. Sustain. Energy Rev. 43: pp. 95–101, (2015).

Mathieu, C.; Eyl-Mazzega, M-A. (Eds.): Biogas and Biomethane in Europe: Lessons from Denmark, Études de l'Ifri, Ifri, Germany and Italy, (2019).

MNES, Annual Report, 1982–1998, Ministry of Non-conventional Energy Sources, New Delhi, (1982).

MNES, Annual Report, 1998–2010, Ministry of Non-conventional Energy Sources, New Delhi, (1998).

Pfeiffer, D.; Thrän, D.: 'One century of bioenergy in Germany: Wildcard and advanced technology'. Chem. Ing. Tech. 90, No. 11, pp. 1676–1698, (2018).

Rajabapaiah, P.; Chanakya, H.N.; Reddy, A.K.N.: 'Interim Report Further Studies on Biogas Technology'. CST, Technical Report, Indian Institute of Science, Bangalore 560012, India, (1981).

Rajabapaiah, P.; Jayakumar, S.; Reddy, A.K.N.: 'Biogas electricity—The Pura village case study'. In Renewable Energy—Sources for Fuels and Electricity (Eds. Johansson, T.B. et al.), Island Press, Washington, DC, 1993, pp. 787–815, (1993).

Ramachandra Rao, T.N.: 'Fermentation Technology—A Personal Journey'. KSCST, Indian Institute of Science, Bangalore, p. 26, (1996).

Ramana, P.V.; Sinha, C.S.: 'Renewable Energy Programmes in India'. Tata Energy Research Institute, New Delhi, (1995).

Reddy, A.K.N.; Reddy, B.S.: 'Substitution of energy carriers for cooking in Bangalore'. Energy, 19, No. 5, pp. 561–571, (1994).

Sathianathan, M.A.: 'Biogas-Achievements and Challenges, Association of Voluntary Agencies in Rural Development'. New Delhi, p. 192, (1989).

Souza, G.M.; Victoria, R; Joly, C.; Verdade, L. (Eds.): Bioenergy & Sustainability: Bridging the Gaps, SCOPE, Paris, Vol. 72, p. 779, (2015). ISBN 978-2-9545557-0-6; http://bioenfapesp.org/scopebioenergy/index.php

Statista: Anzahl der Biogasanlagen in Deutschland in den Jahren 1992 bis 2020. https://de.statista.com/ →Energie & Umwelt, accessed 10.03.2021, (2020).

2 Kinetic Modelling of Biogas Production from Organic Fractions of Municipal Solid Waste and Projection of Energy and Energy Potentials

V. Mozhiarasi, P. M. Benish Rose, S. M. Elavaar Kuzhali, S. Kanyapushpanjali, D. Weichgrebe, and S. V. Srinivasan

2.1 INTRODUCTION

Rapid urbanisation and industrialisation in India have increased the per capita municipal solid waste (MSW) generation with organic matter share of about 40.2 wt.%–51.0 wt.% (Kumar et al. 2017; CPCB, 2013; Speier et al. 2018). Improper collection and segregation of waste are major issues, which prevents the usage of wastes' potentials for bioenergy generation (Singh et al. 2012). It was reported that urban areas in India generated about 68.8 million tons of MSW during 2011, which will have a two-fold increase by 2030 (Annepu, 2012; PIB, 2016). Hence, the organic fractions of MSW need to be segregated and a suitable treatment option needs to be explored.

Although various thermal and biological methods exist for the treatment of solid wastes, thermal treatment technologies are less opted due to the pretreatment cost involved in removing the moisture, since almost all the organic fractions of municipal solid waste (OFMSW) are enriched with a high share of moisture. Hence, treatment through biological options would lead to an effective means of treatment and disposal (Velmurugan and Ramanujam, 2011). Anaerobic digestion (AD) and composting are the widely used biological treatment technologies for handling organic solid wastes. However, composting, although comparatively cheaper option, requires a large area and doesn't give a better output or yield from the process; rather, it gives out a nutrient rich fertiliser. However, the market value of the compost is less than biogas, which can be obtained through AD of organic wastes (Patil and Deshmukh, 2015). In addition, the digestate from the AD process can also be utilised as a liquid/solid fertiliser after sufficient treatments. Thus, the AD process could lead to the eco-friendly conversion of wastes, which therein aids in meeting the sustainable development goals of the United Nations, such as affordable and clean energy (SDG 7), sustainable cities and communities (SDG 11), and climate action (SDG 13) (Song et al. 2019; Li et al. 2016).

AD involves the biological degradation of organic matter by anaerobic bacteria in a series of processes: hydrolysis, acidogenesis, acetogenesis, and methanogenesis (Mir et al. 2016). Hydrolysis involves the breakdown of chemical bonds in the complex organic compounds into simple compounds and is usually a slow process for a solid substrate (Anukam et al. 2019; Van et al. 2020). Acidogenesis is a fermentation process in which the acidogenic bacteria convert the products from hydrolysis into fatty acids, alcohols, hydrogen, and carbon dioxide (Karuppiah and Azariah 2019). The next process is acetogenesis, where the acetogenic bacteria convert the volatile fatty acids and alcohols into acetate and hydrogen (Venkiteshwaran et al. 2015). The last stage is the methanogenesis, where the products from the former stage are converted into methane and carbon dioxide upon

DOI: 10.1201/9781003204435-3

the act of methanogenic bacteria. Thus, an alternative biofuel, methane, could be obtained from wastes, thereby providing dual benefits by serving as an alternative fuel, in addition to the effective and sustainable management of waste.

For the determination of a substrate's suitability for the AD process, biogas potential tests are considered as a viable method, since these tests provide information on the rate and quantity of substrate degradation in addition to its biogas potential (Ware and Power, 2017). The biogas production curve obtained from an AD test could also provide meaningful implications on substrate degradation. The kinetics of biogas production at the varying stages of hydrolysis, acidogeneiss, acetogenesis, and methanogenesis and the shape of the biogas production curves are primarily influenced by substrate characteristics, inhibitory compounds, and microbial community within the AD reactor (Labatut et al. 2011). These curves can be assessed using mathematical modelling of biogas production kinetics.

Mathematical modelling of kinetics of biogas production aids in determining the substrate behaviour in an AD system (Ware and Power, 2017). Further, the kinetic parameters predicted using these models could help to monitor the reactor performance under varying operating conditions (Chala et al. 2019; Pramanik et al. 2019). In a full scale AD system, the kinetic parameters thus obtained from these models could also aid in designing and optimising the reactor parameters while compared to the experimental biogas yields (Teng et al. 2014). Hence, this study focuses on the kinetic modelling of the biogas production, from various individual OFMSW, to predict the kinetic parameters by using first-order kinetic, modified Gompertz, and logistic models.

The biogas potential of OFMSW is an essential component for designing an energy efficient AD plant. Hence, the AD potential of various individual organic fractions of MSW, generated in a smart city, was tested in our previous study (Mozhiarasi et al. 2019b). However, these biogas potentials obtained for individual organic wastes in turn influence the energy, and exergy potentials of the waste. Hence, in this current study, the energy and exergy potentials of various individual OFMSW were projected. Also, biogas potential of the wastes were compared against the calculated biogas yields by following the methods of Baserga (1998), Weiland (2001) and ATV-DVWK-M 363 (2002). This comparison would help to identify biodegradability and inhibitions to determine unexploited energy and exergy potential of wastes, which in turn could help for further process improvements. Further, the effective validation of the models could substitute detailed biogas measurements in the future. Following different calculation methods applied, energy and exergy changes were presented. Finally, kinetic modelling of biogas production was done by following first-order kinetic, modified Gompertz and logistic models.

2.2 MATERIALS AND METHODS

2.2.1 Sampling and Experimental Study

This investigation was conducted at Chennai, one of the largest metropolitan cities in India, with population of about 10.7 million inhabitants. The various individual organic fractions of MSW, such as vegetable market wastes (VMW), fruit market wastes (FRW), banana peduncle (BP), flower market wastes (FLW), goat ruminal contents (GRC), fish market wastes (FMW), slaughterhouse wastes (SHW), chicken wastes (CHW), and canteen food wastes (CFW) were sampled from centralised wholesale markets, and the biogas potential for each were found from AD tests by following VDI 4630 (VDI-Guideline 4630, 2006) standard protocol (Mozhiarasi et al. 2019b). The SBYs of various OFMSW were measured using anaerobic batch reactors of 600 mL capacity. The pressure was measured every day using a manometer (EXTECH 407910), which was converted into normalised volume of biogas (mL) by standard temperature and pressure conversions. The SBY so obtained is presented as millilitre of biogas per gram of volatile solids (VS) of waste added (mL/g VS). All the samples were fed in the AD reactors and maintained at a mesophilic temperature of 37°C in triplicates. The food/microorganisms ratio of 0.5 was maintained with biogas monitoring period of

30 days. Once the reactors were fed with wastes, they were immediately flushed with nitrogen gas to create anaerobic environment.

2.2.2 Energy and Exergy Contents

The theoretical biogas yields were calculated using substrates' carbohydrate, protein, lipid, and fibre contents and, compared with the experimental biogas yield obtained by batch AD tests. The SBY approaches of Baserga, Weiland, and ATV-DVWK-M 363 were used. The digestibility contents of carbohydrate, protein, and lipids were estimated by following DLG (1997) fodder analysis database.

Further, the energy potentials of wastes were arrived based on the gross calorific value of methane as 39.8 MJ/m^3 by following Hahne (2010) and Mozhiarasi et al. (2019a). The amount of exergy was calculated by following Szargut and Egzergia (2007) by considering the standard chemical exergy of methane as $Ex_{ch}(CH4) = 831.2$ (kJ/mol) and expressed dry matter basis as MJ/kg of DM.

2.2.3 Kinetic Modelling

The widely used statistical models such as first-order kinetic model (1), modified Gompertz model (2), and logistic model (3) were used for kinetic modelling of biogas production (Abudi et al. 2016; Duan et al. 2018). Among these models, the first-order kinetic model is a simplest model, which is based on the assumption that the hydrolysis of a substrate is the rate-limiting step in an AD process. Also, this model does not consider the conditions for reaching maximum biological activity and reactor failures (Kafle and Chen 2016). This model has been used by several researchers to obtain hydrolysis kinetics of biogas production from batch AD tests (Li et al. 2018).

The modified Gompertz model has also been widely used for modelling the biogas production kinetics (Abudi et al. 2016; Salehiyoun et al. 2019). Initially, this model was developed for organ growth prediction and human mortality data (Gil et al. 2011). Later, this model was modified by Gibson et al. (1987) to describe cell density by means of lag phase period during bacterial growth and exponential growth rate. This lag phase data obtained from this model could be used to determine the minimum time required by the bacteria for biogas production after microbial acclimatisation (Ware and Power 2017). Also, this model helps to determine the specific growth rate of methanogenic bacteria and the higher the growth rate, ultimately, the higher the biogas production (Szlachta et al. 2018).

The logistic model is also among the widely used models for modelling the kinetics of biogas production (Ghatak and Mahanta, 2017; Pramanik et al. 2019). This model assumes that the rate of methane production is proportional to the quantity of methane production (Chala et al. 2019). This model outputs the lag phase delay, maximum methane production rate, and the predicted methane production potential of a substrate (Ali et al. 2018).

The first-order kinetic model was carried out in Microsoft Excel 2010, whereas the modified Gompertz and logistic models were carried out using IBM SPSS statistical software 20. The kinetic parameters of the models were detailed below in the following equations.

$$G(t) = Gmax\left(1 - \exp(-k.\ t)\right) \tag{2.1}$$

$$G(t) = Gmax.\exp\left\{-\exp(Rmax.\frac{e}{Gmax}(\lambda - t) + 1)\right\} \tag{2.2}$$

$$G(t) = \frac{Gmax}{\left\{1 + \exp(4.\frac{Rm\ (\lambda - t)}{Gmax}) + 2\right\}} \tag{2.3}$$

where $G(t)$ – cumulative methane production (mL/g VS), k – first-order model constant (1/day), t – day, e is Euler's number – exp(1) = 2.7183, R_m – maximum methane production rate (mL/g VS.d), λ – lag phase time (d), G_{max} – ultimate methane production potential (mL/g VS).

2.3 RESULTS AND DISCUSSION

2.3.1 Specific Biogas Yields of Various OFMSW

The Specific biogas yield of various OFMSW i.e., VMW, FRW, FLW, BP, GRC, SHW, FMW, CHW, and CFW are presented in Figure 2.1 as normalized volume (mL_N) of biogas/g VS of substrate added.

The maximum SBY was obtained for FMW as 514.2 mL_N/g VS in concurrence with its highest VS reduction of 61.1 wt.%. Also, FMW showed a steady rise in the biogas production in comparison with other substrates and reached its stead state in about 25 days (Figure 2.1). The second highest SBY was obtained for CHW of 480.3 mL_N/g VS with VS reduction of 53.8 wt.%.

Among the carbohydrate-rich substrates like vegetable, fruit, and flower market wastes, the highest SBY was obtained for FRW of 437.9 mL_N/g VS with VS reduction of 61.5 wt.%, which could be due to the comparatively higher shares of easily degradable matters, i.e., hemicelluloses in fruit wastes (Mozhiarasi et al. 2019b). For CFW, the SBY obtained was 435.1 mL_N/g VS with VS reduction of 59.7 wt.%. For VMW and FLW, the SBY obtained were 428.8 mL_N/g VS and 240.2 mL_N/g VS with VS reduction of 57.5 wt.% and 36.4 wt.% respectively. For SHW, the SBY obtained was 309.3 mL_N/g VS with VS reduction of 44.6 wt.%, and this lower yield could be due to the higher share of hardly degradable leafy and grassy residues in the ruminal contents (Jensen et al. 2016). The SBY of BP was found to be 331.6 mL/g VS with VS reduction of 50.0 wt.% and the lower yield could be attributed to the high fibre content of the banana peduncle wastes. For GRC, the SBY obtained was 277.5 mL_N/g VS with VS reduction 41.3 wt.%. This lower biogas yield of GRC could be attributed to the slow rise in the biogas production (Figure 2.1) during the initial weeks, and later on produced lesser biogas in comparison with other substrates, which could be due to the complex biodegradability of the ruminal shares, similar to SHW.

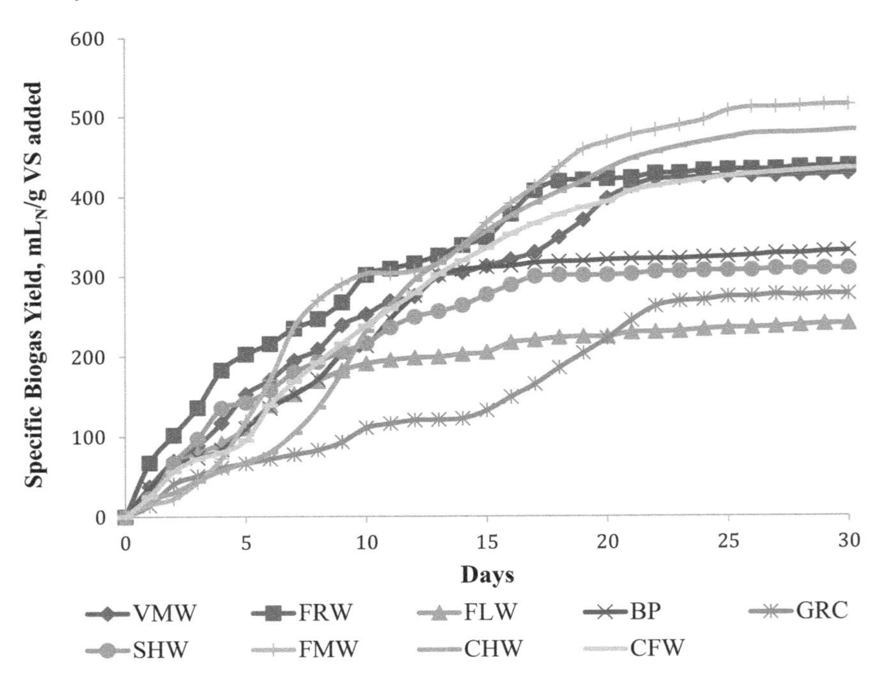

FIGURE 2.1 Specific biogas yields of OFMSW.

2.3.2 Comparison of Measured and Calculated Biogas Yields

The measured biogas yields by BMP tests and the calculated biogas yields obtained by following Baserga (1998), Weiland (2001), and ATV-DVWK-M 363 (2002) are shown in Table 2.1. The mean methane content of the biogas and VS reduction of substrates is also presented in Table 2.1.

From the measured and calculated SBY, it was observed that SBY obtained by Baserga (1998) showed variations in the range of 13.0–22.9% for FMW, CHW, CFW, FRW, and VMW. However, it showed greater differences in the range of 47.6–110.3% for FLW, BP, GRC and SHW, thereby showing strong fluctuations, which could be due to the poor degradability of substrates evident from the obtained low VS reductions. Similarly, calculation approaches by following Weiland (2001) showed differences in the biogas yields of about 8.9–24.2% for VMW, FMW, FMW, CHW, and CFW. For other substrates, such as FLW, BP, GRC, and SHW, this approach showed greater differences, similar to Baserga (1998), of about 54.2–118.4%. By following approaches of ATV-DVWK-M 363 (2002), the estimated biogas yield showed higher yields of about 24.3–39.8% compared to measured yields for VMW, FMW, FMW, CHW, and CFW. However, similar to the previous two approaches, the approach by ATV-DVWK-M 363 (2002) also showed greater differences between the biogas yields of about 71.4–144.0% for FLW, BP, GRC, and SHW. Hence, all three approaches showed higher calculated biogas yields than measured biogas yields, thereby showing the occurrence of strong inhibitions, which could be due to the poor C/N ratio, ammonia inhibition, etc. during mono-digestion of substrates (Yoruklu et al. 2017). For instance, wastes of animal origin such as FMW, GRC, SHW, and CHW are prone to be inhibited mainly by long chain fatty acids, poor biodegradability, and ammonia and sulphide inhibition (Palatsi et al. 2011; Nges et al. 2012; Zhu et al. 2019; Fuchs et al. 2018). The waste of animal manure like GRC is subject to be inhibited by poor biodegradability (Van et al. 2018; Zhang et al. 2014). Likewise, waste of plant origin like VMW, FRW, FLW, and BP is prone to be affected by volatile fatty acid inhibition and poor degradability of ligno-cellulosic matters (Edwiges et al. 2018; Scano et al. 2014; Alkanok et al. 2014; Jadhav et al. 2013). Hence, the greater fluctuation between the measured and calculated biogas yields reveals the need of a co-treatment and pretreatment system for increasing the biogas production from these wastes to cope with the calculated yields using the substrates' composition by following the calculation approaches by Baserga (1998), Weiland (2001), and ATV-DVWK-M 363 (2002).

2.3.3 Energy and Exergy Potentials

The energy and exergy potentials of VMW, FRW, FLW, BP, GRC, SHW, FMW, CHW, and CFW are presented in Table 2.2. Both the measured and calculated energy potentials using the calculation approaches for biogas yields determination by following Baserga (1998), Weiland (2001), and ATV-DVWK-M 363 (2002) approaches are presented in Table 2.2 and expressed on dry matter basis as MJ/kg DM.

The highest energy potential of 10.52 MJ/kg DM was obtained for FMW due to its highest biogas potential, however, the exergy potential remains at 9.89 MJ/kg DM; therefore only 94.0% of the obtained energy is available as exergy. The second highest energy potential was obtained for CHW of about 9.73 MJ/kg DM, out of which 93.9% of the energy content is available as exergy. The energy potentials of 8.76, 8.61, 8.51, 5.44, 5.28, 4.88 and 4.58 MJ/kg DM were obtained for FRW, CFW, VMW, GRC, BP, SHW and FLW respectively. Similarly the exergy potentials remain as 8.23, 8.09, 7.99, 5.11, 4.96, 4.59 and 4.30 MJ/kg DM for FRW, CFW, VMW, GRC, BP, SHW and FLW respectively. The energy potentials obtained from calculation approaches ranged between 7.71–12.95 MJ/kg DM for various wastes. Hence, these differences showed that there is a possibility of increasing the measured biogas yields by addressing the process inhibitions through anaerobic co-digestion with different substrates, thereby balancing the nutrients for stable reactor performances, which in turn increases the biogas yields.

TABLE 2.1
Measured and Calculated Biogas Yields

Substrate	Specific Biogas Yield (SBY) (mL biogas/g VS)				Specific Methane Yield (SMY) (mL methane/g VS)				Mean Methane Content vol.%	VS Reduction wt.%
	BMP Test	Baserga, 1998	Weiland, 2001	ATV-DVWK-M 363, 2002	BMP Test[a]	Baserga, 1998	Weiland, 2001	ATV-DVWK-M 363, 2002		
VMW	428.8	527.0	532.6	599.5	240.0	293.0	304.2	331.1	56.0	57.5
FRW	437.9	520.4	533.4	598.1	239.7	282.5	298.6	322.8	54.7	61.5
FLW	240.2	505.1	524.5	586.1	128.8	268.0	288.3	309.6	53.6	36.4
BP	331.7	489.6	511.5	568.4	175.5	256.4	277.9	295.7	52.9	50.0
GRC	277.5	540.9	547.3	620.4	155.8	301.2	313.5	344.6	56.2	41.3
SHW	309.3	530.0	540.1	607.3	171.0	290.8	305.1	331.3	55.3	44.6
FMW	514.2	581.3	560.3	639.5	309.2	348.0	341.8	380.4	60.1	61.1
CHW	480.3	551.0	558.7	628.7	267.5	304.7	317.7	345.4	55.7	53.8
CFW	435.1	511.8	536.5	598.1	229.7	266.9	290.9	311.0	52.8	59.7

[a] Calculated based on measured SBY from BMP tests and average methane contents of the used calculation approaches.

TABLE 2.2
Energy and Exergy Potentials of OFMSW

	Energy Content (MJ/kg DM)				Exergy Content (MJ/kg DM)			
Substrate	**BMP Test**	**Baserga, 1998**	**Weiland, 2001**	**ATV-DVWK-M 363, 2002**	**BMP Test**	**Baserga, 1998**	**Weiland, 2001**	**ATV-DVWK-M 363, 2002**
VMW	8.51	10.39	10.78	11.73	7.99	9.76	10.13	11.02
FRW	8.76	10.33	10.92	11.80	8.23	9.70	10.25	11.08
FLW	4.58	9.53	10.25	11.01	4.30	8.95	9.63	10.34
BP	5.28	7.71	8.36	8.89	4.96	7.24	7.85	8.35
GRC	5.44	10.52	10.95	12.04	5.11	9.88	10.29	11.31
SHW	4.88	8.30	8.71	9.46	4.59	7.80	8.18	8.88
FMW	10.52	11.85	11.64	12.95	9.89	11.13	10.93	12.16
CHW	9.73	11.08	11.55	12.56	9.14	10.40	10.85	11.80
CFW	8.61	10.01	10.91	11.66	8.09	9.40	10.25	10.96

2.3.4 Kinetic Modelling of Biogas Potentials

The kinetic modelling results obtained from the first-order kinetic model, modified Gompertz model, and logistic model are presented in Table 2.3. The first-order kinetic model parameters presented in Table 2.3 include first-order disintegration rate constant (k), the statistical indicator R^2, and the predicted SBY values. The kinetic parameters of modified Gompertz and logistic models include lag phase (L), maximum methane production rate (R_m), statistical indicator R^2, and the predicted SBY.

TABLE 2.3
Summary of Kinetic Modelling of Biogas Production from VMW, FRW, FLW, BP, and CFW

Parameter		**Units**	**VMW**	**FRW**	**FLW**	**BP**	**CFW**
First order kinetic model							
First-order disintegration rate constant, k		1/d	0.157	0.172	0.159	0.165	0.139
R^2			0.829	0.906	0.890	0.950	0.8738
Methane yield (30 days)	Measured	mL/g VS	428.78	437.88	240.16	331.65	435.13
	Predicted	mL/g VS	424.96	435.33	238.14	329.30	428.43
	Difference	%	0.89	0.58	0.84	0.71	1.54
Modified Gompertz model							
Lag phase (L)		D	0.406	1.417	0.996	0.913	1.212
Maximum methane production rate (R_m)		mL/g VS.d	24.658	27.843	18.744	27.462	27.942
R^2			0.984	0.981	0.985	0.989	0.997
Methane yield (30 days)	Measured	mL/g VS	428.78	437.88	240.16	331.65	435.13
	Predicted	mL/g VS	418.85	432.69	239.25	330.36	427.43
	Difference	%	2.32	1.18	0.38	0.39	1.77
Logistic model							
Lag phase (L)		D	0.664	1.912	1.332	1.205	1.343
Maximum methane production rate (R_m)		mL/g VS.d	22.409	24.733	16.981	26.657	26.428
R^2			0.978	0.975	0.968	0.993	0.994
Methane yield (30 days)	Measured	mL/g VS	428.78	437.88	240.16	331.65	435.13
	Predicted	mL/g VS	423.64	435.50	239.91	331.42	432.11
	Difference	%	1.20	0.54	0.10	0.07	0.69

TABLE 2.4
Summary of Kinetic Modelling of Biogas Production from GRC, SHW, FMW, and CHW

Parameter		Units	GRC	SHW	FMW	CHW
First-order kinetic model						
First-order disintegration rate constant, k		1/d	0.127	0.198	0.151	0.140
R^2			0.665	0.913	0.755	0.8289
Methane yield (30 days)	Measured	mL/g VS	277.48	309.29	514.23	480.29
	Predicted	mL/g VS	271.28	308.47	508.72	474.97
	Difference	%	2.23	0.27	1.07	1.11
Modified Gompertz model						
Lag phase (L)		D	1.889	0.564	1.17	3.571
Maximum methane production rate (R_m)		mL/g VS.d	13.47	23.895	32.19	34.881
R^2			0.939	0.988	0.981	0.997
Methane yield (30 days)	Measured	mL/g VS	277.48	309.29	514.23	480.29
	Predicted	mL/g VS	259.61	307.92	503.98	475.08
	Difference	%	6.44	0.44	1.99	1.08
Logistic model						
Lag phase (L)		D	2.766	0.834	1.096	3.802
Maximum methane production rate (R_m)		mL/g VS.d	13.42	21.543	29.614	33.543
R^2			0.962	0.979	0.969	0.996
Methane yield (30 days)	Measured	mL/g VS	277.48	309.29	514.23	480.29
	Predicted	mL/g VS	267.31	308.87	509.40	479.87
	Difference	%	3.67	0.14	0.94	0.09

From the results of the first-order kinetic model (Table 2.4), it was observed that first-order disintegration constant (k) value ranged 0.127–0.198 for different fractions of OFMSW. The difference between predicted and measured SBY ranged 0.27–2.23%, thereby showing minimum differences. The statistical parameter, R^2, which represents the soundness of the statistical model, ranged 0.665–0.950, thereby showing less accuracy and suitability of the model in prediction.

From the prediction results of logistic model, it was observed that the lag phase varied 0.7–2.8 days for various waste fractions. Further, the maximum methane production rate (R_m) ranged 13.42–33.54 mL/gVS.d. The statistical indicator, R^2 value ranged 0.962–0.996, thereby showing the best fitness of the model. The difference between the measured and predicted SBY ranged 0.07–3.67%.

From the results obtained from the modified Gompertz model, it was found that the lag phase varied 0.4–3.6 days for various OFMSW. The R_m value ranged 13.47 – 34.88 mL/gVS.d. The R^2 value ranged 0.939–0.997, thereby showing the soundness of the model by showing the best fitness similar to the logistic model; this can be confirmed with the least differences between the predicted and measured SBY in the range of 0.38–2.32%.

Further, from the biogas production curves (Figure 2.2), it could be visualised that the curves obtained from the first-order kinetic model showed greater fluctuations during the initial weeks of the incubation period, although this tends to coincide with the experimental biogas yield during the stationary period. However, this could be due to the assumption of this model, as hydrolysis is considered as a rate-limiting step since it may fail for a substrate with low lipid content that could have faster hydrolysis rate (Ware and Power, 2017). This could also be noticed from the curves of FLW and BP, where the first-order kinetic curve coincides with the experimental curves better than the result of the substrates, which could be attributed to the fibrous richness of both wastes, where often hydrolysis will be the rate-limiting step. However, it could be seen that the curves obtained

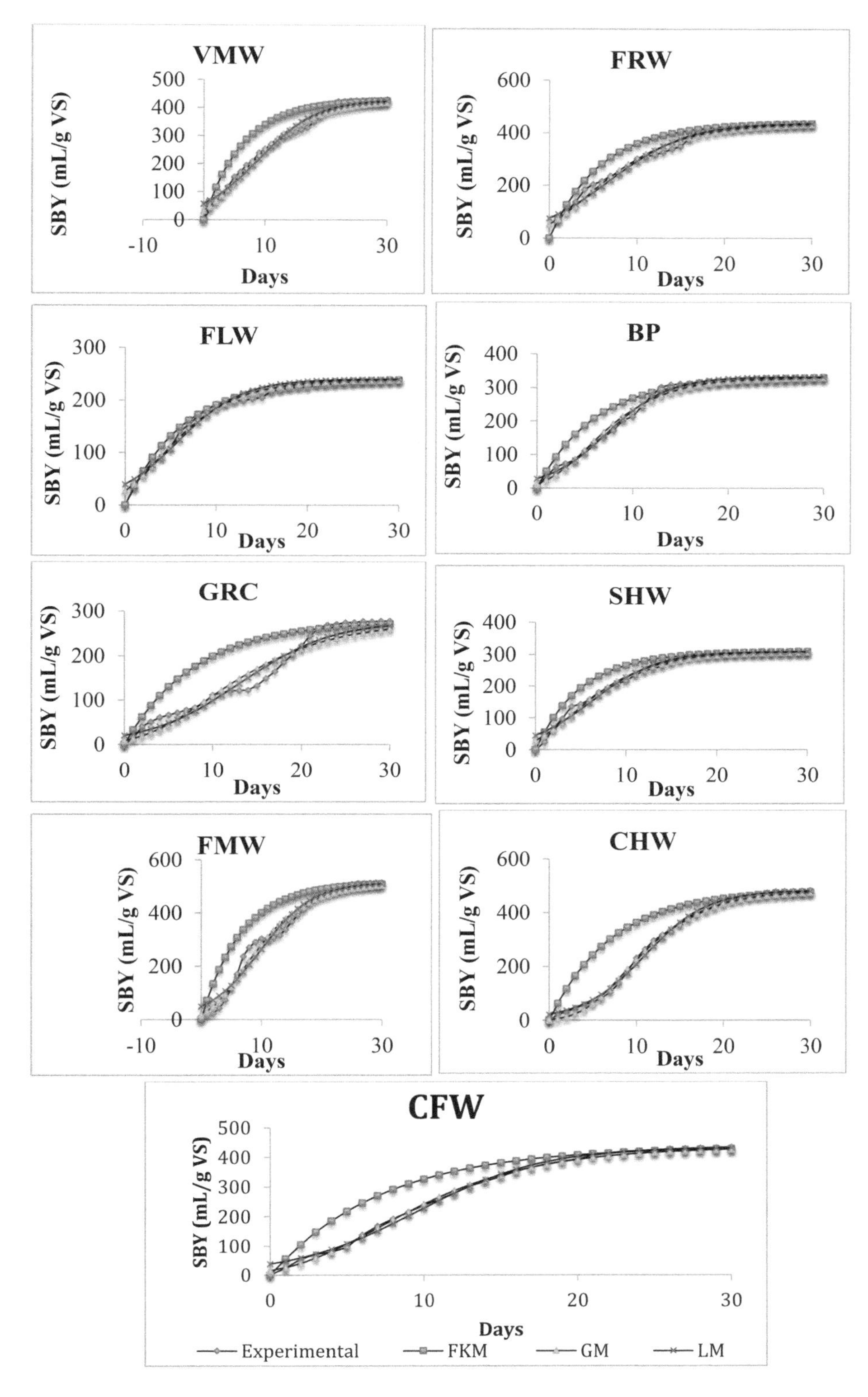

FIGURE 2.2 Comparison between experimental and predicted biogas yields.

from both the modified Gompertz and logistic models coincide well with the experimental biogas profile, during both the incubation and stationary periods.

Hence, while considering the low deviations between the measured and predicted SBY (0.38–2.32%), the R^2 value (0.94–0.99) that represents the accuracy of a model prediction and SBY profiles (Figure 2.2), it could be concluded that the modified Gompertz model reveals its accuracy and best fitness compared to the other two kinetic models for modelling the kinetics of biogas production from organic fractions of MSW.

2.4 CONCLUSION

In this study, the energy and exergy potentials of various bulk generators of organic fractions of MSW were determined. The assessment of energy potentials showed varying energy potentials for various organic fractions of MSW in the range of 4.88–10.52 MJ/kg VS, with the highest being obtained for FMW. However, only 94.0% of the available energy of FMW reveals exergy, which could be improved by reducing process inhibitions. Further, the experimental biogas potentials of various individual OFMSW were compared with the calculated biogas yield obtained by various calculation approaches. The results showed greater differences (13–144%) between the calculated and measured biogas yields while applying calculation approaches. The higher deviations show the presence of process inhibitions during AD of mono-digestion of wastes. This provides a demand for the co-substrate addition and pretreatment for overcoming the process inhibitions. Further kinetic modelling of biogas production was carried out by following first-order kinetic, modified Gompertz and logistic models. The comparison of the statistical indicator, R^2 value and the difference between the measured and predicted yields obtained by three models showed that the modified Gompertz model showed best fit compared with other models.

ACKNOWLEDGEMENTS

This work was supported by the German Federal Ministry of Education and Research and the Indian Department of Science and Technology under the Indo-German Science and Technology Centre (Grant number 01DQ15007A) under the 2+2 Project "RESERVES – Resource and energy reliability by co-digestion of veg-market and slaughterhouse waste". The authors also thank the director of the Central Leather Research Institute for permitting to execute the research work. The authors also acknowledge both Indian (Ramky Enviro Engineers Ltd., India) and German Industrial (Lehmann-UMT GmbH, Germany) partners of this project.

NOMENCLATURE

AD = anaerobic digestion
BP = banana peduncle
CFW = canteen food wastes
CHW = chicken wastes
FLW = flower market wastes
FMW = fish market wastes
FRW = fruit market wastes
G_{max} = ultimate methane production potential
$G(t)$ = cumulative methane production
GRC = goat ruminal contents
k = first-order model constant
λ = lag phase time
mL_N = normalised millilitre, volume of biogas converted to standard temperature and pressure
MSW = municipal solid wastes

$\boldsymbol{R_m}$ = maximum methane production rate
SBY = specific biogas yield
SHW = slaughterhouse wastes
VMW = vegetable market wastes
VS = volatile solids

REFERENCES

Abudi, Z. N., Hu, Z., Sun, N., Xiao, B., Rajaa, N., Liu, C., Guo, D. (2016). Batch anaerobic co-digestion of OFMSW (organic fraction of municipal solid waste), TWAS (thickened waste activated sludge) and RS (rice straw): influence of TWAS and RS pretreatment and mixing ratio. Energy, 107, 131–140. https://doi.org/10.1016/j.energy.2016.03.141.

Ali, M. M., Bilai, B., Dia, N., Youm, I., Ndongo, M. (2018). Modeling the kinetics of methane production from slaughter-house waste and Salvinia Molesta: Batch digester operating at ambient temperature. Energy and Power, 8(3), 61–70. https://doi.org/10.5923/j.ep.20180803.01.

Alkanok, G., Demirel, B., Onay, T. T. (2014). Determination of biogas generation potential as a renewable energy source from supermarket wastes. Waste Management, 34(1):134–140. https://doi.org/10.1016/j.wasman.2013.09.015.

Annepu, R. (2012). Sustainable Solid Waste Management in India. Columbia University, Earth Engineering Center. http://www.seas.columbia.edu/earth/wtert/sofos/Sustainable%20Solid%20Waste%20Management%20in%20India_Final.pdf

Anukam, A., Mohammadi, A., Naqvi, M., Granström, K. (2019). A review of the chemistry of anaerobic digestion: Methods of accelerating and optimizing process efficiency. Processes, 7(8), 1–19. https://doi.org/10.3390/pr7080504.

ATV-DVWK-M363. (2002). Herkunft, Aufbereitung und Verwertung von Biogas. ATV-DVKW, Hennef (ISBN: 3936514119).

Baserga, U. (1998). Landwirtschaftliche Co-Vergärungs-Biogasanlagen, Biogas aus organischen Reststoffen und Energiegras. FAT, Nr. 512.Central Pollution Control Board (CPCB). (2013). Status report on municipal solid waste management. Retrieved from: http://www.cpcb.nic.in/divisionsofheadoffice/pcp/MSW_Report.pdfhttp://pratham.org/images/paper_on_ragpickers.pdf

Chala, B., Oechsner, H., Müller, J. (2019). Introducing temperature as variable parameter into kinetic models for anaerobic fermentation of coffee husk, pulp and mucilage. Applied Sciences, 9(3), 1–15. https://doi.org/10.3390/app9030412.

DLG. (1997). DLG-Futterwerttabellen Wiederkäuer, 7th edn. DLG e.V. DLG-Verlag, Frankfurt.

Duan, N., Ran, X., Li, R., Kougias, P., Zhang, Y., Lin, C., Liu, H. (2018). Performance evaluation of mesophilic anaerobic digestion of chicken manure with algal digestate. Energies, 11(7), 1–11. https://doi.org/10.3390/en11071829.

Edwiges, T., Frare, L. M., Alino, J. H. L., Triolo, J. M., Flotats, X., Costa, M. (2018). Methane potential of fruit and vegetable waste: an evaluation of the semi-continuous anaerobic mono-digestion. Environmental Technology, 1–25. https://doi.org/10.1080/09593330.2018.1515262.

Fuchs, W., Wang, X., Gabauer, W., Ortner, M., Li, Z. (2018). Tackling ammonia inhibition for efficient biogas production from chicken manure: status and technical trends in Europe and China. Renewable and Sustainable Energy Reviews, 97, 186–199. https://doi.org/10.1016/j.rser.2018.08.038.

Ghatak, M. D., Mahanta, P. (2017). Kinetic model development for biogas production from cattle dung. In International Conference on Functional Materials, Characterization, Solid State Physics, Power, Thermal and Combustion Energy. AIP Conference Proceedings 1859, 020010 (pp. 1–7). https://doi.org/10.1063/1.4990163.

Gibson, A. M., Bratchell, N., Roberts, T. A. (1987). The effect of sodium chloride and temperature on the rate and extent of growth of Clostridium botulinum type A in pasteurized pork slurry. Journal of Applied Bacteriology, 62(6), 479–490. https://doi.org/10.1111/j.1365-2672.1987.tb02680.x.

Gil, M. M., Miller, F. A., Brandão, T. R. S., Silva, Cristina, L. M. S. (2011). On the use of the Gompertz model to predict microbial thermal inactivation under isothermal and non-isothermal conditions. Food Engineering Reviews, 3(1), 17–25. https://doi.org/10.1007/s12393-010-9032-2.

Hahne, E. (2010). Technische Thermodynamik: Einführung und Anwendung. Oldenbourg, Boston (ISBN: 3486592319).

Jadhav, A. R., Chitanand, M. P., Shete, H. G. (2013). Flower waste degradation using microbial consortium. IOSR Journal of Agriculture and Veterinary Science, 3(5), 1–4. https://doi.org/10.9790/2380-0350104.

Jensen, P. D., Mehta, C. M., Carney, C., Batstone, D. J. (2016). Recovery of energy and nutrient resources from cattle paunch waste using temperature phased anaerobic digestion. Waste Management, 51, 72–80. https://doi.org/10.1016/j.wasman.2016.02.039.

Kafle, G. K., Chen, L. (2016). Comparison on batch anaerobic digestion of five different livestock manures and prediction of biochemical methane potential (BMP) using different statistical models. Waste Management, 48, 492–502. https://doi. org/10.1016/j.wasman.2015.10.021.

Karuppiah, T., Azariah, E. V. (2019). Biomass pretreatment for enhancement of biogas production. In Anaerobic Digestion. IntechOpen. https://doi.org/10.5772/intechopen.82088.

Kumar, S., Smith, S. R., Fowler, G., Velis, C., Kumar, S. J., Arya, S., Rena, Kumar, R., Cheeseman, C. (2017). Challenges and opportunities associated with waste management in India. Royal Society Open Science, 4 (160764), 1–11. http://dx.doi.org/10.1098/rsos.160764.

Labatut, R. A., Angenent, L. T., Scott, N. R. (2011). Biochemical methane potential and biodegradability of complex organic substrates. Bioresource Technology, 102, 2255–2264, http://dx.doi.org/10.1016/j.biortech.2010.10.035.

Li, P., Li, W., Sun, M., Xu, X., Zhang, B., Sun, Y. (2018). Evaluation of biochemical methane potential and kinetics on the anaerobic digestion of vegetable crop residues. Energies, 12(1), 26. https://doi.org/10.3390/en12010026.

Li, Y., Jin, Y., Li, J., Li, H., Yu, Z. (2016). Effects of thermal pretreatment on the biomethane yield and hydrolysis rate of kitchen waste. Applied Energy, 172, 47–58. https://doi.org/10.1016/j.apenergy.2016.03.080.

Mir, M. A., Hussain, A., Verma, C. (2016). Design considerations and operational performance of anaerobic digester: a review. Cogent Engineering, 3(1), 1–20. https://doi.org/10.1080/23311916.2016.1181696.

Mozhiarasi, V., Speier, C. J., Benish Rose, P. M., Mondal, M. M., Pragadeesh, S., Weichgrebe, D., Srinivasan, S. V. (2019a). Variations in generation of vegetable, fruit and flower market waste and effects on biogas production, exergy and energy contents. Journal of Material Cycles and Waste Management, 3, 1–16. https://doi.org/10.1007/s10163-019-00828-2.

Mozhiarasi, V., Speier, C. J., Michealammal, B. R. P., Shrivastava, R., Rajan, B., Weichgrebe, D., Venkatachalam, S. S. (2019b). Bio-reserves inventory-improving substrate management for anaerobic waste treatment in a fast-growing Indian urban city, Chennai. Environmental Science and Pollution Research, 1–17. https://doi.org/10.1007/s11356-019-07321-1.

Nges, I.A., Mbatia, B., Björnsson, L. (2012). Improved utilization of fish waste by anaerobic digestion following omega-3 fatty acids extraction. Journal of Environmental Management, 110, 159–165. https://doi.org/10.1016/j.jenvman.2012.06.011.

Palatsi, J., Viñas, M., Guivernau, M., Fernandez, B., Flotats, X. (2011). Anaerobic digestion of slaughterhouse waste: main process limitations and microbial community interactions. Bioresource Technology, 102(3), 2219–2227. https://doi.org/10.1016/j.biortech.2010.09.121.

Patil, V. S., Deshmukh, H. V. (2015). Anaerobic digestion of vegetable waste for biogas generation: a review. International Research Journal of Environmental Sciences, 4(6), 80–83.

PIB. (2016). Solid waste management rules revised after 16 years; rules now extend to urban and industrial areas: Javadekar. April 05, 2016. Retrieved from: Press Information Bureau (PIB), Ministry of Environment and Forests, India. http://pib.nic.in/newsite/ PrintRelease.aspx?relid=138591

Pramanik, S. K., Suja, F. B., Porhemmat, M., Pramanik, B. K. (2019). Performance and kinetic model of a single-stage anaerobic digestion system operated at different successive operating stages for the treatment of food waste. Processes, 7(9), 600. https://doi.org/10.3390/pr7090600.

Salehiyoun, A. R., Sharifi, M., Di Maria, F., Zilouei, H., Aghbashlo, M. (2019). Effect of substituting organic fraction of municipal solid waste with fruit and vegetable wastes on anaerobic digestion. Journal of Material Cycles and Waste Management. https://doi.org/10.1007/s10163-019-00887-5.

Scano, E.A., Asquer, C., Pistis, A., Ortu, L., Demontis, V., Cocco, D. (2014). Biogas from anaerobic digestion of fruit and vegetable wastes: experimental results on pilot-scale and preliminary performance evaluation of a full-scale power plant. Energy Conversion and Management, 77, 22–30. https://doi.org/10.1016/j.enconman.2013.09.004

Singh, A., Kuila, A., Adak, S., Bishai, M., Banerjee, R. (2012). Utilization of vegetable wastes for bioenergy generation. Agricultural Research, 1(3), 213–222.

Song, L., Li, D., Fang, H., Cao, X., Liu, R., Niu, Q., Li, Y. Y. (2019). Revealing the correlation of biomethane generation, DOM fluorescence, and microbial community in the mesophilic co-digestion of chicken manure and sheep manure at different mixture ratio. Environmental Science and Pollution Research, 26(19), 19411–19424. https://doi. org/10.1007/s11356-019-05175-1

Speier, C. J., Mondal, M. M., Weichgrebe, D. (2018). Data reliability of solid waste analysis in Asia's newly industrialised countries. International Journal of Environment and Waste Management, 22(1/2/3/4), 124–146. https://doi.org/10.1504/IJEWM.2018.094101

Szargut, J., Egzergia, P. (2007). Oradnikobliczania I stosowania, Widawnictwo Politechniki Shlaskej, Gliwice 2007. Substance. State. Molecular mass. Enthalpy of devaluation. Standard Chemical Exergy.

Szlachta, J., Prask, H., Fugol, M., Luberański, A. (2018). Effect of mechanical pre-treatment of the agricultural substrates on yield of biogas and kinetics of anaerobic digestion. Sustainability, 10(10), 1–16. https://doi.org/10.3390/su10103669

Teng, Z., Hua, J., Wang, C., Lu, X. (2014). Design and optimization principles of biogas reactors in large scale applications. Reactor and Process Design in Sustainable Energy Technology, 99–134. https://doi.org/10.1016/b978-0-444-59566-9.00004-1

Van, D. P., Fujiwara, T., Tho, B. L., Toan, P. P. S., Minh, G. H. (2020). A review of anaerobic digestion systems for biodegradable waste: Configurations, operating parameters, and current trends. Environmental Engineering Research, 25(1), 1–17. https://doi.org/10.4491/eer.2018.334

Van, P., Minh, H. G., Phu, P. S. T., Fujiwara, T. (2018). A new kinetic model for biogas production from co-digestion by batch mode. Global Journal of Environmental Science Management, 4(3), 251–262. https://doi.org/10.22034/ gjesm.2018.03.001

VDI-Guideline 4630 (2006). Fermentation of organic materials. Characterization of substrate, sampling, collection of material data, fermentation tests. Beuth Verlag GmbH, Germany.

Velmurugan, B., Ramanujam, A. (2011). Anaerobic digestion of vegetable wastes for biogas production in a fed-batch reactor. International Journal of Emerging Sciences, 1(3), 478–486. ISSN: 2222-4254

Venkiteshwaran, K., Bocher, B., Maki, J., Zitomer, D. (2015). Relating anaerobic digestion microbial community and process function: supplementary issue: Water microbiology. Microbiology Insights, 8(s2), 37–44. https://doi.org/10.4137/mbi.s33593

Ware, A., Power, N. (2017). Modeling methane production kinetics of complex poultry slaughterhouse wastes using sigmoidal growth functions. Renewable Energy, 104, 50–59. https://doi.org/10.1016/j.renene.2016.11.045

Weiland P (2001). Grundlagen der Methangärung – Biologie und Substrate. In: Biogas als regenerative Energie, Stand und Perspektiven, VDI-Bericht 1620. VDI Verlag, Düsseldorf.

Yoruklu, H. C., Korkmaz, E., Demir, N. M., Ozkaya, B., Demir, A. (2017). The impact of pretreatment and inoculum to substrate ratio on methane potential of organic wastes from various origins. Journal of Material Cycles Waste Management, 20(2), 800–809. https://doi. org/10.1007/s10163-017-0641-1

Zhang, C, Su, H., Baeyens, J., Tan, T. (2014). Reviewing the anaerobic digestion of food waste for biogas production. Renewable and Sustainable Energy Reviews, 38, 383–392. https://doi.org/10.1016/j.rser.2014.05.038

Zhu, J., Wu, S., Shen, J. (2019). Anaerobic co-digestion of poultry litter and wheat straw affected by solids composition, free ammonia and carbon/nitrogen ratio. Journal of Environmental Science and Health, Part A, 54(3), 1–7. https://doi.org/10.1080/10934529.2018.1546494

3 Biomass Supply Chain for Anaerobic Methane Generation

Issues and Concern

Dipal Baruah, Moonmoon Hiloidhari, and Debendra Chandra Baruah

3.1 WORLD ENERGY CRISIS AND EMERGENCE OF BIOMASS AS AN ENERGY SOURCE

The industrial development omnipresent in the world is dependent upon fossil fuels as the primary source of energy. The use of fossil fuels faces the problem of rapidly depleting reserves and its adverse impact on the environment resulting in climate change. This has led to the quest for renewable energy sources, which lessen the burden on the environment with relatively lower greenhouse gas (GHG) emissions (Deng et al., 2012). Also, the dependency on fossil fuels, which has a significant contribution to total energy cost, has inherent variations in cost, resulting in economic and political challenges (Huang et al., 2010). An adequate supply of energy is the prime requirement for the growth and development of human society. There is an evident linkage between availability vis-à-vis consumption of energy and development. The developmental disparities among regions are attributed to many factors, including availability and affordability of energy. Energy poverty is a very serious issue for many Asian, African, and Latin American countries, and it has drawn global attention. Among the Asian countries, China and India's energy concerns are taking centre stage because of their current status and future aspiration for development. Further, with the increase in population and the rise in income and infrastructure developments, most affluent countries' energy demands are rising continuously. Thus, it is imperative to supply and generate additional energy to meet the overall human development goal.

It is reported that the world's primary energy demand will rise by 35% between 2010 and 2035, or on an average of 1.2% annually (IEA, 2012). There have been many studies to predict future demand. The predictions of one such study are presented in Table 3.1. Fossil fuels will continue to be the dominant source of global energy through 2035. It is also evident that conventional fuels are going to be diminished very soon. By the mid-22nd century, only coal will be available as fossil energy. Aside from this energy crisis, global warming attributed to fossil fuel burning is also a prominent threat to humanity's very survival.

In this aspect, bioenergy has been identified as a promising renewable energy option due to its widespread availability, low GHG emission, and low price. This is evident from the fact that biomass is the fourth largest renewable electricity source (8% of total renewable energy generation) globally after hydropower (63%), wind (18%), and solar (9%). In 2018, 523 TWh of biomass-based electricity was generated globally, 70% of which was from solid biomass sources like wood chips, wood pellets, and bagasse; 17% from biogas; and 12% from renewable municipal and industrial waste (IRENA, 2020). Figure 3.1 depicts the increase in biomass-based electricity generation from different sources during 2000–2017. Biomass represents a growing renewable energy source with high growth potential due to its wide availability worldwide (Li et al., 2017). As evident

DOI: 10.1201/9781003204435-4

TABLE 3.1
World Energy Demand under Three Different Scenarios (Mtoe) (IEA, 2012)

	Actual		New Policies[b]		Current Policies[c]		450 Scenarios[d]	
Energy Source	**2000**	**2010**	**2020**	**2035**	**2020**	**2035**	**2020**	**2035**
Coal	2378	3474	4082	4218	4417	5523	3569	2337
Oil	3659	4113	4457	4656	4542	5053	4282	3682
Gas	2073	2740	3266	4106	3341	4380	3078	3293
Nuclear	676	719	898	1138	886	1019	939	1556
Hydro	226	295	388	488	377	460	401	539
Bioenergy[a]	1027	1277	1532	1881	1504	1741	1568	2235
Other renewables	60	112	299	710	265	501	340	1151
Total	10,099	12,730	14,922	17,197	15,332	18,677	14,177	14,793

[a] Includes both traditional and modern uses.

[b] New policies: Existing policies are maintained, and recently announced commitments and plans, including those yet to be formally adopted, are implemented cautiously.

[c] Current policies: Government policies that had been enacted or adopted by mid-2012 continue unchanged.

[d] Four hundred fifty scenarios: Policies are adopted that put the world on a pathway that is consistent with having around a 50% chance of limiting the global increase in average temperature to 2°C in the long term, compared with pre-industrial levels.

from Table 3.1, generation and demand for renewable energy, including biomass energy, will continue to rise. The common biomass feedstock includes sugarcane bagasse, rice husk, rice straw, cotton stalk, coconut shells, soya husk, coffee waste, jute wastes, groundnut shells, and forest-based biomass, including sawdust. A variety of biomasses, including woody biomass and loose biomass such as rice husk, cashew nutshells, areca nuts, and sugarcane residue have been tested for bioenergy generation in India. The modern uses of biomass take advantage of modern

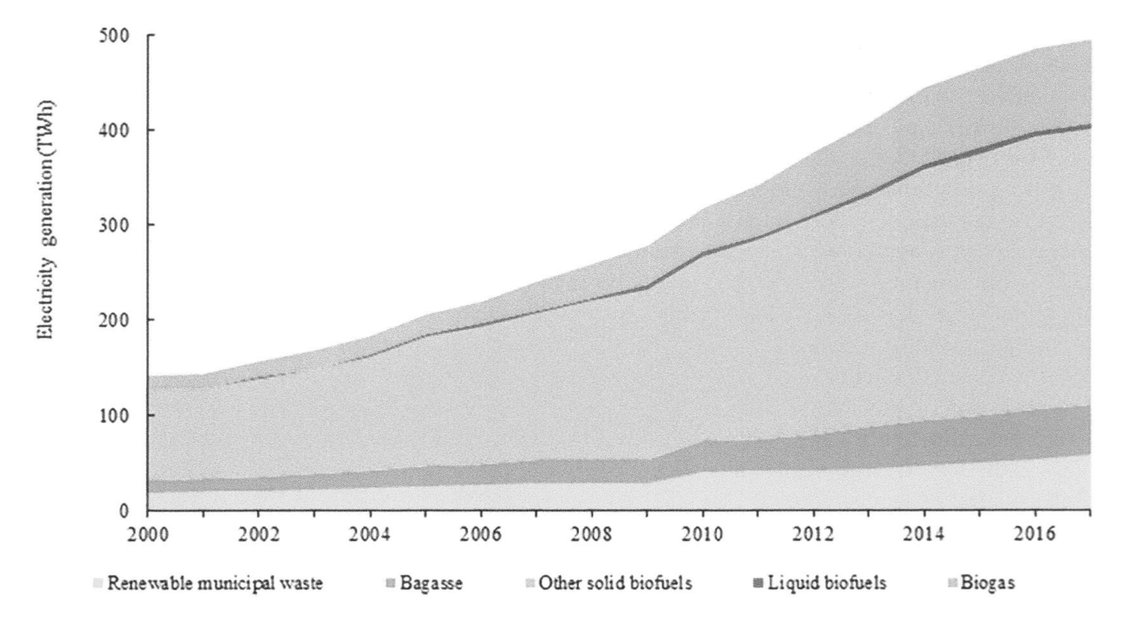

FIGURE 3.1 Yearly, biomass-based electricity generation from different sources worldwide. (Representation based on data of IRENA, 2020.)

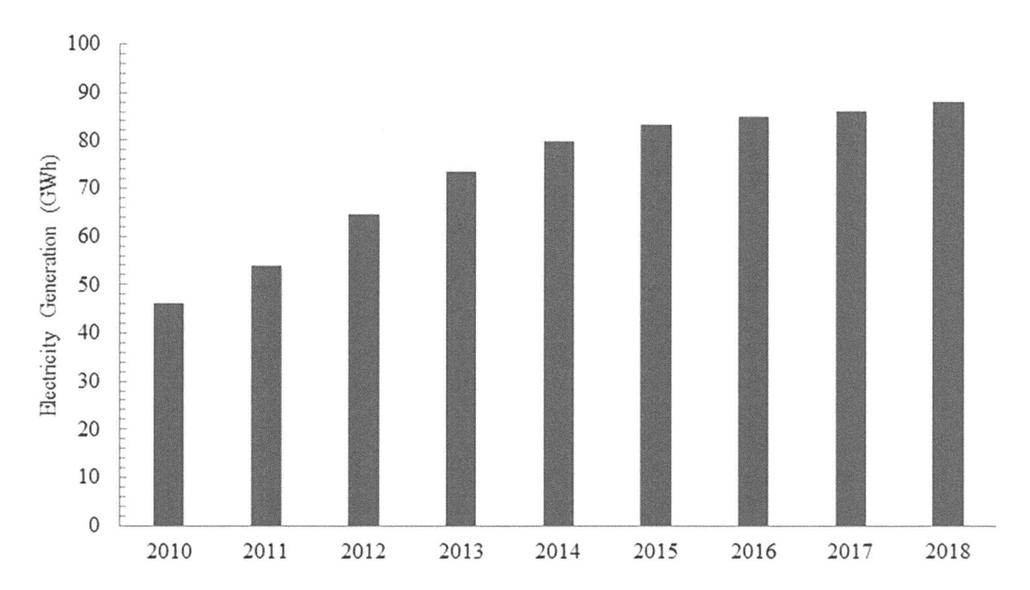

FIGURE 3.2 Biogas-based electricity generation in the world.

biomass conversion technologies such as combustion, pyrolysis, gasification, fermentation, anaerobic digestion (AD) for production of heat and electricity, liquid and gaseous transportation fuel, and biogas for cooking. The use of agricultural residues as a feedstock of biomass energy has been gaining popularity in many countries. Agricultural residues could be a significant source of biomass energy in agriculturally dominant countries like India and China. However, owing to the distribution over large geographical areas, there is inherent heterogeneity in the composition and thermo-chemical properties of different biomass. There is also spatial and temporal variability in the availability of biomass, and its bulkiness and inconvenient form hinder its convenient utilisation. In comparison to fossil fuels, most forms of biomass have low energy density. For example, sub-bituminous coal's energy content is around 20–25 GJ/t (low heat values), whereas for air-dried woody biomass, it is around 12–15 GJ/t (Singh, 2013). Consequently, handling, storage, and transportation of biomass in its raw form become more costly than conventional fuels. Thus, for fruitful utilisation of biomass, it becomes imperative to improve its properties, elevating its handling, storage, and transportation ability. One such method is to convert solid biomass into liquid or gaseous fuels, which can be achieved one of the two ways, biochemical or thermochemical conversion. Within the biochemical route, AD is a significant technology that is being extensively used to generate energy using various substrates, such as sewage sludge, food waste, forestry resource, living stock manure, and wastewater.

It is worth mentioning that biogas-based electricity generation has grown from 46,108 GWh in 2010 to 87,500 GWh in 2016 (IRENA, 2020). Figure 3.2 depicts the growth of this sector, which has grown nearly 90% in eight years.

3.2 AGRICULTURAL BIOMASS AS A SOURCE OF ENERGY

Successful research and development (R&D) has been demonstrated in many parts of the world concerning the use of agricultural residues as a potential feedstock for biomass energy. Matsumura et al. (2005) reported the use of agricultural residue in Japan as an energy resource. Rice straw and rice husk are the two main agricultural residues in Japan. Rice residue could provide 0.47% of Japan's total electricity demand. However, the cost of electricity production from rice straw is double the current cost of electricity. Nevertheless, with the improvement in conversion technology and introducing cost incentives, rice residue-based power generation could be an attractive option in

Japan, and GHG emission reduction achieved through this process can be counted under the Kyoto Protocol. Zeng et al. (2007) reviewed popular, common, and successfully commercialised available technologies in China, such as direct combustion, biogas, straw gasification, and straw briquetting for straw-based biomass energy, including improved cook stove, biogas, straw gasification, and straw briquette. Jenkins et al., 1998 presented a comprehensive discussion on the properties (composition, proximate, ultimate, elemental, heating value, alkali index) of both woody and loose biomass feedstocks relevant to combustion. Biomass fuel, by leaching with water, leads to improvements in ash fusion temperatures.

Physico-chemical properties of some biomass fuels were presented by Zhang and Zhang (1999) and investigated rice straw suitability for gasification with an anaerobic-phased solids digester system. Ammonia is used as a supplemental nitrogen source for rice straw digestion. It is found that a combination of grinding (10-mm length), heating (110°C), and ammonia treatment (2%) resulted in the highest biogas yield, which is 17.5% higher than the biogas yield of untreated whole straw. The pretreatment temperature is critical and has a significant effect on the digestibility of straw. Gadde et al., 2009 investigated rice straw availability as a power-generation source, GHG emission due to its current uses, and GHG saving potential if the straw is utilised for power generation. India, Thailand, and the Philippines produce 97.19, 21.86, and 10.68 Mt of rice straw residue per annum, respectively. In India, 23% of rice straw residue produced remains as surplus or unused. Punjab, Haryana, and Uttar Pradesh are the three major rice straw producing states in India. About 48% and 95% of rice straw residue are open-field burnt in Thailand and the Philippines, respectively. The GHG emissions due to open-field burning of rice straw in India, Thailand, and the Philippines are 0.05%, 0.18%, and 0.56%, respectively. The GHG emission mitigation potential from rice straw–based power would be 0.75%, 1.81%, and 4.31% in India, Thailand, and the Philippines, respectively, compared to the total country GHG emissions. Kim and Dale, 2004 estimated the global annual potential bioethanol production from major crops such as corn, barley, oat, rice, wheat, sorghum, and sugar cane. Overall, total potential bioethanol production from crop residues and wasted crops is 491 GL yr^{-1} (Giga Liters per year), about 16 times higher than the current world ethanol production. The potential bioethanol production could replace 353 GL of gasoline (32% of the global gasoline consumption) when bioethanol is used in E85 fuel for a midsize passenger vehicle. Asia is the largest potential producer of bioethanol from crop residues and wasted crops and could produce up to 291 GL yr^{-1} bioethanol.

Rice straw, wheat straw, and corn stover are the most favourable bioethanol feedstocks in Asia. Globally, rice straw can produce 205 GL of bioethanol annually, which is the largest amount from a single biomass feedstock (Kim and Dale, 2004). Lim et al., (2012) reviewed the key factors of utilising rice husk and rice straw as renewable energy sources. These factors are (i) physical and chemical characteristics that influence the quality of rice biomasses, (ii) various chemical and physical pretreatment techniques that can facilitate handling and transportation of rice straw and husk, and (iii) the state-of-the-art thermo-chemical and bio-chemical technologies to convert rice husk and rice straw into energy. Binod et al. (2010) reviewed the currently available technologies for bioethanol production from rice straw. Bioethanol produced from rice straw can be used as a transportation fuel. Rice straw is abundantly available and is an attractive lignocellulosic material for bioethanol production. Suramaythangkoor and Gheewala (2010) reported that rice straw is a potential source for heat and power generation in Thailand. Although the cost of rice straw for power generation is not competitive with coal, it is comparable with other biomasses. Delivand et al. (2011) evaluated the economic feasibility of rice straw–based combustion projects of various capacities (ranging from 5 MWe to 20 MWe) in Thailand. For an assumed lifespan of 20 years, rice straw–fueled combustion power plants would generate net present values (NPV) of 3.15, 0.94, 2.96, 9.33, and 18.79 million USD for projected 5, 8, 10, 15, and 20 MWe plants, respectively. Furthermore, examining the effects of scale on the cost of generated electricity (COE) over the considered range of capacities indicates that COE varies from 0.0676 USD kWh^{-1} at 20 MWe to 0.0899 USD kWh^{-1} at 5 MWe. Nevertheless, to ensure a secure fuel supply, smaller-scale power plants, such as 8 and 10 MWe, may be more practicable. Hassan et al. (2014) demonstrated electricity generation from rice straw without pretreatment in a two-chambered microbial fuel cell (MFC) inoculated with a mixed culture of cellulose-degrading bacteria (CDB). Their work demonstrated that electricity could be produced from rice straw by exploiting CDB as the biocatalyst. This method provides a promising way to utilise rice straw for bioenergy production. Mussoline et al. (2014) used untreated rice straw in combination with piggery wastewater in a farm-scale biogas system to generate electricity. The authors recommended an overall straw (dry wt.) to wastewater ratio (wet wt.) of 1 to 1.4 to improve gas production and decrease the acclimation period. They also recommended improvements such as continuous leachate recirculation, a more efficient heat exchange system to maintain mesophilic conditions year-round, and periodic addition of fresh wastewater and sludge acclimated to lignocellulosic material to achieve a more sustainable and profitable system.

Hu et al., 2013 investigated the diffusion of rice straw cofiring systems in the Taiwanese power market. They developed a linear complementarily model to simulate the power market equilibrium with cofiring systems in Taiwan. The GIS (Geographical Information System)-based analysis is also used to analyse the geospatial relationships between rice farms and power plants to assess potential biomass for straw power generation. Ranjan et al. (2013) studied the feasibility of using rice straw as a substrate for biobutanol production. Mussoline et al. (2013) reviewed the AD of rice straw. Removal of rice straw from rice fields can reduce GHG emission significantly, as rice fields are regarded as a major source of methane emission. Through the AD process, methane yields from rice straw range from 92 to 280 l/kg of volatile solids. Operating conditions such as pH (6.5–8.0), temperature (35–40°C), and nutrients (C/N ratio of 25–35) are important for optimum digestion of rice straw. Furthermore, pretreatment (i.e., fungi, acid, and alkali solutions) and microbial engineering can increase biogas production.

3.3 ANAEROBIC DIGESTION OF BIOMASS

AD has been accepted as a sustainable and cost-effective technology for energy recovery from waste in the form of biofuel resulting in minimisation of waste (Kothari et al., 2014). As a clean energy source, biogas can substitute fossil fuels (Sreekrishnan et al., 2004). Another added advantage of AD is the generation of nutrient-rich digestate, which can be used as a fertiliser in crop cultivation. Digestate-based fertilisers help improve crop yield and maintain soil health.

AD involves the conversion of biodegradable organic feedstock, such as cattle dung, agro-residue, gardens, kitchens, industry, poultry droppings, night soil, and municipal waste into biogas within a closed chamber (reactor) in a limited supply of oxygen and in the presence of some specific types of micro-organisms. Biogas is a mixture of gases, methane (CH_4) and carbon dioxide (CO_2), traces of hydrogen sulphide (H_2S), nitrogen, ammonia, and moisture. The average calorific value of biogas is about 5000 kcal/m^3. Biogas contains about 55–65% of CH_4, 35–44% of CO_2 and traces of other gases.

The overall process of AD is subdivided into hydrolysis, acidogenic fermentation, hydrogen-producing acetogenesis, and methanogenesis (Figure 3.3). In the hydrolysis process, the extracellular hydrolytic enzyme of acidogens decomposes complex organic substances that bacteria cannot directly utilise into soluble monomers. During acidification, soluble monomers are converted into terminal products (volatile fatty acid, accompanied by cellular material generation) by acidogenic fermentation bacteria. In the next step, hydrolytic products are metabolised into acetic acid with the hydrogen and carbohydrate yield in the presence of hydrogen-producing acetogens. CO_2/H_2 are utilised as substrates to form acetic acid by a small fraction of the homo-acetogenic bacteria. In the methanogenesis process, anaerobic methanogens convert the acidification products (acetic acid, formic acid, CO_2/H_2). This process is represented using Eqs. (3.1)–(3.5) in Figure 3.3. Various mesophilic bacterial species acting in synergy accomplishes the methanogenesis process.

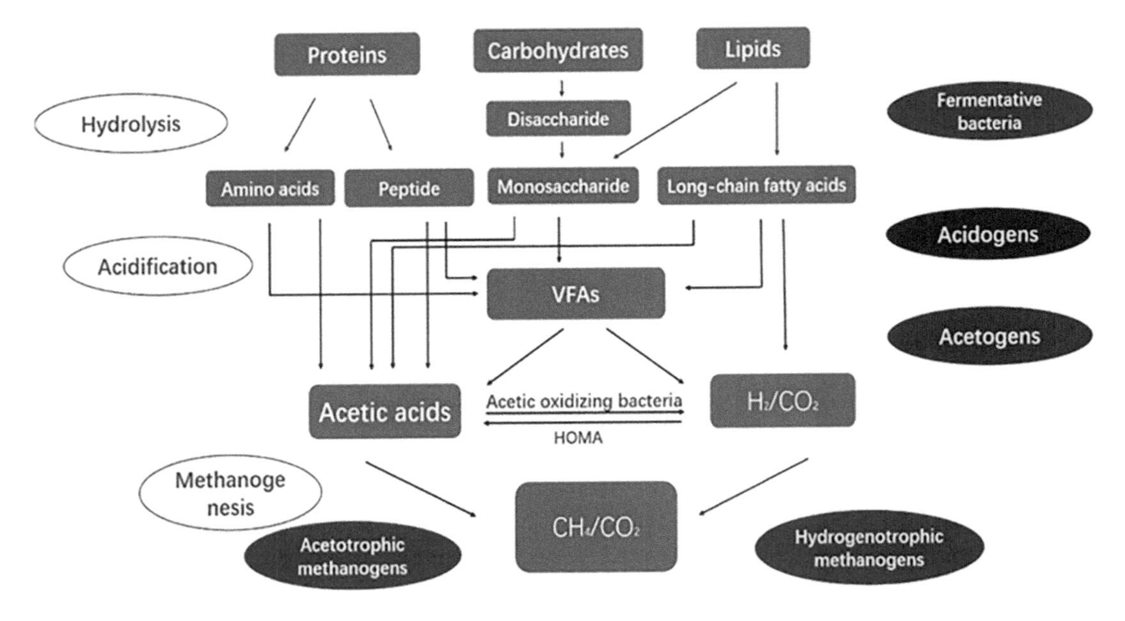

FIGURE 3.3 Flow chart of anaerobic digestion. (Adapted from Li et al., 2019.)

$$CH_3COO^- + H_2O \rightarrow CH_4 + HCO_3^- \tag{3.1}$$

$$HCO_3^- + H^+ \rightarrow CH_4 + 3H_2O \tag{3.2}$$

$$4CH_3OH \rightarrow CO_2 + 2H_2O \tag{3.3}$$

$$4HCOO^- + 2H^+ \rightarrow CH_4 + CO_2 + 2HCO_3^- \tag{3.4}$$

$$4H_2 + CO_2 \rightarrow CH_4 + 2H_2O \tag{3.5}$$

Raw biogas (without any purification) can be used as a cooking fuel like liquified petroleum gas (LPG) or used in for lighting gas lamps after minor treatment. Biogas can be combusted in boilers to generate heat, internal or external combustion engines to produce electricity, combined heat and power (CHP) plants to produce both heat and electricity, and trigeneration systems to provide cooling via absorption chillers, in addition to heat and electricity. Biogas has been used to substitute diesel up to 80% in diesel engines. Also, 100% replacement of diesel can be achieved by using 100% biogas engines. Purified biogas can be upgraded up to 98% purity of CH_4 content. This purified and upgraded biogas can be used as a clean and green fuel for transportation or bottled in cylinders at high pressure (around 250 bar) and called compressed biogas (CBG). CBG can be used as vehicle fuel in gas-powered vehicles; in place of natural gas in industrial, commercial, and domestic uses; or pumped into gas grids to substitute natural gas supplied to households and businesses. The CO_2 fraction of biogas may be extracted for commercial use, such as feedstock in greenhouses, or it can be reconverted into fuels. There is also a possibility of converting biogas into higher-value products such as bio-plastics or bio-chemicals.

Impurities in biogas include water, hydrogen sulphide, nitrogen, oxygen, ammonia, siloxanes, and particles (Masebinu et al., 2014). The impurities cause corrosion and degrade biogas's energy quality, therefore they have to be removed. Biomethanation is a process of enriching CH_4 by removing CO_2 and other impurities from biogas, making it suitable for use as equivalent to natural gas (Lorenzi et al., 2019).

3.4 PURIFICATION AND UPGRADING OF BIOGAS

Adsorption, absorption, membrane separation, and cryogenic separation are commercially available biogas purification and upgrading technologies.

3.4.1 Absorption

The separation of CO_2 by absorption is based on the solubility of various gas components in a liquid scrubbing solution (Persson et al., 2006). When water is used as a solvent for absorption, it is known as water scrubbing (Bauer et al., 2013). An organic solvent solution, like polyethylene glycol, is also used to remove CO_2 (Nie et al., 2013). Chemical absorption, such as amine scrubbing, is used for CO_2 removal (Bauer et al., 2013).

3.4.2 Adsorption

Adsorption is based on the adsorption of various gas components of biogas on a solid surface under high pressure. Pressure swing absorption (PSA) is a standard method (Grande, 2012). Adsorption of CO_2 on a solid surface and its removal by changing pressure (high to low and low to high) is known as pressure swing adsorption. CO_2 is separated by adsorption at high pressure, and the resultant gas is rich in CH_4 (Masebinu et al., 2014).

3.4.3 Membrane Separation

The process is based on the principle that it is permeable for CO_2, water, NH_3, O_2, and N_2 and not permeable to CH_4. Typical membranes for biogas upgrading are made of polysulfone, polyimide, or polydimethylsiloxane (Molino et al., 2013; Sridhar et al., 2007). The membrane also removes acid gases and water vapour to improve gas quality.

3.4.4 Cryogenic Separation

Cryogenic separation uses the difference of temperature to separate different gas species (Awe et al., 2017). CO_2 is separated from CH_4 at low temperature with continuous sublimation (gas to solid) and desublimation of CO_2 (Jonsson and Westman, 2011).

3.5 ENVIRONMENTAL PROFILE OF BIOMETHANE TECHNOLOGIES

The environmental performance of biogas production or upgrading technologies can be assessed through life cycle assessment (LCA). The LCA is a tool used to assess and compare the environmental impacts of products or services throughout their entire life cycle, from the extraction of material, transport, production, use, and end of life treatment from raw materials extraction to production to consumption processes (European Commission, 2010; Lyng and Brekke, 2019). A number of LCA studies have been done on biogas production systems worldwide, generating useful information for policymakers and researchers to improve efficiency and decrease GHG emissions and other environmental impacts (Hijazi et al., 2016).

Some LCA works on biomethane purification, and upgrading technologies are presented in Table 3.2.

3.6 APPLICATION AND STATUS OF BIOGAS TECHNOLOGIES IN INDIA

Micro digesters have been reported to be playing a crucial role in developing countries, especially in rural areas. In these areas, micro digesters have been found to be an integral part of farming, waste management, and energy security (WBA, 2019). Around 50 million micro-scale digesters are in

TABLE 3.2
Selected LCA Works on Biomethane Purification and Upgrading Technologies

Author	Biogas Upgrading Technology	Results and Conclusions
Starr et al. (2012)	HPWS, AwR, and BABIU	AwR upgrading technology has the highest environmental impacts in all impact categories except GWP due to high energy demand for alkaline reactants production. The GWP is negative for AwR and BABIU, but positive for HPWS. The BABIU technology is the most environmentally friendly when compared with HPWS, AwR, BABIU, OPS, PSA, MS and CSe.
Florio et al. (2019)	MS, CSe, PSA, CS, HPWS	Membrane upgrading technology is slightly more environmentally friendly than the others. Using biogas digestate as organic fertiliser can further enhance the environmental benefit.
Castellani et al. (2018)	Hydrate-based biogas upgrading	The carbon and energy footprints of biosynthetic methane are estimated to be 1.875 kg CO_2 eq Nm^{-3} and 40.57 MJ Nm^{-3}. Conventional grid-energy consumption for the upgrading process contributes about 70% of the carbon footprint. Using solar photovoltaic-based electrical energy for methane upgrading can reduce the carbon footprint and energy footprint to 0.7081 kg CO_2 eq Nm^{-3} and 28.55 MJ Nm^{-3}.
Lorenzi et al. (2019)	SOECs and HPWS	The source of electricity for upgrading and carbon content of electricity is the critical factor in determining the environmental burden of biogas upgrading.
Ferreira et al. (2019)	Photosynthetic algae biogas upgrading units	Biomass dewatering and drying account for the highest energy consumption and CO_2 emissions, followed by energy consumption for agitation of algal bacterial broth at high-rate algal pond.
Navas-Anguita et al. (2019)	Fischer–Tropsch synthesis and dry reforming process	DRM biodiesel has high GWP, AP, and EP than fossil diesel. ODP and CED_{nr} of DRM biodiesel are lower than fossil diesel. Although DRM biodiesel is not yet a competitive alternative from an environmental perspective, reducing methane leakage and heat demand at the biogas-production phase and reducing NH_3 and NO emissions in the biogas-to-liquid plant can improve the environmental performance.
Lombardi and Francini (2020)	HPWS, AS, PCS, PSA and MP	AS technology has the lowest GWP due to low methane leakage and minimal electricity consumption. If only upgrading part is considered, AS technology results in 13%, 1%, 23%, and 3% less GWP than HPWS, PCS, PSA and MP, respectively. Environmental performance of the upgrading technologies is strongly influenced by the amount and type of electricity consumption.
Ardolino et al. (2018)	MS	Biomethane as a transportation fuel can reduce GHGs emissions and non-renewable energy consumptions.
Leonzio (2016)	CA	The environmental impact of the process using a KOH aqueous solution is lower, followed by chemical absorption with MEA and NaOH aqueous solution.
Collet et al. (2017)	P2G	Electricity consumption is a significant contributor to GHG emissions in P2G technologies. Continuous P2G process generates more GHGs than direct injection. Intermittent P2G operation using renewable electricity can reduce GHG emissions substantially. Reducing electricity consumption in the electrolysis process and integrating renewable credits from CO_2 valorisation can make P2G competitive.

(Continued)

TABLE 3.2 *(Continued)*
Selected LCA Works on Biomethane Purification and Upgrading Technologies

Author	Biogas Upgrading Technology	Results and Conclusions
Zhang et al. (2017)	P2G	Type of electricity and sources of CO_2 influence the environmental performance of P2G. Electricity from fossil fuels for the P2G process causes higher GHG emissions than conventional technologies.

Abbreviations: AP: Acidification Potential; AS: Amine Scrubbing; AWR: Alkaline with Regeneration; BABIU: Bottom Ash Upgrading; CA: Chemical Adsorption; CEDnr: Cumulative Non-Renewable Energy Demand; CS: Chemical Scrubbing; CSe: Cryogenic Separation; DRM biodiesel: Synthetic biodiesel; EP: Eutrophication Potential; GWP: Global Warming Potential; HPWS: High Pressure Water Scrubbing; MEA: Mono Ethanol Amine; MP: Membrane Permeation; MS: Membrane Separation; ODP: Ozone Depletion Potential; OPS: Organic Physical Scrubbing; P2G: Power to Gas; PCS: Potassium Carbonate Scrubbing; PSA: Pressure Swing Adsorption; SOEC: Oxide Electrolyser Cells.

operation worldwide (IRENA, 2020), with 5 million in operation in India (MNRE, 2020). In the rest of Asia, Africa, and South America, around 700,000 biogas plants have been installed (SNV, 2020).

In India, the Ministry of New and Renewable Energy (MNRE) promotes the installation of biogas plants by implementing two schemes under off-grid/distributed and decentralised renewable power through the New National Biogas and Organic Manure Programme for biogas plant size ranging from 1 m^3 to 25 m^3 per day and Biogas Power Generation (Off-grid) and Thermal energy application Programme, for setting up biogas plants in size range of 30–2500 m^3 per day, for corresponding power generation capacity range of 3–250 kW from biogas or raw biogas for thermal energy/cooling applications (MNRE, 2020). The Ministry's earlier schemes were the National Project on Biogas Development, National Biogas and Manure Management Programme (NBMMP). Based on these schemes and an estimated initial potential of 12.3 million biogas plants, a cumulative total of 50.28 Lakh family/small-sized biogas plants had been set up in the country by 2018–2019. The estimated potential and cumulative total numbers of family size/small biogas plants installed in different States and Union Territories of India as of 2019 was made by MNRE in 2020 (MNRE, 2020), estimating the potential number of biogas plants at 12,339,300 and cumulative achievement on number of biogas plants up until 2018–2019 as of March 31, 2019, is 5,028,340, with a big gap from the estimated number.

Biogas from micro-scale digesters (mostly used in stoves for cooking or heating) has contributed significantly to displacing high-emissions solid fuels like firewood and charcoal. Around 10 million biogas stoves were used in India, with a total production of nearly 2×10^6 m^3, in the year 2016 (Murdock et al., 2019). In India, MNRE has approved different models of biogas plants (available from 1 M m^3 per day capacities) for fixed dome design and floating Gasholder KVIC design-type biogas plants (Table 3.3).

Biogas-based electricity generation is an established technology and adopted by many countries worldwide to supplement their electricity requirements. The process involves a CHP engine with provision for recovery and use of some of the waste heat. A CHP engine can be operated by linking it to biogas-generating anaerobic reactor. However, for the economic operation of a CHP engine, there is a minimum size requirement. There has been research on parametric influence on biogas plants' efficiency while trying to maximise efficiency and income, specifically by increasing the utilisation of heat. Another developing area is tri-generation, which aims at generating electricity, heat, and cooling when required. In respect of scale digester–based electricity generation, 10.5 GW was the installed capacity in Europe with 17,783 plants in 2017. Within Europe, Germany is the leader with 10,971 installed plants, followed by Italy (1,655), France (742), Switzerland (632), and

TABLE 3.3
MNRE Approved Models of Biogas Plants in India

S No.	Biogas Plant Models	Specifications/Ministry's Letter Nos. and Date of Approval.
1	Fixed dome/gasholder biogas plants • Deenbandhu fixed dome model with brick masonry construction. • Deenbandhu Ferro-cement model with in -situ technique. • Prefabricated RCC fixed dome model. • Solid-state Deenbandhu fixed dome.	• Ministry's letter No. 13-10/96-BG dated January 10, 2002. • Code of Practices (Second Revision, IS 9478: 1989 of the BIS, New Delhi. • Ministry's letter No. 13-11/99-BG dated March 5, 1999 • Ministry's O.M No. 13-5/2016-BG (NBMMP) dated November 7, 2016
2.	Floating dome design biogas plants • KVIC floating steel metal gasholder with brick masonry digester • KVIC floating gasholder–type plant with Ferro-cement digester. • KVIC design biogas plant with fibreglass reinforced plastic (FRP) and digester. • Pragati model biogas plants. • KVIC design type digester with floating gasholder made up of HDPE/PVC/FRP/RCC etc. material plant	• Code of Practices (Second Revision), IS 9478:1989 of the BIS, New Delhi. • Code of practice IS-12986:1990 of BIS, New Delhi. Specifications of FRP Gasholder should be as per IS-12986:1990 • Code of Practices (Second Revision), IS 9478:1989 of the BIS, New Delhi. • MNRE O.M No. 18-1/2014-BE(NBMMP) dated November 26, 2014
3.	Prefabricated model biogas plants • Prefabricated reinforced cement concrete (RCC) digester with KVIC floating drum/gasholder.	• Ministry's O.M. NO. 18-1/2014 BE (NBMMP), dated November 26, 2014
4.	Bag-type biogas plants. (Flexi model)	• Ministry's letter No. 7-39/89-BG dated July 14, 1995

the United Kingdom (613) (EBA, 2018). In the United States, around 2,200 anaerobic digesters operate with an installed capacity of 977 MW (WBA, 2018). Canada has around 196 MW installed capacity with nearly 180 digesters (Biogas Association, 2020).

In India, MNRE promotes decentralised biogas energy–based power generation, specifically in the small capacity range (3 kW to 250 kW) and thermal energy for heating and cooling from the biogas produced from biogas plants of 30–2500 m^3 size. Under the Biogas Power (Off-grid) Programme (BPP)/Biogas-based Power Generation and Thermal Application Programme (BPGPT) schemes implemented by MNRE, the focus has been on setting up medium-sized biogas plants for electricity generation and/or using biogas for various thermal and cooling applications. By 2018–2019, a total of 389 power projects were commissioned in India. The State/Union Territory achievements under India's biogas power program up to 2018–2019 are summarised in Table 3.4.

3.7 CONCLUSION

Biogas is an increasingly popular renewable energy source. Biogas can be generated from a wide range of biomass feedstocks, such as crop residues, livestock manure, forest residues, household and municipal organic waste, and industrial wastewater. The physico-chemical properties, heating value, pretreatment process, and conversion technology vary with different biomass sources. Biogas mainly consists of methane (50–75%) and carbon dioxide (25–50%), and trace amounts of hydrogen sulphides, hydrogen, ammonia, water and other gases such as oxygen and nitrogen. Biogas is not only renewable and sustainable, it also causes a reduction of GHG emissions and is a source of

TABLE 3.4
State Cumulative Achievements under Biogas Power (Off-Grid) Programme in India through 2018–2019

		Installed		
S. No	State	Nos.	Capacity (m³)	Capacity (kW)
1	Andhra Pradesh	34	4320	481
2	Gujrat	2	285	30
3	Haryana	3	2540	155
4	Karnataka	70	15670	1581.5
5	Maharashtra	68	11690	1257.5
6	Punjab	41	9980	1035
7	Rajasthan	2	120	15
8	Tamil Nadu	52	30360	2853.5
9	Uttarakhand	17	1070	124
10	Uttar Pradesh	30	4400	591
11	Madhya Pradesh	6	735	70
12	Kerala	36	1010	118
13	West Bengal	1	340	60
14	Odisha	2	60	10
15	Telangana	25	5410	574
	Total	389	87990	8951.5

organic fertiliser. Biogas can be upgraded to biomethane, a methane-rich energy source, for applications equivalent to fossil natural gas. However, CO_2 and other impurities have to be removed from raw biogas for its use as a vehicular fuel or as natural gas. Cleaning, purification, and upgrading of biogas are essential to utilise it as modern industrial and vehicular fuel. Several biogas upgrading technologies are available, with each having its own merits and demerits. Technologies that are currently matured and available at commercial markets include adsorption, absorption, membrane separation, and cryogenic separation. The performance, efficiency, energy, and material input/output flows for these upgrading technologies vary, and, consequently, so do their environmental impact in terms of global warming potential and other environmental indicators like eutrophication, acidification, ozone layer depletion, and energy consumption. LCA enables the understanding of biogas upgrading technologies' environmental performance and helps to identify emissions hotspots for better management of the upgrading processes. The available LCA studies of biogas upgrading technologies have noted that the consumption of electricity in the biogas upgrading phase and leakage of methane in the biogas production phase are the two main contributors to the overall environmental impact categories. Future developments to increase the environmental performance of biomethane production and upgrading includes: (i) increasing the share of zero- or low-carbon renewable energy as an electricity source at the upgrading phase, (ii) reducing methane leakage from the biogas production, storage, and pipeline transfer, (iii) proper management of digestate (e.g., as organic fertiliser), and (iv) minimising nitrogen fertiliser use for biomass feedstock production and cultivation.

REFERENCES

Ardolino, F., Parrillo, F. and Arena, U., 2018. Biowaste-to-biomethane or biowaste-to-energy? An LCA study on anaerobic digestion of organic waste. Journal of Cleaner Production, 174, pp. 462–476.

Awe, O.W., Zhao, Y., Nzihou, A., Minh, D.P. and Lyczko, N., 2017. A review of biogas utilisation, purification and upgrading technologies. Waste and Biomass Valorisation, 8(2), pp. 267–283.

Bauer, F., Persson, T., Hulteberg, C. and Tamm, D., 2013. Biogas upgrading–technology overview, comparison and perspectives for the future. Biofuels, Bioproducts and Biorefining, 7(5), pp. 499–511.

Binod, P., Sindhu, R., Singhania, R.R., Vikram, S., Devi, L., Nagalakshmi, S., Kurien, N., Sukumaran, R.K. and Pandey, A., 2010. Bioethanol production from rice straw: an overview. Bioresource Technology, 101(13), pp. 4767–4774.

Biogas Association, 2020. Canadian Biogas Association. Biogas Projects in Canada. Retrieved November 21, 2020, from https://biogasassociation.ca/index.php/about_biogas/projects_canada

Castellani, B., Rinaldi, S., Bonamente, E., Nicolini, A., Rossi, F. and Cotana, F., 2018. Carbon and energy footprint of the hydrate-based biogas upgrading process integrated with CO_2 valorisation. Science of the Total Environment, 615, pp. 404–411.

Collet, P., Flottes, E., Favre, A., Raynal, L., Pierre, H., Capela, S. and Peregrina, C., 2017. Techno-economic and Life Cycle Assessment of methane production via biogas upgrading and power to gas technology. Applied Energy, 192, pp. 282–295.

Delivand, M.K., Barz, M., Gheewala, S.H. and Sajjakulnukit, B., 2011. Economic feasibility assessment of rice straw utilisation for electricity-generating through combustion in Thailand. Applied Energy, 88, pp. 651–3658.

Deng, Y.Y., Blok, K. and van der Leun, K., 2012. Transition to a fully sustainable global energy system. Energy Strategy Reviews, 1(2), pp. 109–121.

EBA, 2018. EBA Statistical Report, 2017. European Biogas Association. Annual Report.

European Commission, 2010. Joint Research Centre—Institute for Environment and Sustainability. ILCD Handbook. International Reference Life Cycle Data System. Framework and Requirements for Life Cycle Impact Assessment Models and Indicators.

Ferreira, A.F., Toledo-Cervantes, A., de Godos, I., Gouveia, L. and Munõz, R., 2019. Life cycle assessment of pilot and real scale photosynthetic biogas upgrading units. Algal Research, 44, p. 101668.

Florio, C., Fiorentino, G., Corcelli, F., Ulgiati, S., Dumontet, S., Güsewell, J. and Eltrop, L., 2019. A life cycle assessment of biomethane production from waste feedstock through different upgrading technologies. Energies, 12(4), p. 718.

Gadde, B., Menke, C. and Wassmann, R., 2009. Rice straw as a renewable energy source in India, Thailand, and the Philippines: Overall potential and limitations for energy contribution and greenhouse gas mitigation. Biomass and Bioenergy, 33(11), pp. 1532–1546.

Grande, C.A., 2012. Advances in pressure swing adsorption for gas separation. International Scholarly Research Notices, 2012.

Hassan, S.H., El-Rab, S.M.G., Rahimnejad, M., Ghasemi, M., Joo, J.H., Sik-Ok, Y., Kim, I.S. and Oh, S.E., 2014. Electricity generation from rice straw using a microbial fuel cell. International Journal of Hydrogen Energy, 39(17), pp. 9490–9496.

Hijazi, O., Munro, S., Zerhusen, B. and Effenberger, M., 2016. Review of life cycle assessment for biogas production in Europe. Renewable and Sustainable Energy Reviews, 54, pp. 1291–1300.

Hu, M.C., Huang, A.L. and Wen, T.H., 2013. GIS-based biomass resource utilization for rice straw cofiring in the Taiwanese power market. Energy, 55, pp. 354–360.

Huang, A.Q., Crow, M.L., Heydt, G.T., Zheng, J.P. and Dale, S.J., 2010. The future renewable electric energy delivery and management (FREEDM) system: the energy internet. Proceedings of the IEEE, 99(1), pp. 133–148.

International Energy Agency (IEA), 2012. World Energy Outlook 2012, International Energy Agency, France.

IRENA, 2020. Renewable Energy Statistics, 2020.

Jenkins, B., Baxter, L.L., Miles Jr, T.R. and Miles, T.R., 1998. Combustion properties of biomass. Fuel Processing Technology, 54(1-3), pp. 17–46.

Jonsson, S. and Westman, J., 2011. Cryogenic biogas upgrading using plate heat exchangers (Master's thesis).

Kim, S. and Dale, B.E., 2004. Global potential bioethanol production from wasted crops and crop residues. Biomass and Bioenergy, 26(4), pp. 361–375.

Kothari, R., Pandey, A.K., Kumar, S., Tyagi, V.V. and Tyagi, S.K., 2014. Different aspects of dry anaerobic digestion for bio-energy: An overview. Renewable and Sustainable Energy Reviews, 39, pp. 174–195.

Leonzio, G., 2016. Upgrading of biogas to bio-methane with chemical absorption process: simulation and environmental impact. Journal of Cleaner Production, 131, pp. 364–375.

Li, Y., Chen, Y. and Wu, J., 2019. Enhancement of methane production in anaerobic digestion process: A review. Applied Energy, 240, pp. 120–137.

Li, Y., Rezgui, Y. and Zhu, H., 2017. District heating and cooling optimisation and enhancement–Towards integration of renewables, storage and smart grid. Renewable and Sustainable Energy Reviews, 72, pp. 281–294.

Lim, J.S., Manan, Z.A., Alwi, S.R.W. and Hashim, H., 2012. A review on utilisation of biomass from rice industry as a source of renewable energy. Renewable and Sustainable Energy Reviews, 16(5), pp. 3084–3094.

Lombardi, L. and Francini, G., 2020. Techno-economic and environmental assessment of the leading biogas upgrading technologies. Renewable Energy, 156, pp. 440–458.

Lorenzi, G., Gorgoroni, M., Silva, C. and Santarelli, M., 2019. Life cycle assessment of biogas upgrading routes. Energy Procedia, 158, pp. 2012–2018.

Lyng, K.A. and Brekke, A., 2019. Environmental life cycle assessment of biogas as a fuel for transport compared with alternative fuels. Energies, 12(3), p. 532.

Masebinu, S.O., Aboyade, A. and Muzenda, E., 2014. Enrichment of biogas for use as vehicular fuel: a review of the upgrading techniques. International Journal of Research in Chemical, Metallurgical and Civil Engineering, 1(1), pp. 88–97.

Matsumura, Y., Minowa, T. and Yamamoto, H., 2005. Amount, availability, and potential use of rice straw (agricultural residue) biomass as an energy resource in Japan. Biomass and Bioenergy, 29(5), pp. 347–354.

MNRE, 2020. Bio Energy: Current Status. Ministry of New and Renewable Energy, Government of India. Retrieved November 21, 2020, from https://mnre.gov.in/bio-energy/current-status

Molino, A., Nanna, F., Ding, Y., Bikson, B. and Braccio, G., 2013. Biomethane production by anaerobic digestion of organic waste. Fuel, 103, pp. 1003–1009.

Murdock, H.E., Gibb, D., André, T., Appavou, F., Brown, A., Epp, B., Kondev, B., McCrone, A., Musolino, E., Ranalder, L. and Sawin, J.L., 2019. Renewables 2019 global status report.

Mussoline, W., Esposito, G., Giordano, A. and Lens, P., 2013. The anaerobic digestion of rice straw: a review. Critical Reviews in Environmental Science and Technology, 43(9), pp. 895–915.

Mussoline, W., Esposito, G., Lens, P., Garuti, G. and Giordano, A., 2014. Electrical energy production and operational strategies from a farm-scale anaerobic batch reactor loaded with rice straw and piggery wastewater. Renewable Energy, 62, pp. 399–406.

Navas-Anguita, Z., Cruz, P.L., Martin-Gamboa, M., Iribarren, D. and Dufour, J., 2019. Simulation and life cycle assessment of synthetic fuels produced via biogas dry reforming and Fischer-Tropsch synthesis. Fuel, 235, pp. 1492–1500.

Nie, H., Jiang, H., Chong, D., Wu, Q., Xu, C. and Zhou, H., 2013. Comparison of water scrubbing and propylene carbonate absorption for biogas upgrading process. Energy & Fuels, 27(6), pp. 3239–3245.

Persson, M., Jönsson, O. and Wellinger, A., 2006. Biogas upgrading to vehicle fuel standards and grid injection. In IEA Bioenergy Task, 37, pp. 1–34.

Ranjan, A., Khanna, S. and Moholkar, V.S., 2013. Feasibility of rice straw as alternate substrate for biobutanol production. Applied Energy, 103, pp. 32–38.

Singh, B.P. ed., 2013. Biofuel Crops: Production, Physiology and Genetics. CABI.

SNV, 2020. Biogas. SNV Netherlands Development Organisation. Retrieved November 21, 2020, from https://snv.org/sector/energy/topic/biogas

Sreekrishnan, T.R., Kohli, S. and Rana, V., 2004. Enhancement of biogas production from solid substrates using different techniques––a review. Bioresource Technology, 95(1), pp. 1–10.

Sridhar, S., Smitha, B. and Aminabhavi, T.M., 2007. Separation of carbon dioxide from natural gas mixtures through polymeric membranes—a review. Separation, Purification Reviews, 36(2), pp. 113–174.

Starr, K., Gabarrell, X., Villalba, G., Talens, L. and Lombardi, L., 2012. Life cycle assessment of biogas upgrading technologies. Waste Management, 32(5), pp. 991–999.

Suramaythangkoor, T. and Gheewala, S.H., 2010. Potential alternatives of heat and power technology application using rice straw in Thailand. Applied Energy, 87(1), pp. 128–133.

WBA global bioenergy statistics, 2019. World Bioenergy Association.

World Biogas Association, 2018. WBA Market report 2017-USA.

Zeng, X., Ma, Y. and Ma, L., 2007. Utilization of straw in biomass energy in China. Renewable and Sustainable Energy Reviews, 11(5), pp. 976–987.

Zhang, R. and Zhang, Z., 1999. Biogasification of rice straw with an anaerobic-phased solids digester system. Bioresource Technology, 68(3), pp. 235–245.

Zhang, X., Bauer, C., Mutel, C.L. and Volkart, K., 2017. Life cycle assessment of power-to-gas: approaches, system variations and their environmental implications. Applied Energy, 190, pp. 326–338.

4 Bioenergy in Germany – Status and Outlook[1]

Michael Nelles, Romann Glowacki, Ingo Hartmann, Volker Lenz, Jan Liebetrau, Franziska Müller-Langer, Satya Narra, and Daniela Thrän

4.1 CURRENT ENERGETIC USE OF BIOMASS IN GERMANY

4.1.1 Primary Energy Consumption

Bioenergy is the most important renewable energy source in Germany. In 2018, it represented 8.5% of the total primary energy consumption in Germany. The other renewables accounted for 5.5% (AGEB 2019). Biomass and the biogenic share of waste together accounted for over 60% of renewable energy. Percentages of different types of natural resources used in electricity generation are shown in Figure 4.1.

Compared to the previous year, the share of renewable energies in the German primary energy mix increased by 1.1%. This was mainly due to the expansion in the photovoltaic and wind sectors, as well as extreme weather conditions, with 26% more hours of sunshine than in 2017 and very good wind conditions (AGEB 2019). However, due to the drought, 18% less energy was generated from hydropower. The supply of bioenergy remained almost unchanged at -0.7%. This shows an important characteristic of providing energy independent of the weather and, above all, in a plannable manner. This strength will play an important role in the future energy system; predictability and reliability are important criteria for energy pricing. The associated benefits for users are manifold and range from grid stabilisation and security of supply in the electricity and heat sector to detailed planning of production processes.

These positive features must be evaluated against the background of German energy policy. The German government's climate protection goals are to halve primary energy consumption in Germany by 2050 (compared to 2008) and to achieve a 60% share of renewables in final energy consumption. The share of renewable electricity should be between 80% and 95%.

4.1.2 Bioenergy in the Heating and Cooling Sector

Over 86% of renewable heat in 2018 was provided from biomass, including the biogenic share of waste (UBA 2019). This share is slightly lower than in the previous year. In contrast, solar thermal energy, geothermal energy, and environmental heat (heat pumps) recorded relatively large increases, as shown in Figure 4.2. The annual fluctuations in the share of renewables in final energy consumption are due to weather conditions.

A distinction must be made between the provision of thermal energy for heating residential buildings; the commercial, trade, and service sectors; and industrially used heat (process energy). In the industrial sector in particular, the integrated provision of process energy (steam, heat, electricity) is

[1] **Note:** This chapter is based on a German presentation at the 13th Rostocker Bioenergy Conference 2019: Nelles, M.; Glowacki, R.; Hartmann, I.; Lenz, V.; Liebetrau, J.; Müller-Langer, F.; Narra S.; Thrän, D. (2019). Bioenergie heute und was kann/muss die energetische Biomasseverwertung bis 2030 bzw. 2050 leisten! In: Nelles, M. (Hrsg.) 13. Rostocker Bioenergieforum: Tagungsband. (87). ISBN: 978-3-86009-487-7. S. 15–29.

DOI: 10.1201/9781003204435-5

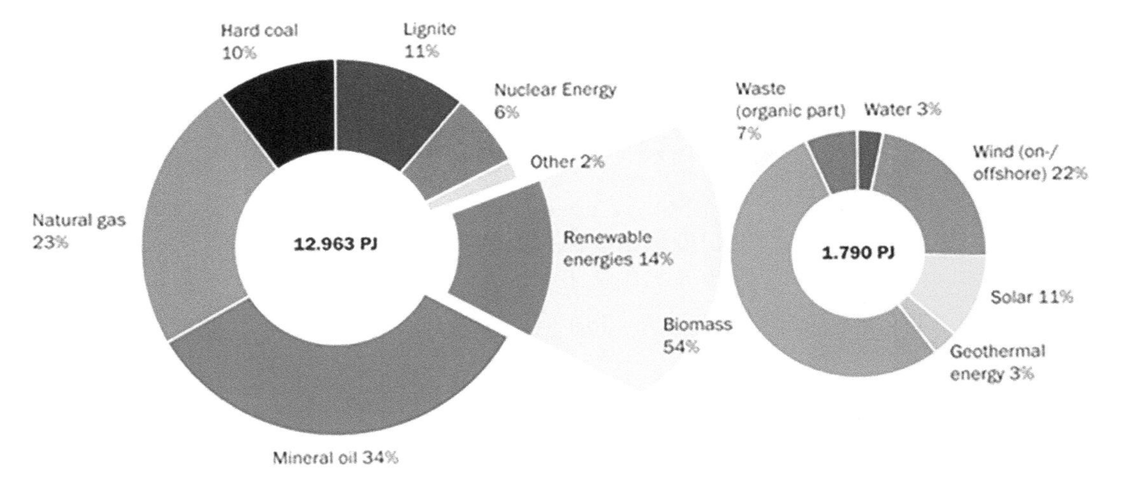

FIGURE 4.1 Presentation DBFZ. (Arbeitsgemeinschaft Energiebilanzen e.V.: Energy consumption in Germany in 2018. Figures as of February 2019.)

a decisive production factor that will gain in importance in view of a growing and politically driven bioeconomy.

According to the German Energy Wood and Pellet Association, about 2.2 million tonnes of wood pellets were sold in the domestic heating sector, and the number of pellet heating systems increased by 3% to 33,000 (UBA 2019). Heat recovery from nearly 9000 biogas plants is also considerable. All heat applications are decentralised.

4.1.2.1 Biogenic Solid Fuels

The majority of biogenic solid fuels are used to provide heating and hot water in private households and the commercial, trade, and service sectors (approx. 48%, as seen in Figure 4.2). The 2.2 million tonnes of wood pellets are mainly derived from by-products of the sawmill industry, producing

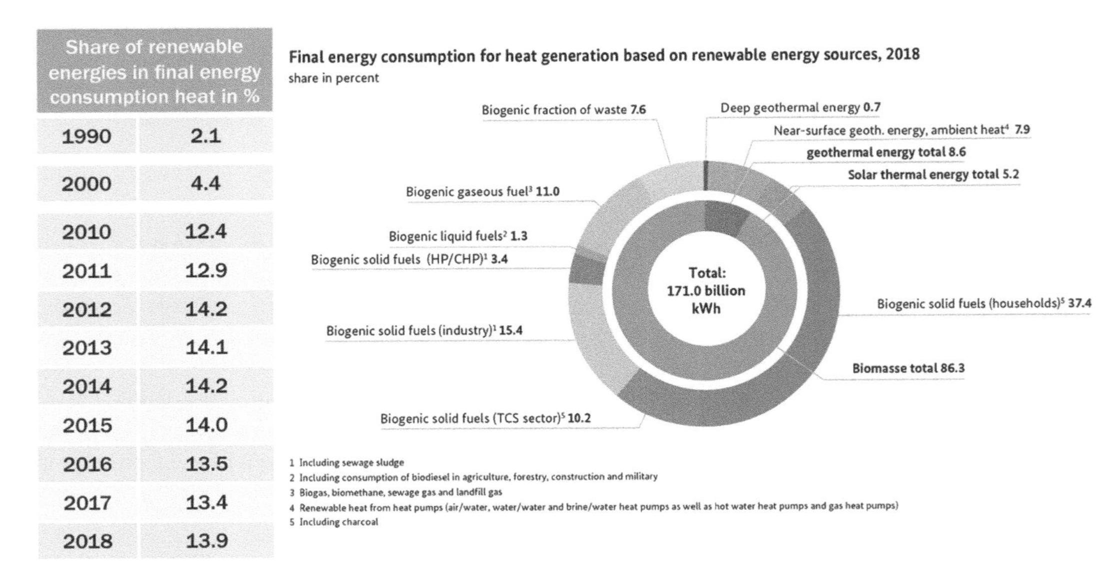

Share of renewable energies in final energy consumption heat in %	
1990	2.1
2000	4.4
2010	12.4
2011	12.9
2012	14.2
2013	14.1
2014	14.2
2015	14.0
2016	13.5
2017	13.4
2018	13.9

FIGURE 4.2 Final energy consumption of heat from renewable energies in 2018. (Modified from Federal Environment Agency (UBA)/Federal Ministry for Economic Affairs and Energy "Renewable energy sources in figures", p. 17.)

bark-free sawdust. This is an example of a value-added combination of material and energetic use of biomass, in terms of the bioeconomy. By far the largest energy source is logs, which account for approximately 80% of the total fuel input (see Rönsch et al. 2015). In some cases, single-room wood stoves produce considerable emissions, which is an important aspect to consider in the future development of energetic biomass use in private households. The amount of biomass used must also be taken into account with consideration to the future raw material supply; in the "Rohstoffmonitoring Holz" report, Mantau determines that almost 28 million solid cubic metres of firewood were used in 2014 (Mantau et al. 2018).

More than 15% of the heat generated from biogenic solid fuels is attributable to the industrial sector. These are mainly integrated industrial power and heating plants, which primarily provide process energy based on wood fuels. In 2016, just under 49% of waste wood was used as fuel, while bark, sawmill by-products, and other industrial waste wood together accounted for almost 20% (Mantau et al. 2018). Comparable differentiated figures for the energy supply from by-products and residues (not solid fuels) of other sectors, such as the food or chemical industry, are currently not available.

4.1.2.2 Biogas

In 2018, the supply of heating and cooling from biogas and biomethane reached 11% of the final energy consumption of heat from renewable energies (see Figure 4.2). Evaluations of operator surveys by the German Biomass Research Centre (DBFZ) show that, after deducting the heat demand of the biogas plant, the externally available heat is provided to one or more uses (Denysenko et al. 2019). Important uses include the heating of residential buildings and hot water production or stables in the immediate vicinity of the biogas plant or for drying processes (for approx. 37% of the plants, see Figure 4.3).

Rensberg et al. found that the importance of heat supply via heat networks (local and district heating) until 2016, as well as heat utilisation from biogas in drying processes, has continuously increased and reflects the increased economic relevance of heat utilisation from biogas.

4.1.3 Bioenergy in the Transport Sector

Bio-based fuels will account for 88% of final energy consumption of renewable energies in the transport sector in 2018 (UBA 2019). The Working Group on Renewable Energy Statistics points to an increase in sales of biodiesel by about 5% over the same period the previous year to about 2.2 million tonnes. Changes in the Federal Emission Control Act regarding the calculation of greenhouse gas (GHG) quotas and changed accounting regulations for biodiesel led to an increase in sales (UBA 2019). With the conversion of the energy-based biofuel quota to a GHG avoidance quota, the GHG balance became an increasingly relevant competitive criterion for

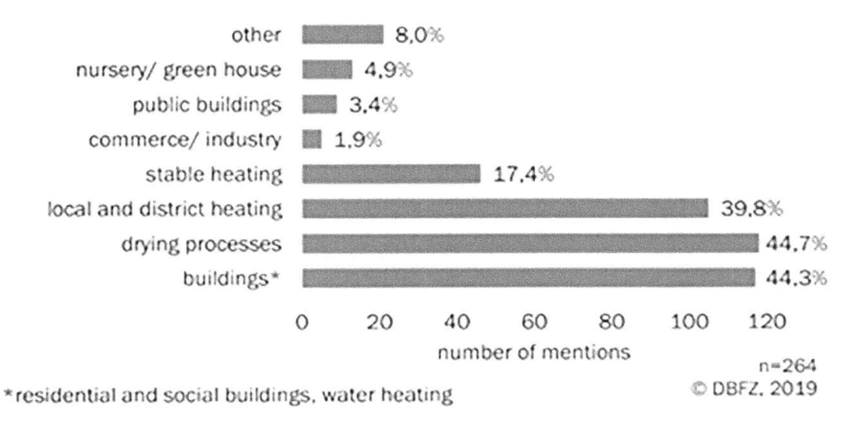

FIGURE 4.3 Rensberg et al. (2019): Type of external heat use, absolute number of nominations and relative frequency in relation to the sample (n = 372) (reference year 2017); multiple nominations possible.

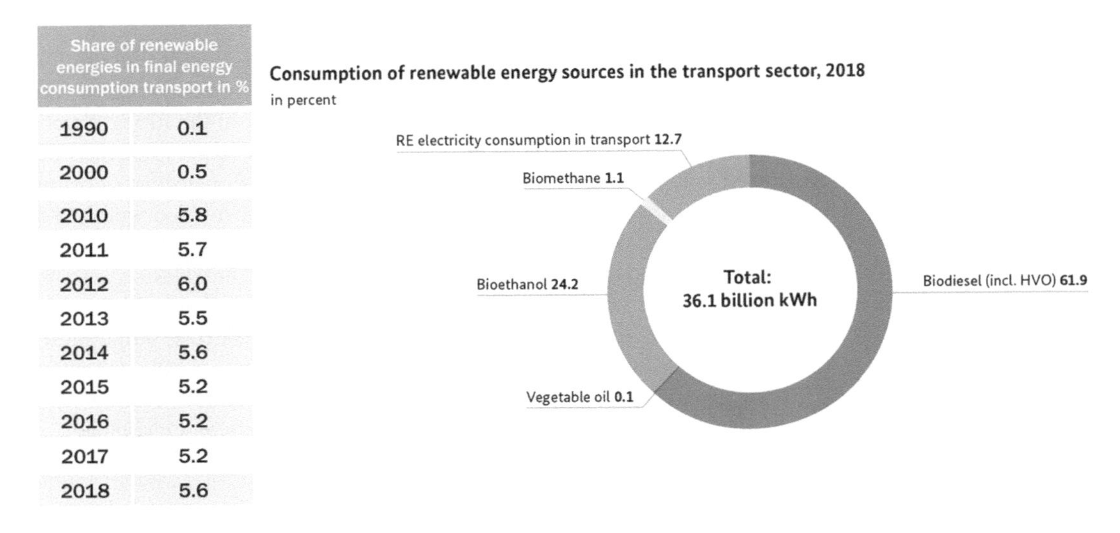

FIGURE 4.4 Final energy consumption from renewable energies in the transport sector in 2018, figures provisional. (Modified from Federal Environment Agency [UBA] Federal Ministry for Economic Affairs and Energy "Renewable energy sources in figures", p. 21.)

biofuels (3.5% from 2015 and 4% from 2017). The average GHG balance of the biofuels used within this quota, as recognised in the certification process, increased from 70% in 2015 to over 80% in 2017. This has resulted in savings of approximately 7.7 million tonnes of CO_2, which represents approximately 4% of total transport emissions. Despite the increase in the GHG quota to 4%, the amount of biofuel used (energy equivalent) has remained largely the same overall due to the parallel increase in average specific GHG avoidance. The main fuels used are biodiesel/FAME, bioethanol, and, to a much lesser extent, HVO/HEFA and biomethane (see Lenz et al. 2019).

Electromobility is recording considerable growth: in 2018, the number of electric cars rose from 53,861 to 83,175 vehicles, and the number of plug-in hybrids from around 42,500 to 67,000 (see Figure 4.4). In relation to the total fleet in Germany of around 47 million passenger cars, however, these are only the first small steps towards establishing electromobility. A total of 12% of final energy consumption in the transport sector is provided by renewable electricity, with rail-bound transport accounting for a large share.

Despite this, the transport sector faces particular challenges in enabling sustainable and climate-friendly mobility. This is being achieved while the number of vehicles and total mileage is increasing, leading to rising overall energy consumption. An important framework for climate protection and renewable energies in transport is provided by the Renewable Energies Directive II (RED II), adopted at the EU level, the concrete national implementation of which is still pending in Germany.[2] It is already foreseeable that with the existing measures and direct implementation of RED II, Germany will miss the climate targets of 40–42% GHG reduction by 2030.

4.1.4 Bioenergy in the Electricity Sector

The share of renewable electricity has been rising continuously for many years. In 2017, wind drove the growth, and in 2018, above-average solar radiation and new wind energy capacities were added to the grid (see Figure 4.5). The drought led to a drastic decline in electricity generation from hydropower (–18%). At just under 23%, electricity generation from biomass was roughly at the same level as the previous year (UBA 2019).

[2] (See also the article "The importance of biomass-based biofuels in the Renewable Energies Directive (RED II) as a contribution to climate protection in transport").

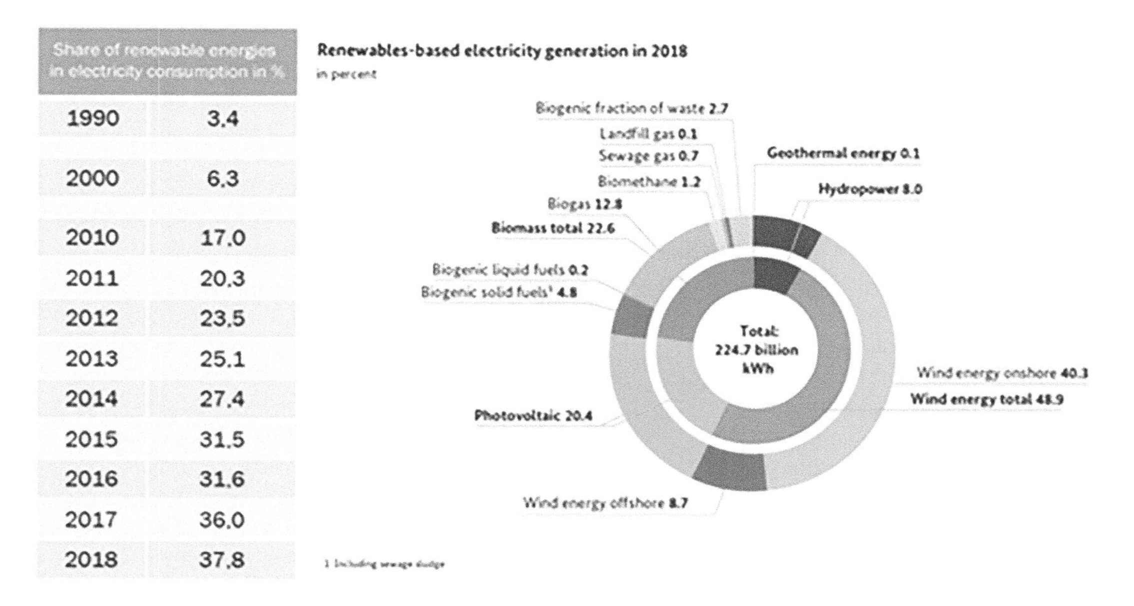

Share of renewable energies in electricity consumption in %	
1990	3.4
2000	6.3
2010	17.0
2011	20.3
2012	23.5
2013	25.1
2014	27.4
2015	31.5
2016	31.6
2017	36.0
2018	37.8

FIGURE 4.5 Electricity generation from renewable energies in 2018, provisional figures. (Modified from Federal Environment Agency (UBA)/Federal Ministry for Economic Affairs and Energy "Renewable energy sources in figures", p. 12.)

4.1.4.1 Biogas

In the biogas sector, a capacity of 411 MW was added, more than 90% of which was built over existing sites. At approximately 80%, the majority of the plants are marketed directly.

As Denysenko et al. described in 2019, the amendment of the EEG in 2012 introduced the flexibility premium for plants generating electricity from biogas (EEG 2012, §33i) and continued with the EEG 2014 in the form of the flexibility surcharge (§54 EEG 2014). Incentives were created to make additional installed electrical capacity available for demand-oriented electricity generation. This can be achieved by installing one or more additional combined heat and power (CHP) units (superstructure) or by replacing an old CHP unit with a larger one. The flexibility bonus is intended to partially refinance the necessary investments for flexible plant operation. Additional income can be generated by shifting electricity production to high-price periods, which can generate revenues above the monthly average price on the exchange for short-term electricity market products (EPEX SE).

In relation to the installed electrical capacity, around 15% of biogas plants currently receive the EEG fixed remuneration, while the remaining 85% market the electricity generated directly on the exchange. Since 2014, there have been signs of an increase in the flexibility of existing biogas plants. In the course of the amendment of the EEG 2014, there was a sharp increase in applications for the flexibility premium, especially in June/July 2014, which can be interpreted as a pull-forward effect due to the great uncertainty surrounding the EEG 2014.

"As of 09/2018, some 2,678 biogas plants with a total capacity of 1,715 MWel for which the flexibility bonus was granted were allocated on the basis of data from the Federal Network Agency and transmission system operators based on evaluations by the DBFZ. In addition, about 155 biomethane CHP plants with a capacity of 161 MWel were operated flexibly. Thus, a total of more than 2,900 biogas and biomethane CHPs with a total installed capacity of more than 2.02 GWel received flexibility premiums or supplements" (Denysenko et al. 2019).

4.1.4.2 Waste Digestion Plants

By the end of 2018, 136 waste digestion plants were operated in Germany (Lenz et al. 2019). This includes fermentation plants in which biowaste and green waste from separate collection is used, as well as plants in which commercial organic waste (food, leftovers from canteen kitchens, canteens and catering, fats and flotates), waste from the food industry, or other organic waste is used. About 90 waste digestion plants use bio and green waste from separate collection – with very different proportions of the total input. Evaluations of the DBFZ database show that around 70 plants are fermentation plants under §27a EEG 2012/§45 EEG 2014/§43 EEG 2017. In 2018, two new digestion plants based on biowaste from separate collection were commissioned. The treatment of the green and biowaste quantities of around 7.4 million t (fresh mass) is currently carried out largely in composting plants. A small proportion of the biowaste collected separately (approx. 20%) is used for digestion and biogas production, increasingly in combined plants (composting plants with digestion stage), as Scholwin et al. explained in 2018 (Denysenko et al. 2019). Figure 4.6 shows the locations of the waste digestion plants, differentiated by operating status and substrate input. Since the introduction of separate support for the digestion of biowaste when the EEG came into force in 2012, around 30 waste digestion plants have gone into operation in Germany. Around half of the new waste digestion plants commissioned since 2012 have been integrated into existing composting plants as so-called upstream plants (upstream digestion of biowaste and green waste before composting). By the end of 2018, more than 50 digestion plants with downstream composting were in operation in Germany (Denysenko et al. 2019).

The targeted redirection from pure composting plants to composting plants with a digestion stage (requirements of TA Luft) for the combined material and energy recovery of the residues

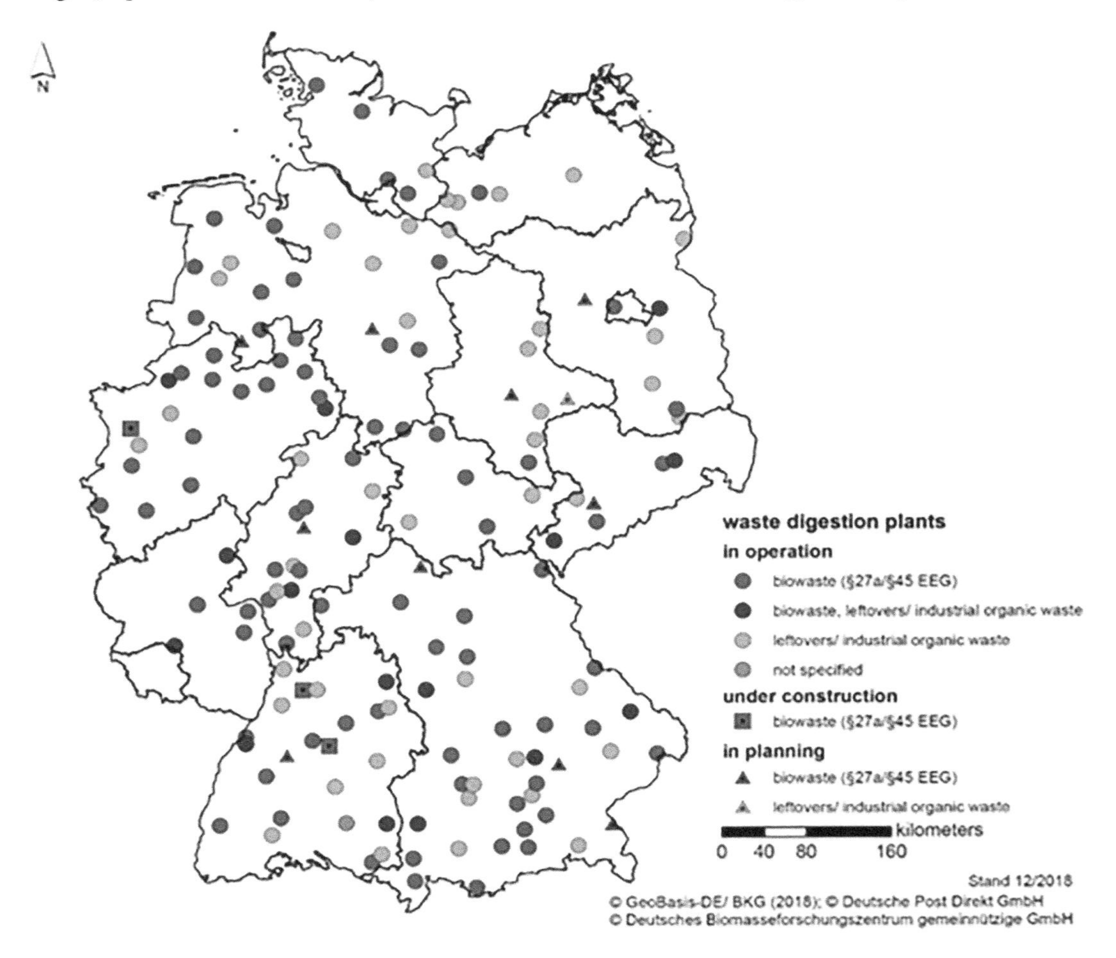

FIGURE 4.6 Denysenko et al. (2019) – Bio-waste digestion plants in Germany by operating status and substrate input.

allows a more efficient use of the residues and savings of GHG emissions. Selected flows of the landscaping and organic materials that have so far been used in pure composting plants should also be used in biogas plants in the future. Further potential results from the consistent implementation of the collection and recycling of separately collected biowaste quantities and, in the longer term, from the collection of the organic portion in residual waste (Denysenko et al. 2019).

4.1.4.3 Solid Fuels

Integrated plants from the sawmill industry were awarded contracts in the last EEG tendering round, as reported by the Holzzentralblatt on May 5, 2019. The German Sawmill and Wood Industry Association, for example, considers the instrument to be an important connection perspective for existing EEG plants. However, the demands for flexibility are described as obstructive. The sawmill industry alone currently operates approximately 2000 wood combustion plants for heat (necessary for the provision of process energy), partly with electricity generation in CHP. In addition, there are about 70 waste wood (heating) power plants in Germany (HZB 2019).

4.2 ENERGETIC BIOMASS USE WORLDWIDE

The latest estimates assume a bioenergy production worldwide of approximately 51 EJ in 2015. Eighty-seven percent of this is wood-based, 10% from agriculture, and 3% from the biogenic fraction of household waste (Klepper, Thrän 2019). More than two-thirds of the energy use is accounted for by traditional forms of use in open fireplaces for heating and cooking, with extremely low efficiencies. The flow chart shown in Figure 4.7 shows the biomass flows in the year 2000. Of the total biomass harvested by humans, 50% comes from cultivated plants (arable farming), approximately 33% from grazing grass eaten by animals, and 16% from wood harvesting. About 60% of the total

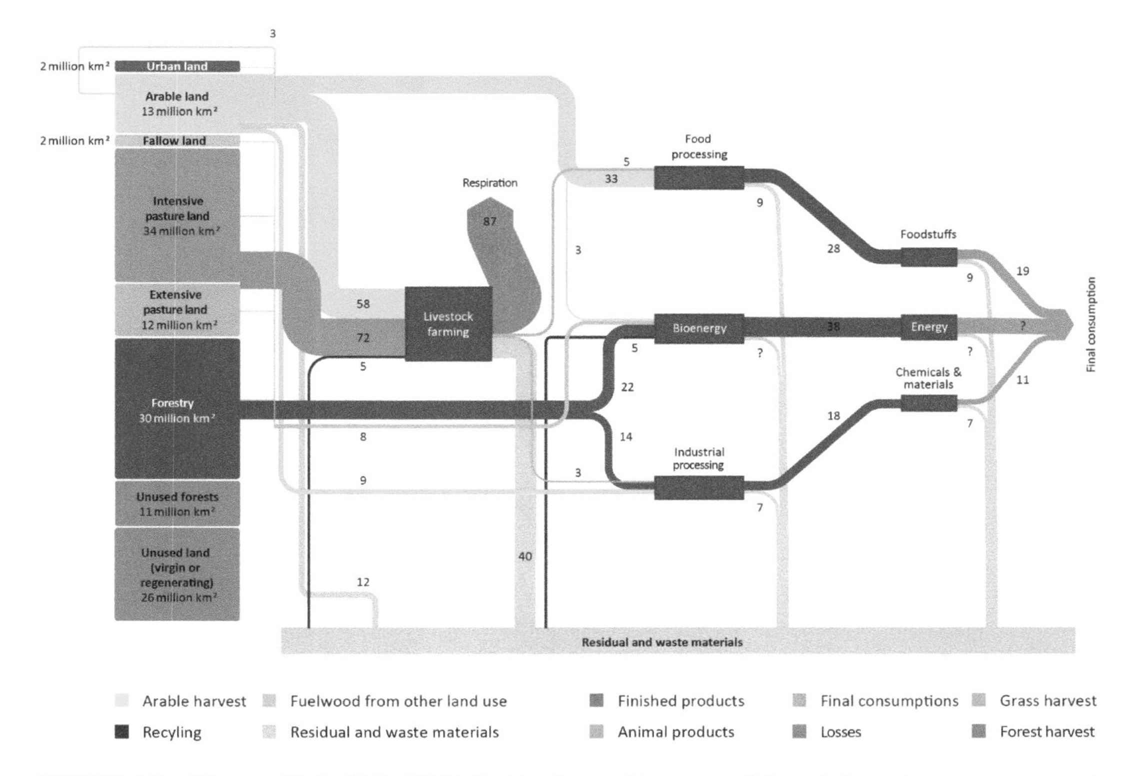

FIGURE 4.7 Klepper, Thrän 2019: ESYS Position Paper: Biomass: striking a balance between energy and climate policies, p. 24: Flow diagram of global harvested biomass flows in exajoules/year for 2000: the left-hand column illustrates the use of global land area.

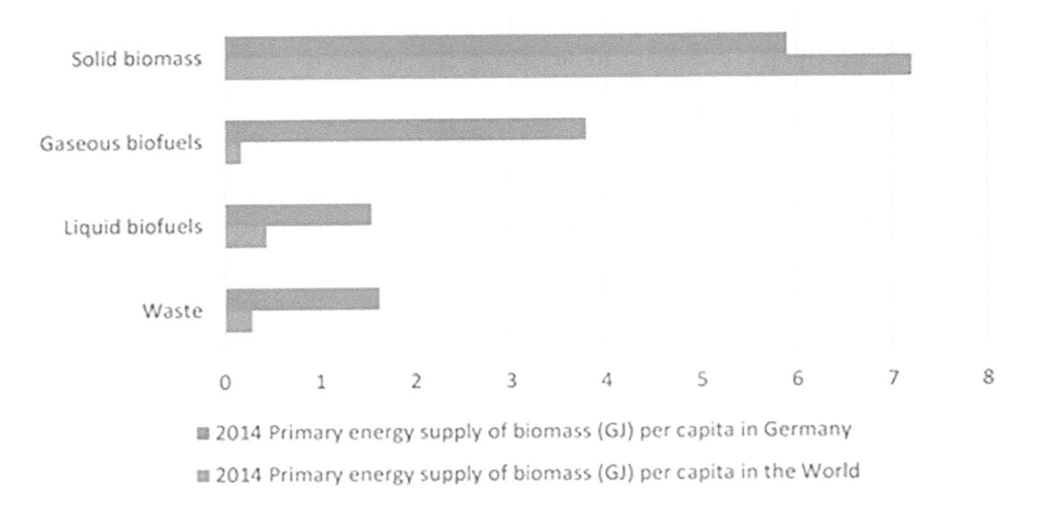

FIGURE 4.8 Comparison of primary energy consumption per capita globally and in Germany in 2014. The figures for gaseous bioenergy sources also include sewage gas. (Pfeiffer, Thrän 2018.)

biomass (130 EJ) is used as animal feed, of which 3 EJ (less than 2.5%) is available for human consumption. Most of it is consumed during the conversion into animal proteins. About 10% of the cultivated biomass is used directly as food.

From a global perspective, most of the biomass harvested is therefore currently used or burned with low efficiency levels in livestock farming.

Figure 4.8 shows the per capita primary energy consumption of Germany in a global comparison. The relatively high use shares of biogas and liquid bioenergy sources, as well as the recycling of waste, should be emphasised. From a global perspective, there is still considerable potential here that can be exploited.

A substantial increase in global biomass production appears unlikely. The analysis of Leopoldina/Acatech "Biomass in the field of tension between energy and climate policy" (Klepper, Thrän 2019) compares different studies on the estimation of the potential for arable land, meadows and pasture land, forests, and residues. It concludes that without serious environmental impacts, a further increase in global biomass production appears unrealistic. Impacts of climate change on the productivity of arable land, including forests, or the growing world population are not yet considered. The current discussion on the extinction of species due to currently practiced cultivation methods and the decreasing biodiversity also makes a significant increase seem unlikely. Intact ecosystems, such as near-natural managed forests, also require functioning destructor systems that mineralise residual forest wood, circulate nutrients, and feed important food chains.

4.3 REQUIREMENTS FOR FUTURE ENERGETIC USE OF BIOMASS

The comprehensive study "Milestones 2030 – Elements and milestones for the development of a viable and sustainable bioenergy strategy" (Thrän, Pfeiffer 2015) provides an outlook beyond 2020. Based on modelling, it analyses various scenarios of bioenergy use in the areas of resources and markets, resources and land use, on life cycle assessment, ecological effects, regional and international effects, and feedback. The more recent analysis of Leopoldina/Acatech, "Biomass in the field of tension between energy and climate policy" (Klepper, Thrän 2019) also presents the development options for the energetic use of biomass in 2050 in a differentiated way. Based on these two studies and the status quo presented in the previous section, essential requirements and possible development directions for bioenergy use are briefly summarised below:

- **Use of scarce biomass resources and highest possible efficiency**
  A further increase in global biomass production appears hardly realistic. The growing world population and the transition to a bioeconomy, which is being pursued in many developed countries, will significantly increase the demand for biomass. In addition, sustainability requirements are increasingly being imposed in order to keep ecosystems – and thus our biomass production systems – functioning. **All paths of biomass use must be used with the highest possible efficiency.** This applies in particular to the current form of traditional bioenergy use in developing countries, where a large proportion of the solid fuels used worldwide for cooking and heating are consumed with minimum efficiencies of around 10%. As Klepper and Thrän stated in 2019, by far the largest part of the biomass produced is used as animal feed in livestock farming. This potential is used to produce animal proteins and food. A meat-reduced orientation of food supply, especially in developed countries, could contribute to a considerable release of additional biomass potential. Between the production of a vegetarian and a meat calorie lies the efficiency factor 10.
  A comprehensive orientation of biomass utilisation towards utilisation cascades in the targeted bioeconomy will lead to a **much more comprehensive energy utilisation of biogenic residual and by-product flows**. More efficient technologies for separation, recycling, and energy supply are required.
- **Integration into the energy system and the bioeconomy**
  The **better linkage and integration of material and energetically utilised material flows in coupled production** and the associated technologies will be of decisive importance. The energetic use of biomass must be equally integrated into the future energy system and the bioeconomy. The strengths of bioenergy as an energy source that can be planned and stored must be used intelligently (Dotzauer et al. 2019). Competitive uses must be eliminated or reduced. Sectors for which no meaningful renewable alternatives have emerged yet are primarily in the field of process energy (industry), transport, especially aviation and shipping, and long-distance heavy goods traffic, as well as in the heating of larger buildings or buildings that are difficult to insulate (see also Klepper, Thrän 2019).
  The synergies and integrative approaches resulting from the interaction of the individual renewable products have hardly been considered so far. Smart bioenergy not only plays an important role in the energy turnaround in the electricity sector, but also in networking with the transport sector and the material use of biomass. To this end, the field of sustainable SynBioPTX (Synthetic Biobased Power to chemical products) products from biomass and electricity-based source materials for material and energy recovery opens up new perspectives for successful competition, nationally as well as internationally (see Müller-Langer in Naumann 2019).
  Bioenergy technologies can be integrated in the agricultural sector in the closing of nutrient cycles.
- **Minimal environmental impact: emissions and sustainability criteria**
  The energetic use of biomass must be "emission free" in the long term. This requires the development of appropriate combustion and separation technologies, including the necessary catalysts. Methane emissions must be avoided completely due to the much higher GHG effect – in the fermentation process as well as when operating engines or turbines with biogas/biomethane.
  Strict sustainability criteria must apply equally to all forms of biomass use. This includes material use as well as the production of food and bioenergy. Appropriate cultivation systems must ensure that, among other things, biodiversity and soil fertility are increased.
- **Comprehensive use of all by-products of bioenergy plants**
  Integrated bioenergy technologies enable a wide range of system services in the energy system, in the bioeconomy, the circular economy, and to achieve climate protection goals.

In addition to the previously mentioned options for cycle closure, grid stabilisation through demand-oriented energy supply, bioenergy can provide green CO_2. This can be used in the chemical industry or permanently removed from the atmosphere to achieve negative emissions. In the long term, these technologies will be able to contribute as a further option to achieving the climate protection targets set. However, economic feasibility and social acceptance must be examined in detail (see Klepper, Thrän 2019).

REFERENCES

AGEB 2019 – Arbeitsgemeinschaft Energiebilanzen e.V. (2019): Energieverbrauch in Deutschland im Jahr 2018. Stand Februar 2019.

Denysenko et al. 2019 – Denysenko, V., Rensberg, N., Liebetrau, J., Nelles, M., Jaqueline Daniel-Gromke, J. (2019): Innovationskongress Osnabrück: Aktuelle Entwicklungen bei der Erzeugung und Nutzung von Biogas. Mai 2019.

Dotzauer et al. 2019 – Dotzauer, M., Pfeiffer, D., Lauer, M., Pohl, M., Mauky, E., Bär, K., Sonnleitner, M., Zörner, W., Hudde, J., Schwarz, B., Faßauer, B., Dahmen, M., Rieke, C., Herbert, J., Thrän, D. (2019): How to measure flexibility – Performance indicators for demand driven power generation from biogas plants. Renew. Energy 134, 135–146.

HZB 2019 – Holzzentralblatt vom 3. Mai 2019: "Sägewerker nutzen Chancen des EEG".

Klepper, G./Thrän, D. 2019 - Biomasse im Spannungsfeld zwischen Energie- und Klimapolitik. Potenziale – Technologien – Zielkonflikte (Schriftenreihe Energiesysteme der Zukunft), München

Lenz et al. 2019 - Lenz, V., Naumann, K., Denysenko, V., Daniel-Gromke, J., Rensberg, N., et al. (2019): *Erneuerbare Energien. In: BWK: Das Energie-Fachmagazin 05/2019, im Druck.*

Mantau et al. 2018 – Prof. Dr. Udo Mantau (INFRO e. K.) unter Mitwirkung von Przemko Döring (INFRO e. K.); Dr. Holger Weimar, Sebastian Glasenapp, Dr. Dominik Jochem, Klaus Zimmermann (Thünen-Institut): Rohstoffmonitoring Holz – Erwartungen und Möglichkeiten. Hrsg. FNR 2018.

Naumann 2019 – Naumann, K., Schröder, J., Oehmichen, K., Etzold, H., Müller-Langer, F., Remmele, E., Thuneke, K., Raksha, T., Schmidt, P. (2019): Monitoring Biokraftstoffsektor. 4. überarbeitete und erweiterte Auflage. Leipzig: DBFZ (DBFZ-Report Nr. 11). ISBN 978-3-946629-36-8.

Pfeiffer, Thrän 2018 – Pfeiffer, D., Thrän, D. (2018): One Century of Bioenergy in Germany: Wildcard and Advanced Technology. In: Chemie, Ingeniuer, Technik. Wiley-VCH Verlag GmbH & Co. KGaA.

Rensberg et al. 2019 –Rensberg, N., Denysenko, V., Daniel-Gromke, J. (2019): Wärmenutzung von Biogasanlagen. Report im Rahmen des Projektes "Optionen für Biogas-Bestandsanlagen bis 2030 aus ökonomischer und energiewirtschaftlicher Sicht" (FKZ 37EV 16 111 0), im Druck.

Rönsch et al. 2015 – Rönsch, C., Sauter, P., Bienert, K., Schmidt-Baum, T., Thrän, D. (2015): Biomasse zur Wärmeerzeugung – Methoden zur Quantifizierung des Brennstoffeinsatzes. DBFZ Report Nr. 24. 2015.

Scholwin et al. 2018 – Scholwin, F., Grope, J., Clinkscales, A., Daniel-Gromke, J., Rensberg, N., Denysenko, V., Stinner, W., Richter, F., Raussen, T., Kern, M., Turk, T., Reinhold, G. (2018): Aktuelle Entwicklung und Perspekti-ven der Biogasproduktion aus Bioabfall und Gülle. FKZ 37EV 17 104 0.

Thrän, Pfeiffer 2015 (Hrsg.) – Thrän, Daniela (DBFZ/UFZ) Arendt, Oliver; Ponitka, Jens; Braun, Julian (DBFZ) Millinger, Markus (UFZ) Wolf, Verena; Banse, Martin (TI) Schaldach, Rüdiger; Schüngel, Jan (CESR) Gärtner, Sven; Rettenmaier, Nils (IFEU) Hünecke, Katja; Hennenberg, Klaus (Öko-Institut) Wern, Bernhard; Baur, Frank (IZES) Fritsche, Uwe; Gress, Hans-Werner (IINAS): Meilensteine 2030 Elemente und Meilensteine für die Entwicklung einer tragfähigen und nachhaltigen Bioenergiestrategie. Endbericht zu FKZ 03KB065, FKZ 03MAP230 Herausgegeben von Daniela Thrän, Diana Pfeiffer.

UBA 2019 – Umweltbundesamt (UBA) Hrsg./Arbeitsgruppe Erneuerbare Energien-Statistik (AGEE-Stat) 2019: Erneuerbare Energien in Deutschland. Daten zur Entwicklung im Jahr 2018. März 2019.

5 Feasibility Study of Commercialised Self-Circulating Biogas Generators

A Circular Economy Approach

Varadha V. P. Bhuvana, B. I. Devanand, Nandhini S. Selva, T. Thavasivamanikandan, and V. Kirubakaran

5.1 OBSERVATIONS

Though great number of biogas plants are in operation in India, the research on the feasibility study of Commercialised Self-Circulating Biogas Generators is scarcely available. Hence an analytical experimental study using 20-l floating dome digester available in the laboratory for comparative study was carried out.

5.2 INTRODUCTION

A biogas plant converts biomass into usable energy. Itis one of the most reliable and highly encouraging renewable energy techniques that could be easily installed and can operate in almost any geographical conditions with slight design modifications.. Biogas plants use biodegradable organic materials, especially waste such as cattle dung, garden and kitchen waste, poultry waste, night soil, municipal solid waste (MSW), and agricultural residues, as a raw material to produce biogas. Biogas is a mixture of gas that consists of methane (CH_4) as a primary element, carbon dioxide (CO_2), traces of hydrogen sulfide (H_2S), and moisture. This technology is a highly reliable option that provides a sustainable solution to meeting energy needs with a carbon-neutral option. The population increases daily, and daily waste generated by each individual person has also been increased.

It is predicted that world waste production will be approximately 27 billion tonnes per year by 2050, and that one-third of the waste will be produced from India and China, due to their rapid population growth. At present, the population of India is around 1,380,004,385, with a population density of 464 /km^2; in India, an individual can generate approximately 170–620 g of MSW per day. According to the current statistics, India has the second highest number of biogas plants. The present availability of biomass in India is estimated at about 500 million metric tonnes per year.

The sponsored study conducted by the Ministry of New and Renewable Energy has estimated surplus biomass availability at about 120–150 million metric tonnes per year, covering agricultural and forestry residues, which is an energy potential of about 18,000 MW. Biomass from agricultural residues or energy crops is estimated at about 141 Mha of arable land, producing over 700 MT of estimated biomass surplus, approximately 150 MT by 2020; the urban population, likely to be 550 million, would generate greater than 100 MTA of MSW. With the increased population, the rapid growth of the economy, and industrialisation and urbanisation, India faces difficulties in MSW management. In this study, we analyse the possibility of MSW as a feedstock in an anaerobic digester.

Waste management plays a vital role in the nation's development. There is various research and numerous studies in the area of extraction of energy from waste. The waste to energy conversion methods are promising technology, as the yields are beneficial both environmentally and economically.

DOI: 10.1201/9781003204435-6

The present feasibility study focusses on the assessment of the availability of biomass and its scope for biogas production. It also focusses on the role of community participation in providing biomass for biogas production. The aim of this work is to evaluate the opportunity of the production of biogas from the locally available biomass through self-circulating biogas plants. Here we include the type of substrates, quantity, and quality of available biomass; desirable environmental and economic conditions; problems associated with the collection, management, and distribution of decentralised biomass; and market. This study has been conducted with the following objectives:

- Energy component: biogas generation and distribution, electricity generation
- Agricultural component: production of organic fertiliser
- Social component: improvement of energy security, better livelihood, job opportunities
- Environment component: solution to waste-related problems, carbon emission reduction.

5.3 FACTORS HINDERING THE DISTRIBUTION OF BIOGAS PLANTS

Although it is understood that biogas technology has great potential to meet two basic rural requirements – energy and manure – its wide scale penetration is not without problems. Biogas technology diffusion in India has experienced a number of setbacks, mainly due to the non-operation of large proportion of the plants installed. Though a remarkable feature of the biogas plant technology is its operational simplicity, a host of chemical, microbiological, engineering, and socio-economic problems have to be tackled to ensure large-scale adaptation of these plants. Serious limitations include the unavailability of sufficient feedstock, as well as defects in construction and microbiological failure in plants. The initial cost of biogas installation is prohibitively high and needs sound economic status of the potential users.

The number of cattle owned by the household also determines the size of the plant to be installed. Cattle dung shortages may arise from a decrease in the number of cattle, division of cattle among family members, and installation of large plant, either by accident or by design. Initially, full subsidies were given to many "marginal" farmers to procure biogas plants in order to promote biogas technology. Installations were carried out giving little importance to the technical aspects and with very little knowledge of the macro- and micro-economics, as well as socio-cultural aspects of introducing anaerobic digesters in rural areas. Even the smallest sized plant needs three to four cattle to support them. Short supply of dung eventually led to underfeeding and consequent failure of many plants. Often, delays of one year in obtaining the subsidy are common. Absence of competent bank staff and loan officers with fair knowledge on biogas plants acts as a constraint. In some areas of the country, water scarcity imposes further constraints on the viability of biogas technology, particularly in the arid and semi-arid regions. A water source is needed in the vicinity to enable the supply of enough water to dilute the fresh dung.

Immature technology of plants until the beginning of the 1980s and a diffusion strategy that was only minimally developed and did not recognize the importance of user training and follow-up services until much later acted as stumbling blocks. Lack of servicing facilities, spare parts, repair and maintenance facilities, and inadequate access to technical information also posed problems. Sometimes, the plants are faulty in their construction or develop problems that lead to their non-functioning.

Biogas plant implementation is a globally accepted technique, and biogas technology has huge potential to meet the two basic rural requirements of energy and manure, its wide-scale implementation doing so in a carbon-neutral way. In India, biogas technology diffusion experienced serious setbacks due to the lack of technological advancement, lack of public awareness, problems in the raw materials supply chain, output commercialisation problems, the initial cost of biogas installation, and the economic status of the potential users. The problems discussed may be overcome through effective management, technological selection, and adequate support services. Constant monitoring and evaluation are some other basic requirements for the successful promotion of the technology.

Coordination among the various agencies and benefit monitoring are required to promote and sustain the technology. This feasibility study proposes a model of biogas production and distribution supply chain. Here the supply chain starts and ends at the cluster of people in the selected region. The modelling is done by finding the optimal flow of biodegradable waste and biogas end-product distribution through a number of processes. Furthermore, the model includes public sector, private sector, and people participation.

5.4 METHODOLOGY

The following layout shows our proposed plan for a cluster, as seen in Figure 5.1. In this design layout, we almost nullified the problems faced by the existing digester plants in India by making this as a proper business module that would satisfy both the investors and the local community in a healthy way. Thereby we can satisfy the public–private–people partnership (4P) approach. Biodigester techniques have still not gotten much acceptance in broader society. The primary reason is related to the biomass supply chain. Biomass availability is not certain; biomass availability from the agriculture sector is available after the harvesting period, only 2–3 months in a year. Another problem related to the availability of segregated wastes and difficulties related to biomass transportation. In India, Karnataka, Andhra Pradesh, and Maharashtra lead the pack in establishing biomass-based power projects; ironically, states having an agriculture-based economy do not properly utilise the biomass energy potential.

India is one of the top emerging economies in the world, but 300 million Indians are struggling to meet their energy requirements. Apart from this, unemployment and environmental problems are causing serious setbacks to India's development. Our intention is to address these issues by making use of the available resources and existing technologies. Biogas plants are reliable. They can install and operate in almost any geographical conditions with slight design modifications; they convert biomass into useful energy. Here we are trying to introduce the concept of waste as a source of income. Waste is directly collected from the people, as well as from the local

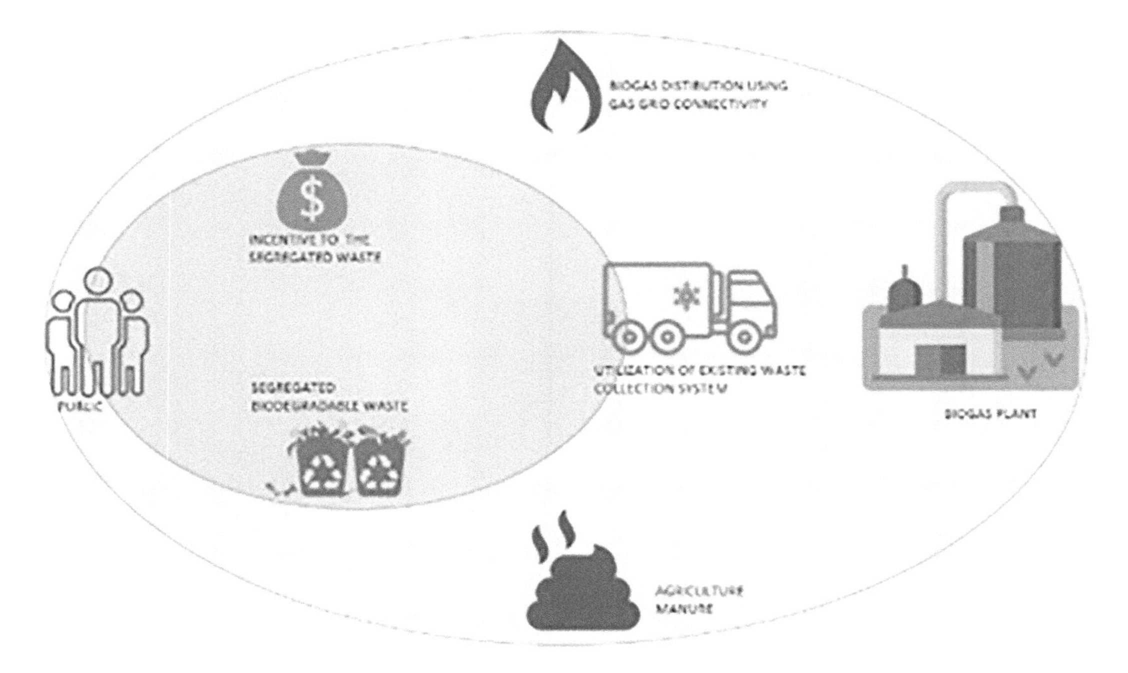

FIGURE 5.1 Schematic representation of plan.

community, based upon energy needs. We directly approach people by avoiding intermediaries and paying them based on waste.

5.4.1 Waste Collection

Though the generation of waste is common, collecting all in a common platform is a difficult task; for desired output, we need an appropriate level of input feed to the digester without delay in the feedstock supply. Most of the existing biogas plants facing problems related to availability of feedstock and unavailability of segregated waste. By segregation and collection of waste at the source, we can thereby ensure waste recycling and resource recovery. The installation of waste-to-compost and bio-methanation plants helps to reduce the load of landfill sites. The proper segregation and waste recovery give a solution for the environmental problems and land area requirements. The biodegradable component of India's solid waste is estimated at almost 50%, as shown in Figure 5.2, and bio-methanation is a suitable solution for processing biodegradable waste. The available statistics show that India has huge potential for bioenergy.

In this approach, we suggest the collection of segregated biodegradable waste at doorsteps by paying based on segregated waste quantity and quality. Here we suggest the waste as alternative means of income, and this approach will help to attract more households to this network. It also ensures that less waste reaches landfills. Here we planned to group the local community to collect the waste by paying them based upon the composition of waste they could offer. Some of the waste can be directly processed and the other waste requires some pretreatment steps, so based upon the plant requirement, collectors who are collecting waste can be paid accordingly. We can also encourage startups and private agencies to collect waste in the selected cluster at this stage. Through the suggested direct approach, and thereby avoiding the maximum intermediaries, participation can increase, with some acceptance from the local community. Waste collection from the public and institutions would follow the same approach. This would then require linking the existing waste transportation facility and mode of transportation for waste by making agreements with a public authority. By making agreements with farmers' clusters, the availability of agriculture residue for the biogas production can be assured.

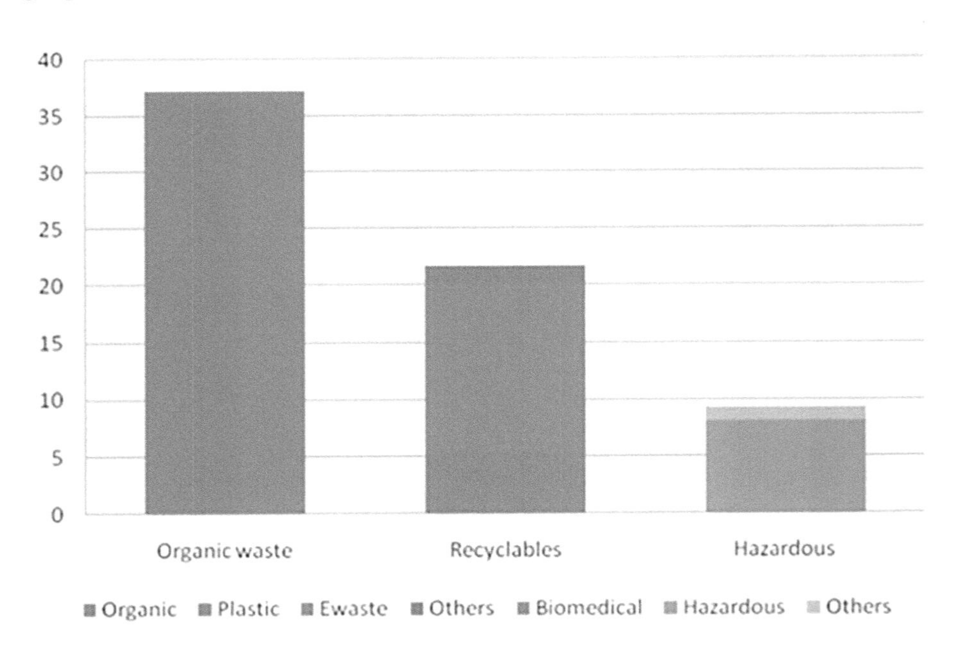

FIGURE 5.2 Waste composition of India, in million metric tonnes per annum. (PIB 2016.)

5.4.2 Waste Processing

Processing of waste means the physical alteration of waste to make it suited for its technology-adopted treatment. This helps in achieving the possible benefits from every functional element of the waste management system. After collecting the required quantity of household, agricultural, and other biodegradable waste, the second step is waste processing. Here, the waste needs to be grouped according to the biochemical properties of biogas production, the carbon-nitrogen ratio, and so on. This part of the process includes participation of the people by providing job opportunities and sources of income. Raw materials are collected from various sources, such as livestock, crop residue, food waste, and other biodegradable wastes, which require different types of pretreatments before being used as digester feed. Residue from the agricultural sector such as hay, spent straw, corn, and plant stubble need to be shredded in order to make use as digester feed, which also helps to increase bacterial activity. The succulent plant material yields more than dried matter, and some materials like brush and weeds needs semi-drying. Pretreatment processes like shedding and grinding increase the surface area available for microbial attack and speedup the digestion process. In the crushing process, raw material is crushed, then mixed with a proper proposition of water and taken to a digester unit, as seen in Figure 5.3.

5.4.2.1 Anaerobic Digestion Process

Anaerobic digestion (AD) is the process in which organic matter such as animal or food waste is broken down, producing biogas and biofertiliser. It is a complex process, which includes hydrolysis, acidogenesis, acetogenesis, and methanation. The feedstock can be biodegradable, such as agricultural residue, kitchen waste, degradable MSW, and slaughterhouse waste. AD is a controlled process involving microbial decomposition of organic matter in absence of oxygen (Municipal Solid Waste Management Manual, 2018, India). AD involves the process of decaying biodegradable material in the absence of any terminal electron acceptors like sulphate, nitrate, and oxygen. AD consists of a number of complex biochemical reactions carried out by several types of microorganisms that survive in oxygen-free conditions. Biogas is produced during this process. Biogas is composed primarily of CH_4, with CO_2 being the other of the major gases, and trace amounts of H_2S, nitrogen

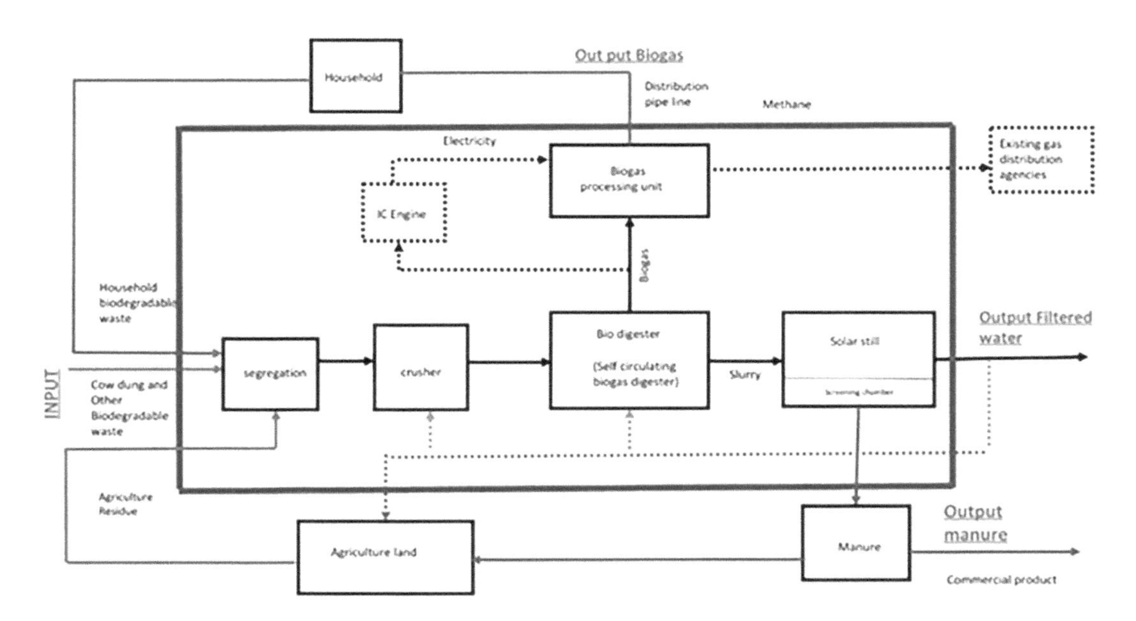

FIGURE 5.3 Plant layout.

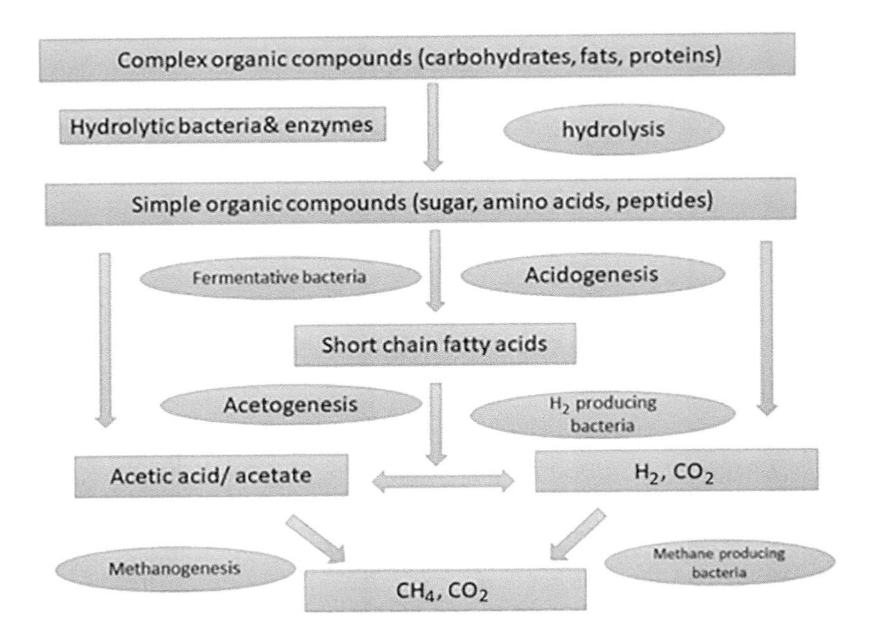

FIGURE 5.4 Chemical processes involved in anaerobic digestion.

(N_2), hydrogen (H_2), and water vapours. The material left after the AD process is called digestate, which is rich in nutrients. Digestate is a wet mixture that can be separated into a solid and a liquid.

The breakdown of organic material inside an anaerobic digester involves four stages of chemical processes, which convert the biogas feed into usable biogases. The first stage of the chemical processes is the hydrolysis stage, which sees complex matter, such as carbohydrates and proteins, broken down into sugars and amino acids. These are normally long-chain chemical compounds, but hydrolysis breaks them down into single molecules. It is a slow process, and this stage is a rate-limiting stage in the AD process. In the acidogenesis stage, acidogenic bacteria transform the products of hydrolysis into short-chain volatile acids (acetic acid, propionic acid, and butyric acid), ketones, alcohols, H_2, and CO_2. Acetic acid, H_2 and CO_2 are directly utilised by methanogenic bacteria. This stage is affected by a diverse group of bacteria, which are capable of bringing down the pH inside the digester to four. The third phase of AD is acetogenesis. In this process, the products of the acidogenic stage (organic acids greater than two carbon atoms, alcohols, and fermentation products) are converted to acetate, H_2, and CO_2 by acetogenic bacteria. The terminal stage of AD is the biological process of methanogenesis, in which methanogens create CH_4 from the final products of acetogenesis, as well as from some of the intermediate products from hydrolysis and acidogenesis.

At the end of these processes, we get biogas, as well as a nutrient-rich fertiliser as a by-product of the process. Figure 5.4 demonstrates steps in the biochemical reactions and microbes involved in the overall digestion process, which occurs in the absence of oxygen inside a sealed, oxygen-free tank called an anaerobic digester.

5.4.2.2 Solar Still

A solar still is a green energy solution that uses the natural energy of the sun to purify water. In this layout the use of a solar still is proposed for drying the slurry from the biogas digester. The solar still process uses the solar energy instead of fossil fuels to gain the energy needed for purification. A solar still is made up of two water troughs and a piece of glass running across the top of the water containers. The liquid slurry is placed in one of the troughs, while the other remains empty. The glass is placed across the top at an angle, angling directly down into the empty trough. The bottom of the trough containing slurry is usually painted black to help absorb the energy from the sun. The slurry dries easily, in a passive way, through this technique, thereby solving the problems related to the liquid slurry.

5.4.2.3 End Products

At the end of the process, two kinds of products are obtained as outputs: biogas and biofertiliser. In this, the biogas that is produced from the digester unit can be taken to the biogas processing unit to produce CH_4. Here we add a direct current engine as a supplementary source that can also be powered by the methane gas to generate electricity; 1 kg of methane will produce 1 KWh of electricity, so based upon the plant's requirements, this supplementary source can run the complete plant, making it a self-sustaining module. The slurry from the biodigester, which is highly rich in nutrient content, is then treated with a solar still to remove the excess water content and packed as rich organic manure.

5.4.2.4 Digester Design Modification

There are several biogas plant designs that have been introduced worldwide. More than 30 digester designs are suggested in India alone by incorporating certain features implemented on digester or in the gasholder, based on the geographical requirement. The simplest design for a biogas digester is a single tank, which is kept completely air- and water-tight. The digester receives organic feed, which decomposes inside a digestive chamber and is converted into biogas. Anaerobic conditions made inside the digester allow for the organic matter to be converted into biogas. In this feasibility study, some modifications are suggested for the conventional digester in order to reduce the hydraulic retention time and yield higher biogas production. The comparative study of conventional and designed modified digester were carried out practically in the bioenergy laboratory of Gandhigram Rural Institute, India.

5.4.2.5 Modified Self-Circulating Anaerobic Digester

Some modification to the digester is suggested for reducing the hydraulic retention time (HRT) and increasing the gas production rate (see Figures 5.5 and 5.6). HRT is the average duration that a given volume of feed stays inside the digester; it is one of the most important design parameters

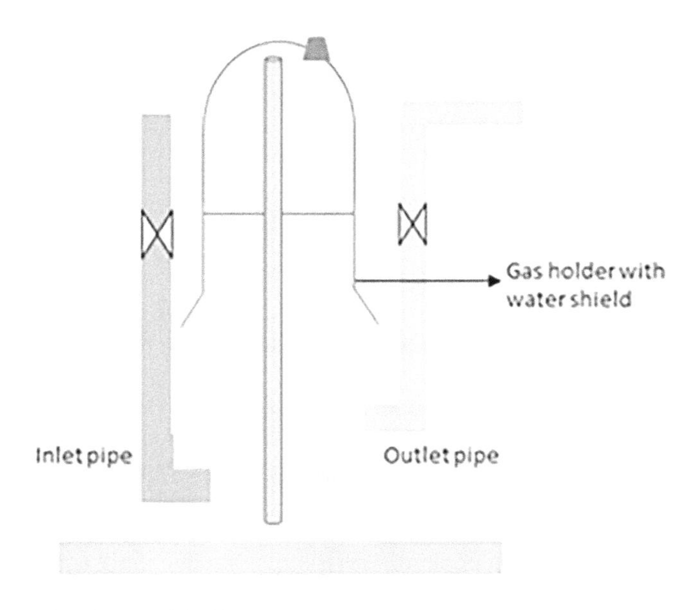

FIGURE 5.5 Conventional anaerobic digester.

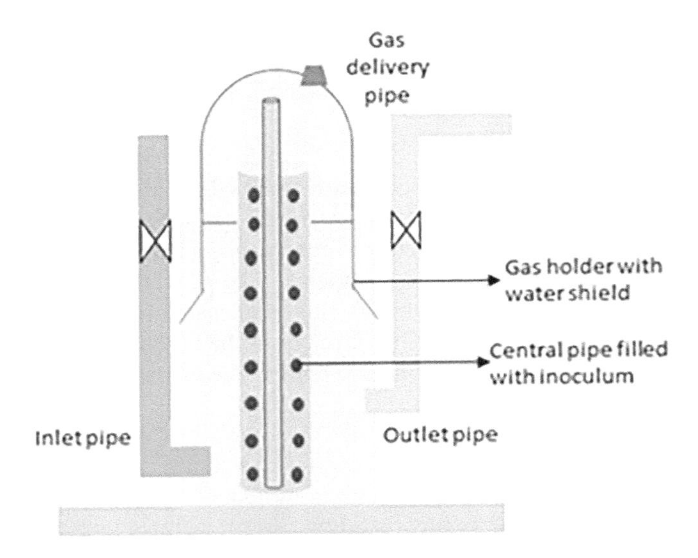

FIGURE 5.6 Design modified anaerobic digester.

in the economics of digester. The suggested digester chamber is redesigned with the concept of self-circulation. The main principle of the self-circulation occurring inside the modified anaerobic digester is density difference. The higher-density organic matter decomposes and gets lower in density. There is then a flow between high-density organic matter and the low-density degraded slurry. This process is effectively carried out by the passively modified central pipe. The inoculum inside this central pipe increases the degradation rate and makes the velocity of the domain higher. This allows the anaerobic digester to carry out the self-circulating process efficiently and increases the gas production rate.

5.4.3 Inoculum

The inoculum used in this project was goat dung. To make use of this inoculum effectively, it was poured into broken brick pieces. This brick/goat dung mixture was soaked for five days to adsorb bacteria from the dung to brick pieces, contained in an air-tight chamber during soaking period. After five days, the inoculum was filled in the passive central pipe to carry out the self-circulation. The components and concepts used in the process will be briefly described in the following sections.

5.4.4 Central Pipe

The central pipe is the main passive modification made in a floating dome anaerobic digester. This central pipe has a diameter of 11 cm and a height of 60 cm. For every 8 cm in height, a 1 cm hole was drilled on the pipe's surface. Holes have been provided in all directions. The pipe is made of polyvinyl chloride material with a thickness of 3 mm. The pipe is placed centrally inside the digester and supported by the dome holder. The soaked bricks were filled inside this pipe, which also provided mechanical support. The pipe was lifted 3 cm by means of the projected tri-stand at the bottom side; this 3 cm gap ensures a path for flow of organic matter. An internal view of the modified digester can be seen in Figure 5.7.

The inlet temperature, outlet temperature, and the gas production were analysed at frequent intervals. The produced biogas composition was analysed at regular intervals using the online biogas analyser, seen in Figure 5.8.

FIGURE 5.7 Inside view of modified digester.

5.5 OBSERVATIONS

We conducted the analytical experimental study using 20 l floating dome digesters in a laboratory.

For the comparative study, the digester was fed with composition of cow dung and water in 1:1 ratio at daily intervals. The biomethane analysis can be seen in Figure 5.9.

The modified digester result shows less HRT when compared to the conventional digester. The anaerobic digester used in our experiment shows the production rate set to 3 days; later it was 7 days. This design eliminates the need for active stirring and instead introduces the self-circulating component. The microbe density is higher in the central pipe, which increases the production rate and reduces microbe loss through slurry. We can incorporate the same design modification to residential-, commercial-, and industrial-scale anaerobic digesters. Through using the modified digester, we attained a higher gas production rate of 65% of CH_4 concentration in raw biogas, as seen in Table 5.1, 72 hours after feed. This design modified digester uses a passive modification technique with no additional energy source needed to decrease the HRT. In the case of commercial plants, the passive structure can be included, which costs significantly less compared to the digester

FIGURE 5.8 Raw biogas analysis.

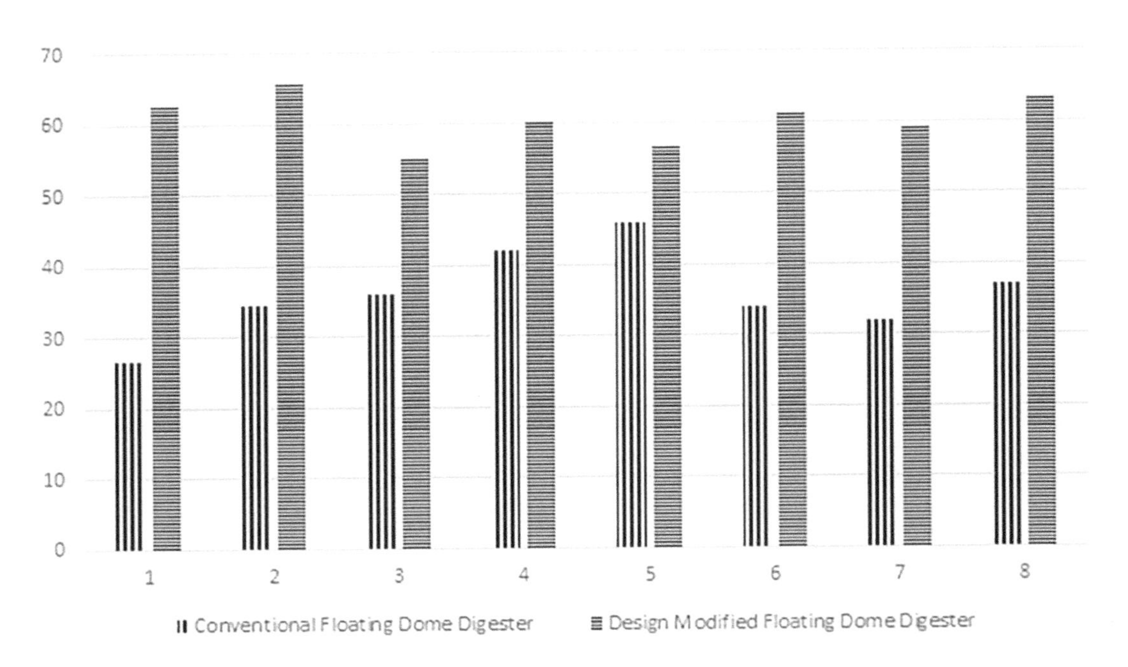

FIGURE 5.9 Biomethane analysis.

structure construction cost. Existing anaerobic digesters can be retrofitted with this modification, in addition to introducing the modification into anaerobic digester manufacturing. The central pipe can also be built using the same material used to build the anaerobic chamber. Biomethane analysis for the modified digester can be seen in Figure 5.10, and for the conventional digester in Figure 5.11.

5.6 FEASIBILITY ANALYSIS OF SELF-CIRCULATING DIGESTERS

Supply chain design is one of the most fundamental and economically dependable modules that help us to project the success and future of the project by analysing the dynamics of order. Here the complete network is framed based on both producer and consumer needs. The feasibility report has been shortened by considering a single district, Dindigul, a small city in Tamil Nadu with 10.362 °N, 77.9596 °E coordinates. The population distribution of this city would be around 25,500,000

TABLE 5.1
Comparison of Raw Biogas Components

Conventional Floating Dome Digester			Design Modified Floating Dome Digester		
Frequency 1	Interval 1 (days)	Methane Concentration (%)	Frequency 2	Interval 2 (days)	Methane Concentration (%)
10-05-2020	7	26.7	12-01-2020	3	62.6
10-12-2020	7	34.5	12-04-2020	3	61.78
10-19-2020	7	36	12-07-2020	3	55
10-26-2020	7	42	12-10-2020	3	60
11-02-2020	7	46	12-13-2020	3	56.76
11-09-2020	7	34	12-16-2020	3	61.4
11-16-2020	7	32	12-19-2020	3	59.3
11-23-2020	7	37	12-22-2020	3	60

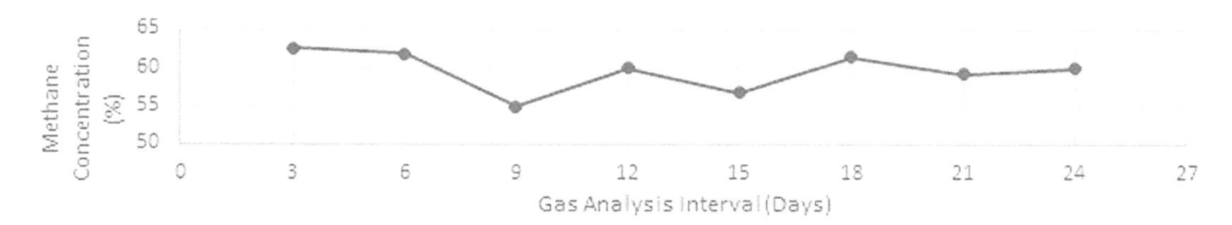

FIGURE 5.10 Biomethane analysis for self-circulating anaerobic digester.

(according to 2011 census), with per capita energy consumption at around 1687.9 kW h based on 2015 data, with 6– 7% annual rise in demand.

The available studies and statistics show huge increases in the population and waste generation in the Dindigul district. Table 5.2 shows the relation between population and the waste generation rate. The waste generated in the district mostly goes to the open dump yard, without proper treatment and segregation, though more than half of the waste generated in the district is biodegradable. In this feasibility check, we analyse the energy potential of the biodegradable waste in a particular city with gas grid integration throughout and attempt to make it self-sufficient in energy production.

Households and industries in the district mainly use liquid petroleum gas (LPG) for cooking purposes. The consumption of LPG for the past few years shows a large increase in consumption; in 2020, the consumption of LPG was about 27.41 million tonnes, about 4.3% higher than 2019, with approximately 20 kg per capita LPG annual consumption. Reviewing this energy demand pattern, we bring the idea of a centralised biogas digester unit in a different locality, a district with integrated gas gridlines. The current trend shows a drastic shift to LPG, and a massive increase in the LPG usage for cooking and other purposes.

The report published by International Institute for Population Sciences in 2016 shows that more than 80% of urban populations and more than 50% of rural populations use LPG and other clean fuel for cooking. In the Dindigul district, 66.2% of total households use only clean energy sources for cooking, as seen in Table 5.3. There is an increase in gas agencies in both urban and rural areas; at present, 38 gas agencies are operating in the district. The current data published in the district fact sheet of Dindigul shows that more than 600,000 LPG consumers are in the Dindigul district alone (Table 5.4).

The domestic price of an LPG cylinder is about Rs. 785 ± 10. The price of a cylinder includes the refilling cost, labour cost, and transportation costs. A sudden rise in the price of petroleum oils clearly projects the surge in price of various products needed for daily living, including LPG cylinders. Bringing this gas grid to a district with a population of about 2,160,000 can help to avoid an unnecessary rise in the cost of the gas; since this gas is manufactured through waste obtained from the people with in the district, it can be supplied at a fixed cost with lot more independence. By doing so, we could attain a proper circular economy with regard to cooking gas.

The initial cost involved in constructing a gasoline grid for a 12-in pipeline would be about 22,078,920 per mile, which is roughly around 1.60934 km. This price may vary according to the

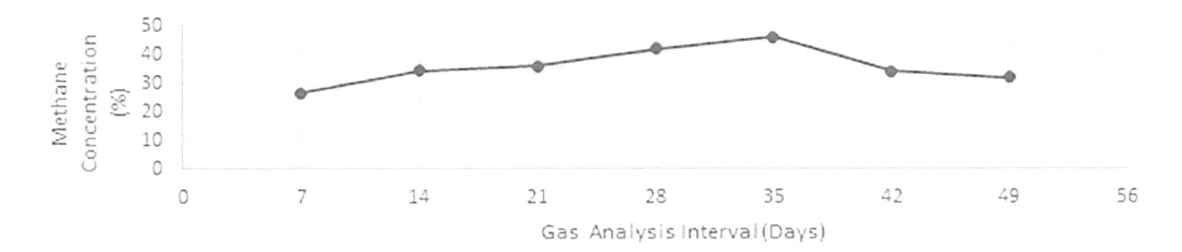

FIGURE 5.11 Biomethane analysis for conventional anaerobic digester.

TABLE 5.2
Projection of Population and Generation of Solid Waste

Year	Population	Following Year Population	Rate of Increase in Generation of Waste	Estimated Quantity in MT/Day	Est. Qty of Organic Waste in MT/Day	Est. Qty of Inert Waste to be Land Filled in MT/Year	Cumulative Total of Inert Waste in MT
2014	213,350	215,729	0.415	88.5	57.52	23.98	23.98
2015	215,729	216,110	0.425	91.68	60.50	24.18	24.18
2016	216,110	217,225	0.430	92.29	60.91	24.38	24.38
2017	217,225	218,350	0.435	94.49	63.30	24.19	24.19
2018	218,350	219,250	0.480	104.8	70.21	28.59	28.59
2019	219,250	200,190	0.480	105.89	70.94	27.95	27.95
2020	219,819	213,122	0.480	105.51	70.69	27.82	27.82
2021	213,122	221,165	0.495	105.49	71.73	27.76	27.76
2022	221,165	226,390	0.500	110.58	76.30	28.00	28.00
2023	226,390	237,210	0.505	114.32	77.73	29.59	29.59

Source: City Sanitation Plan.

TABLE 5.3
Households Using Clean Fuel for Cooking (%)

Sector	Percentage
Urban	82.2 %
Rural	55.6 %
Total	66.2 %
Actual Population	2,538,397
Male	1,174,406
Female	1,357,868

Source: District Fact Sheet, Dindigul, Tamil Nadu.

TABLE 5.4
No. of LPG Connections 2018–2019

Sl. No.	Name of the Company	No. of Agencies	No. of Customers		
			Single	Double	Total
1	Indian Oil Corporation Ltd	17	122,805	107,882	230,687
2	Bharat Petroleum Corp. Ltd.	09	93,311	82,478	175,789
3	Hindustan Petroleum Corp, Ltd.	12	113,921	82,380	196,301
District total		38	330,037	272,740	602,777

diameter of the pipeline. For a city like Dindigul, with 46.9 km^2 area and a scattered population distribution, it may take a significant initial investment to complete the construction, in addition to regular maintenance, but it could give provide a guarantee where power is concerned.

In this study, we are investigating the possibility of energy potential of available biodegradable waste to satisfy energy needs for cooking in the district. The current available data shows that organic waste of 200 kg can produce one LPG cylinder daily. In the case of Dindigul, the generation of organic waste estimated around 71.73 MT/day, making the annual biodegradable waste generation in Dindigul around 26,181.45 MT/year. Using 1 kg of biodegradable waste, we can produce 0.75 m^3 to 1.62 m^3 of biogas, though this may vary according to the digester design and other factors, such as waste characteristics or pH. The biogas potential of biodegradable waste in the district is calculated at around 42,413,949–196,360,875.5 m^3 of biogas annually. By using the suggested self-circulating biogas digester, the gas production rate can improve to an even higher range. In 2020, the consumption of LPG was about 27.41 million tonnes, which is equivalent to 27,410,000 m^3. One domestic LPG cylinder is equivalent to 28 m^3 of biogas. The availability and necessity show that available biodegradable waste potential is more than enough to satisfy the energy for cooking purposes in the Dindigul district.

We are suggesting waste as a source of income and 4P, as well as the design modification of the existing digester and effective utilisation of the raw materials to maximum energy conversion from the waste and achieve the gas requirements of the locality. When we analyse the waste data, we can see that the waste generated has 60% of biodegradable waste. Agro-residues, which have rich cellulosic parts, can also be collected separately from farmers in the district.

As the waste serves as a source of income, the community will also support this 4P model. This district topography supports both high-altitude and low-altitude vegetation. Bio-methanation is the best possible way to treat that waste. MSW management is still in a developing phase in Dindigul. By introducing 4P into the solid waste management principles of the district, it will create a circular model that is also profitable.

Instead of a common centralised plan, we suggest some decentralised biogas units in each taluk. In the case of the Dindigul district, there are eight taluks. The entire utility gas grid line can be designed by connecting these eight biogas plant units and end-users with real-time monitoring systems. By combining these units with the same gas grid network, we can also solve the gas shortage problems, thereby satisfying consumer requirements. Using this decentralised approach, we can also reduce waste collection and transportation problems.

The product outcome of the anaerobic digestor is biogas and digestate. The biogas is then fed into the utility gas grid and the digestate is passed to the solar still. The dried digestate is then powdered into dust and packed as organic manure for agricultural purposes. This will be an additional profit for this plant. Utility gridline infrastructure has to be built in this district, which takes time and is expensive, but once the grid is built, the distribution of the biogas through the city is feasible. The output biogas can be upgraded and filtered to increase the CH_4 concentration.

Biomass has huge energy potential. The suggested circular approach to use the regionally obtainable biomass as usable energy with a self-circulating biogas generator will decrease waste management issues, pollution, greenhouse gas emissions, heating, and, therefore, the use of fossil fuels. The implementation of the 4P approach through waste as a source of income can solve the problems related to the biomass supply chain. The conversion of biomass waste into biogas is one of the most economically viable options for highly populated and developing countries like India. Compared to other technologies for the treatment of MSW, the economic aspects, energy savings, and ecological advantages make anaerobic processing a more attractive choice.

5.7 ECONOMIC ASPECTS

The model seeks to find the optimal way of producing biogas, such that the biogas plant projects become economically feasible by maximising profit.

5.7.1 Commercialisation of End Product

To make this model a successful one, a proper business model is necessary, since most of the digester plants installed today are searching for their customers online and through other sources, but fail to make a proper supply chain. The raw material obtained will also act as the most suitable platform to sell the product more profitably. The suggested model is going to benefit the local people within the district of the project. The gas can be supplied by installing a gas grid with a utility meter so that pure methane gas can be provided at a reasonable cost at all times. The fertiliser produced is also highly nutritious and good plant growth, which can be packaged and sold.

5.7.2 Business Credibility

India is a second largest biogas-installed country, but it still has a long way to go in this field. Both the public and private biogas digester units seem like a failing model because of an inability to take the product to their consumer. We studied a number of digester plants installed in and around Dindigul to understand the problem completely. From that, we came to the conclusion that consumers can be reached through the gas grid, procuring the raw material from consumers, that can then be sold back to them as the end product. The conclusion is reaching the customer through the gas grid. Here we can easily convince the people around the plant, since we take the raw material from them, establishing a relationship; through this we are also able to create a customer base within a short span of time.

Biogas generation also depends upon the raw material that obtained, that is 10 kg of green grass is enough to produce 1 kg of biogas, but for cow dung, 40 kg of dung is needed to produce 1 kg of biogas; if a cow could produce approximately 10–20 kg dung/day, that is equal to approximately 0.25–0.5 kg of biogas. The average cooking fuel need for a single person is 0.1–0.3 kg of biomass, making this sufficient to cook food for four people. If this is used to run a direct current engine, it will produce 0.5 KWh of electricity. Dindigul generated 10 tonnes of biodegradable waste per day according to the Dindigul district data book 2010. If at least a quarter of this is used for biogas generation, an accessible and highly reliable gas connection can be provided to the whole city.

5.7.3 Government Schemes That Support This Project

National Biogas and Organic Manure Programme and Biogas based Power Generation (Off-grid) and Thermal energy applications Programme are two existing government schemes in India related to biogas production and supply. These programmes can help to find a solution for the capital cost for installation of the proposed system and to solve the problems related to gas distribution in the area.

5.8 FUTURE WORK

The production of biogas is a simple and natural method of bioconversion. A comparison of the present worth, future worth, and rate of return of the plant shows that it is an economically feasible business solution. Apart from that model helping to solve the present energy crises, environmental problems are addressed through sustainable utilisation of resources by establishing a proper 4P model. We are facing an energy crisis due to diminishing fossil fuels and their price rise, which causes a serious threat to both industry and consumers. That is why alternate fuel technology is in high demand, as seen through the use of hybrid cars that run through battery and gas. A gas-powered diesel engine could act as a generator and fuel cell technology can provide power. If our waste collection and processing method is success, it will become a profitable business model and a solution for many other problems that we are facing today. In the future, we can expand the unit as per the availability of feeds and can make use of the biogas as an alternate fuel. We can also expand to electricity generation, portable electricity genset, and the automobile industry, as well as other forms of transportation.

5.9 CONCLUSION

Biomass has huge energy potential. The suggested circular approach of using regionally obtainable biomass to convert into usable energy with self-circulating biogas generators will decrease waste management issues, pollution, greenhouse gas emissions, heating, and, therefore, the use of fossil fuels. The implementation of the 4P approach through waste as a source of income can solve the problems related to the biomass supply chain. The conversion of biomass waste into biogas is one of the most economically viable options for highly populated and developing countries like India. Compared to other technologies for the treatment of MSW, the economic aspects, energy savings, and ecological advantages make anaerobic processing the more attractive choice.

ACKNOWLEDGEMENT

The authors acknowledge the support received from the Ministry of New and Renewable Energy under One Time Grant for the laboratory's upgradation.

REFERENCES

1. Ahammad, S.Z.; Sreekrishnan, T.R. Biogas: An evolutionary perspective in the Indian context. In Green Fuels Technology; Green Energy and Technology; Springer: Cham, Switzerland, 2016; pp. 431–443, ISBN 978-3-319-30203-4. https://doi.org/10.1007/978-3-319-30205-8_17
2. Angelidaki, I.; Treu, L.; Tsapekos, P.; Luo, G.; Campanaro, S.; Wenzel, H.; Kougias, P.G. Biogas upgrading and utilization: Current status and perspectives. Biotechnol. Adv. 2018, 36, 452–466. https://doi.org/10.1016/j.biortech.2016.05.026
3. Bos, K.; Chaplin, D.; Mamun, A. Benefits and challenges of expanding grid electricity in Africa: A review of rigorous evidence on household impacts in developing countries. Energy Sustain. Dev. 2018, 44, 64–77. https://doi.org/10.1016/j.esd.2018.02.007
4. Cheng, S.; Li, Z.; Mang, H.-P.; Huba, E.-M.; Gao, R.; Wang, X. Development and application of prefabricated biogas digesters in developing countries. Renew. Sustain. Energy Rev. 2014, 34, 387–400. https://doi.org/10.1016/j.rser.2014.03.035
5. Kemausuor, F.; Bolwig, S.; Miller, S. Modelling the socio-economic impacts of modern bioenergy in rural communities in Ghana. Sustain. Energy Technol. Assess. 2016, 14, 9–20. https://doi.org/10.1016/j.seta.2016.01.007
6. Kinyua, M.N.; Rowse, L.E.; Ergas, S.J. Review of small-scale tubular anaerobic digesters treating livestock waste in the developing world. Renew. Sustain. Energy Rev. 2016, 58, 896–910. https://doi.org/10.1016/j.rser.2015.12.324
7. Lebuhn, M.; Munk, B.; Effenberger, M. Agricultural biogas production in Germany—From practice to microbiology basics. Energy Sustain. Soc. 2014, 4, 10. https://doi.org/10.1186/2192-0567-4-10
8. Mittal, S.; Ahlgren, E.O.; Shukla, P.R. Barriers to biogas dissemination in India: A review. Energy Policy 2018, 112, 361–370. https://doi.org/10.3390/en5082911
9. Mshandete, A.M.; Parawira, W. Biogas technology research in selected sub-Saharan African countries—A review. Afr. J. Biotechnol. 2009, 8, 116–125.
10. Rajendran, K.; Aslanzadeh, S.; Taherzadeh, M.J.; Rajendran, K.; Aslanzadeh, S.; Taherzadeh, M.J. Household biogas digesters—A review. Energies 2012, 5, 2911–2942. https://doi.org/10.3390/en5082911
11. Renewable Energy Network of the 21st Century. Renewables 2018 Global Status Report. 2018. Available online: http://www.ren21.net/status-of-renewables/global-status-report/ (accessed on 12 June 2018).
12. Roubík, H.; Mazancová, J.; Le Dinh, P.; Dinh Van, D.; Banout, J.Biogas quality across small-scale biogas plants: A case of central Vietnam. Energies 2018, 11, 1794.
13. Sehgal, K. Current state and future prospects of global biogas industry. In Biogas: Biofuel and Biorefinery Technologies; Springer: Cham, Switzerland, 2018; pp. 449–472, ISBN 978-3-319-77334-6.
14. UN Climate Change Annual Report. 2018. United Nations Framework Convention on Climate Change (UNFCCC).
15. Year End Review. 2018. MNRE, Ministry of New and Renewable Energy, Government of India.

Section II

Biomethane Research and Development

6 Improvement in Biogas Production by Direct Interspecies Electron Transfer Technique

Current Aspects and Diverse Prospects

Ravi Bajaj,[1] Kapil Garg,[1] Sagarika Panigrahi,[1] and Brajesh Kumar Dubey

6.1 INTRODUCTION

Anaerobic digestion, a process for the production of biogas from wastewater and organic waste, has received more attention in past few decades. Anaerobic digestion is one of the most widely used techniques for producing methane (Panigrahi et al., 2020, 2019; Panigrahi and Dubey, 2019a, 2019b). The last step of anaerobic digestion, methanogenesis, involves conversion of acetate, ethanol, carbon dioxide (CO_2) and hydrogen (H_2) to methane by methanogenic bacteria or methanoarchaea in anaerobic condition. Methanogens are the microbes that ensure the progress of the reaction by producing methane (Thauer, 1998). They consume organic reactants like CO_2, H_2, ethanol, and acetic acid (De Bok et al., 2004; Thauer, 1998). Complex organic compounds are biodegraded to ethanol, propionate etc. first, then these alcohols, lipids, and fatty acids react with methanogens to build syntrophic associations, which results in production of methane (De Bok et al., 2004; Shin et al., 2010). Acetoclastic methanogens exploit acetate, while H_2 is used up by hydrogenotrophic methanogens via reducing CO_2. It could be said that hydrogenotrophic methanogens help in feasibility of this reaction thermodynamically by maintaining low partial pressure of H_2 and making the reaction H_2 driven, therefore helping in fermentation of reduced organic substrates (De Bok et al., 2004). Direct interspecies electron transfer (DIET) involves electron transfer between syntrophic microbial activities. Volatile fatty acids produced from acidogenesis such as formate act as electron carriers in some methanogenic environments (Boone et al., 1989).

In previous studies (Morita et al., 2011; Rotaru et al., 2014a), it was found that, for electron transfer between fermentative bacteria and methanogens, the transfer of hydrogen and formate plays an important role. The reaction is improved thermodynamically and metabolically, organic compounds are consumed, and methane is formed more efficiently in this recently discovered cell-to-cell electron transfer (Cheng and Call, 2016). This accelerates the rate of conversion of organic waste to methane (Morita et al., 2011). Electrophilic methanogens are bacteria that accept electrons from other species directly (Lovley, 2011). In previous studies (Zhao et al., 2015, 2016b), it has been observed that DIET in an anerobic digestion is affected by different factors, such as organic loading rate, hydraulic retention time, and substrate type.

[1] Authors contributed equally

DOI: 10.1201/9781003204435-8

Species such as *Geobacter* and *Pseudomonas* were found to be able to generate electricity in electrochemical microbial systems via extracellular electron transfer. Hence such species were recognised as exoelectrogens, or electrochemically active bacteria (Chang et al., 2006; Logan, 2009). However, the capacities of other fermentative bacteria such as *Syntrophomonas*, *Streptococcus*, *Bacteroides*, and *Sporanaerobacter* are still unknown and yet to be investigated (Dang et al., 2017). An electrical connection between interspecies is mandatory in DIET. *Geobactor* have been found to be making biologically wired connections to methanogens with the help of filamentous protein appendages known as electrically conductive pili or microbial nanowire (Rotaru et al., 2014b, 2014a). Studies have shown that DIET ability can be induced within those bacteria that could not create pili like *Geobacter* spp by the addition of conductive material such as granular activated carbon (GAC), carbon cloth, biochar, and iron nanoparticles. (Chen et al., 2014; Liu et al., 2012). Syntrophic partners can stick to the surface of these conductive materials and can be utilised as electrical conduits for electron exchange. The conductive materials can improve the performance of microbes by reducing their energy utilisation in synthetisation of conductive pili (Zhao et al., 2015). Promoting DIET in anaerobic digestion can automatically boost methane production rates (Dang et al., 2016; Liu et al., 2012; Rotaru et al., 2014a; Zhao et al., 2015). The kinetic analysis showed that the electron transfer obtained by DIET is 8.6 times higher than that of conventional electron transfer through bacteria (Storck et al., 2016). Therefore, higher organic loading rate (OLR) can be handled by DIET-active digesters than that of conventional ones (Dang et al., 2016; Zhao et al., 2016a).

Recently, a number of papers have been published on the improvement of methane production by DIET from anaerobic digestion of organic substrate. However, a review article compiling the research articles has not been published. This review article covers the role of conductive materials in DIET, and the effect of substrate characteristics in DIET is reviewed.

6.2 CONDUCTIVE MATERIALS

The effect of conductive material on enhancement of methane production is dependent upon the digester configuration. The conductive materials perform better electron transfer in up-flow anaerobic digesters compared to continuous stirred tank reactors. The better performance in up-flow anaerobic digesters could be related to aggregation of feedstock. The effects of conductive materials, including biological materials such as GAC, iron nanoparticles, carbon-based materials, and hydrochar have been used to improve methane production rates (Rotaru et al., 2014a).

6.2.1 Granular Activated Carbon

GAC has been exercised as a purifying agent for a long time, because its highly porous surface area provides a larger surface area for the absorption of contaminants. Recently it has been used in bioreactors to improve solid retention time (Singh and Prerna, 2009; Tao et al., 2017). Different types of microorganisms have been found to be accepting electrons from GAC after its deployment for different biocathodes in microbial fuel cells (Wei et al., 2011). The microscopic analysis of GAC shows its surface attracted both the species, to which they were tightly attached. Also, no aggregates were formed for electron transfer through biological connective. The role of conductive pili was seen to relieve once DIET-active sites developed on the solid surface of additive GAC (Zhao et al., 2015). GAC can be used in continuous stirred tank reactors, up-flow anaerobic digesters, and sequential batch anaerobic digesters handling sewage sludge, the organic fraction of municipal solid waste, food waste, etc. The application of GAC in a continuous anaerobic digester can improve the methane production by 1.8 times over the control (Lee et al., 2016a; Lee et al., 2016b). In another study for anaerobic digestion of lignocellulosic biomass, it was observed that addition of GAC improved the methane production rate by 3.7 times (Dang et al., 2017). However, limited information has been provided on optimisation of GAC loading to improve methane production rate. In recent reports, it has been reported that 5 $g/L_{reactor}$ GAC could be optimum among loading range from 0.5 to 5 g/L;

some literature also used higher GAC dose ranging from 10 to 50 $g/L_{reactor}$ (Dang et al., 2017; Liu et al., 2012). More studies need to be conducted to optimise GAC loading for improving methane production rate.

6.2.2 Biochar

Biochar, a high carbon content organic matter, is processed by thermal treatment of organic substrate (Ahmad et al., 2014). It is used in different sectors such as gardening, filtration, and fuel cell and building materials. However, it has been mostly utilised as an adsorbent for removal of contaminants, such as antibiotics, pentachlorophenol, tylosin, atrazine, and pyrimethanil. (Ahmad et al., 2014). The use of biochar is comparable with GAC in anaerobic digestion systems (Liu et al., 2012).

6.2.3 Iron Nanoparticles

Demand for nanoparticles is increasing because of their diverse application. They are used as a primary colourant in glass and ceramics, as a catalyst in medical industry, and many other applications. (Wu et al., 2015). Their application in the field of anaerobic digestion is their use to facilitate electron transfer from exoelectrogenic bacteria to electrotrophic methanogen. In DIET, the addition of iron nanoparticles such as haematite and magnetite improves the production rate of methane (Kato et al., 2012). In most cases, conductive materials provide a path for electron transfer from extracellular electron transfer (EET) bacteria to electrotrophic methanogen, but in case of iron nanoparticles, bacteria attaches to conductive pili and works in place of c-type cytochromes for DIET (Figure 6.1) (Cruz Viggi et al., 2014). Hence, directing DIET in absence of pili may not be possible in the case of iron nanoparticles. The major problem with iron nanoparticles is high cost, which limits their use.

6.2.4 Stainless Steel

Li et al. (2017) investigated the addition of stainless steel in an up-flow anaerobic sludge bed (UASB) digester fed with sulphate-rich acetate medium. It is observed that the use of 0.5 g/L stainless steel increases methane production 4.5 times. High conductivity, affordable price, high chemical resistance, and mechanical strength make stainless steel more acceptable than carbon-based conductive materials. However, the adhesion force of iron nanoparticles with biomass is smaller (Sarathi and Nahm, 2013), which limits the use of stainless steel in the case of complex organic matter. The use of iron nanoparticles to improve methane production is under research.

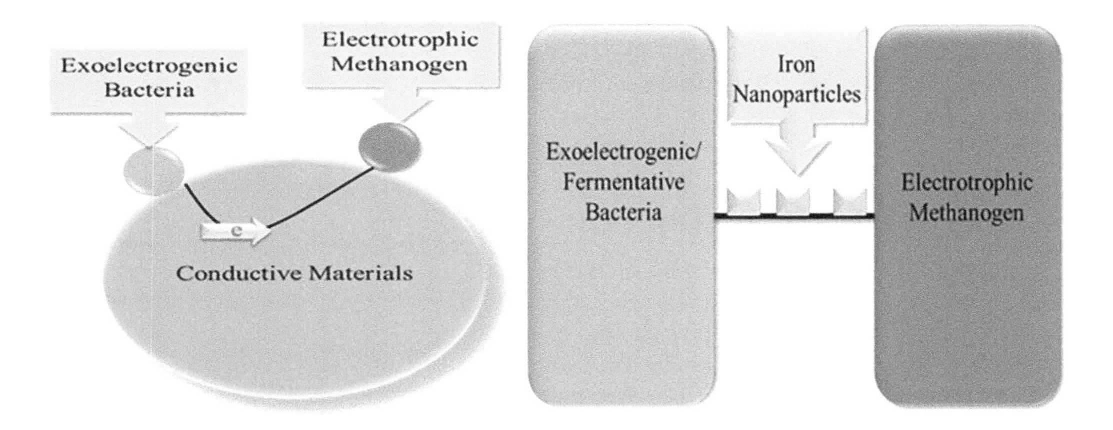

FIGURE 6.1 Methods adopted by different conductive materials versus iron nanoparticles for promoting direct interspecies electron transfer.

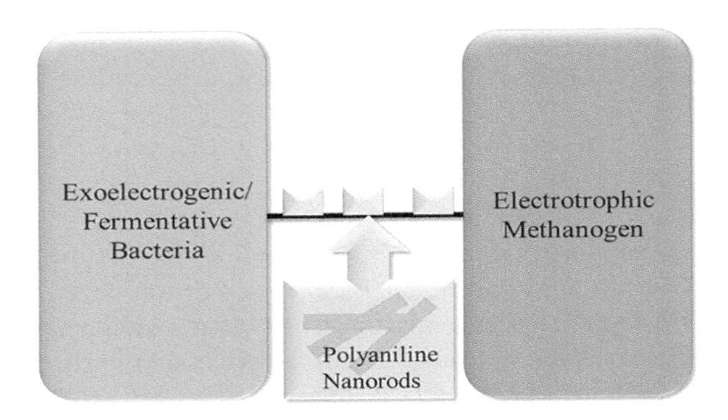

FIGURE 6.2 The mechanism of branching structure of polyaniline in electron transfer.

6.2.5 Carbon-Based Nanomaterials

Carbon-based nanomaterials include materials with high conductivity and large surface area like graphene, carbon nanotubes, and fullerenes (Zhang et al., 2013). They promote the process of DIET in anaerobic digestion. Tian et al. (2017) studied the effect of these materials on methanogenesis. The results showed that short-term exposure to graphene increases the rate of methane production, whereas long-term exposure inhibits the reaction. Zhang and Lu (2016) concluded that dose of about 5 g/L carbon-based nanomaterials increases the gas production rate by 50%. In the study of Li et al. (2015) it was concluded that mesophilic methanogenic communities secret extracellular polymeric substances, which brings them closer to single-walled carbon nanotubes. A similar process was observed in magnetite nanoparticle-supplemented digester (Cruz Viggi et al., 2014). There may be a similarity between iron nanoparticles and carbon-based nanoparticles that needs further study, however the high cost of carbon nanotubes limits their use.

6.2.6 Conductive Polymeric Material

Use of polymers such as polyaniline and polypyrrole improve DIET and increase the methane production rate because of their high conductive nature (Sarathi and Nahm, 2013). In the study of Hu et al. (2017) the addition of 0.6 g/L polyaniline nanorods in batch anaerobic digesters almost doubled the methane produce over control. It is an efficient alternative conductive material for DIET between exoelectrogenic bacteria and electrotrophic methanogen. The study of the process under a microscope showed that polyaniline rods are uniformly diffused in the digester sludge. As the polyaniline is positively charged, it attaches itself to the negatively charged microbial cell membrane (Yong et al., 2012). The branching structure of polyaniline can act like conductive pili (Figure 6.2) and helps in the electron transfer from EET bacteria to electrotrophic methanogens. Hence, polyaniline can be used to increase methane production rate in the absence of conductive pili.

6.3 ROLE OF ELECTRICAL CONDUCTIVITY IN DIET

In the study of Zhao et al. (2016a), the range of ohmic conductivity of multi-species aggregates was from 0.2 to 36.7 μS/cm. Moreover, it was observed that the decrease in the conductivity of aggregate does not affect the rate of DIET, which showed that during electron transfer the reaction was only influenced by the conductivity of conductive materials, not by the conductivity of any inorganic metals or minerals (Morita et al., 2011). Shrestha and Rotaru (2014) observed a relation between the availability of *Geobacter* and conductivity of a UASB aggregate, whereas the presence of other types of methanogens and bacteria did not contribute to the conductivity of aggregate. The study of Eaktasang et al. (2016) indicated that some fermentative and sulphate-reducing bacteria help in the formation of conductive pili, which makes electron transfer more effective. In the study of Malvankar et al. (2012), it was observed that the properties of conductive pili formed by *Geobacter* and other potential DIET-active species depend on external factors like pH, temperature, etc. This field requires further detailed investigation. In the study of Liu et al. (2012) the addition of highly conductive materials like GAC (3000 ± 327 μS/cm) improves DIET kinetics. The conductivity of various material used for improving DIET is significantly higher than that of biochar (2.11–4.41 μS/cm). However, biochar was found to be as effective as GAC in improving methane-producing kinetics in co-cultures of *Geobacter metallireducens* and *Methanosarcina barkeri* (Chen et al., 2014). Despite conductivity differences between semi-conductive haematite and conductive magnetite, the addition of both shows comparable results in DIET-stimulated methanogenic environments (Kato et al., 2012). There may be the concept of threshold conductivity that exists after which individual conductivity does not much affect the methane production rate.

6.4 IMPORTANCE OF SUBSTRATE CHARACTERISTICS IN DIET

Studies on various electron donors have been conducted to study their effect on the methane production kinetics. However, most of these studies have been done on simple organic acids, alcohol, and very few on the complex organic wastes (such as animal food or yard waste). Mixed-culture studies identified *Geobacter* as a major EET bacteria with methanogens (Lin et al., 2017). In the conversion of ethanol to methane, the mechanism of DIET involved is well observed from the studies of co-culture samples (Liu et al., 2012). However, ethanol is not always the intermediate in conversion during fermentation in anaerobic digestion; reduced organic compounds such as propionate and butyrate can be major intermediates in some cases (De Bok et al., 2004). A recent study has shown that propionate and butyrate cannot metabolise in the co-culture of *G. metallireducens* and *Methanosarcina barkeri*, even after the addition of various conductive materials (Wang et al., 2016). Some studies have shown the presence of conductive materials supports the conversion of propionate and butyrate to methane (Cruz Viggi et al., 2014); however, the path adopted for the conversion of propionate and butyrate to methane was not clear in the studies. In the study of Zhao et al. (2016a), the methanogenic communities enriched by ethanol in the long term could degrade propionate and butyrate. Some *Geobacter* species can metabolise butyrate (Prakash et al., 2010). In the case of complex organic waste such as animal food, fermentative microorganisms would mainly participate in DIET in the presence of conductive materials (Dang et al., 2017). In the study of Dang et al. (2016), fermentative *Sporanaerobacter* species follow different pathways for the conversion of complex organic waste to acetate and carbon dioxide along with direct electron transfer to *Methanosarcina* species. More studies are required in the field of DIET to make it more effective for complex organic compounds.

6.5 EFFECT OF ORGANIC LOADING RATES

OLR is an important parameter that affects the activity of methanogens and their kinetics (Ferguson et al., 2016). The operating OLR depends on the parameter called hydraulic retention time. There must be an optimum OLR for every process that depends on the substrate characteristics, such as biodegradability, C/N ratio, etc. In the case of high OLRs, fermentation rate is more than methanogenesis rate, which leads to instability of the process and can promote the reverse reaction (Ferguson et al., 2016). The adaptation of the DIET method increases the methane production rate, which makes the process happen even in the case of high OLRs (Zhao et al., 2016a). Zhao et al. (2015) observed the effect of different OLRs ranged from 4.11 to 12.33 kg COD/m^3/d in continuous UASB digesters added with organic additives like biochar or graphite. At each OLR, the digester showed a 30–40% lower methane production rate in the absence of conductive materials. Dang et al. (2016) found the effect of different OLRs ranged from 1.6 to 10.3 kg COD/m^3/d in digesters fed with complex organic matter in the form of commercial dog food. The digester with GAC did not show accumulation of volatile fatty acids and alcohols at high OLR. However, in a controlled digester, the above circumstances decreased methane production because of the acidification of the digester as the process proceeded. This observation has shown that DIET would be more effective at higher OLRs when a digester is fed with complex feedstock.

6.6 CONCLUSION

DIET is an emerging method that improves the efficiency of anaerobic digestion. The process of EET can be made more effective and less energy consuming by the use of conductive additives. DIET increases the methane production rate even in the case of complex organic waste.

REFERENCES

Ahmad, M., Rajapaksha, A.U., Lim, J.E., Zhang, M., Bolan, N., Mohan, D., Vithanage, M., Lee, S.S., Ok, Y.S., 2014. Biochar as a sorbent for contaminant management in soil and water: a review. Chemosphere. 99, 19–33.

Boone, D.R., Johnson, R.L., Liu, Y., 1989. Diffusion of the interspecies electron carriers H_2 and formate in methanogenic ecosystems and its implications in the measurement of K_m for H_2 or formate uptake. Appl. Environ. Microbiol. 55, 1735–1741.

Chang, I.-S., Moon, H.-S., Bretschger, O., Jang, J.-K., Park, H.-I., Nealson, K.H., Kim, B.-H., 2006. Electrochemically active bacteria (EAB) and mediator-less microbial fuel cells. J. Microbiol. Biotechnol. 16, 163–177.

Chen, S., Rotaru, A.-E., Liu, F., Philips, J., Woodard, T.L., Nevin, K.P., Lovley, D.R., 2014. Carbon cloth stimulates direct interspecies electron transfer in syntrophic co-cultures. Bioresour. Technol. 173, 82–86.

Cheng, Q., Call, D.F., 2016. Hardwiring microbes via direct interspecies electron transfer: mechanisms and applications. Environ. Sci. Process. Impacts. 18, 968–980.

Cruz Viggi, C., Rossetti, S., Fazi, S., Paiano, P., Majone, M., Aulenta, F., 2014. Magnetite particles triggering a faster and more robust syntrophic pathway of methanogenic propionate degradation. Environ. Sci. Technol. 48, 7536–7543.

Dang, Y., Holmes, D.E., Zhao, Z., Woodard, T.L., Zhang, Y., Sun, D., Wang, L.-Y., Nevin, K.P., Lovley, D.R., 2016. Enhancing anaerobic digestion of complex organic waste with carbon-based conductive materials. Bioresour. Technol. 220, 516–522.

Dang, Y., Sun, D., Woodard, T.L., Wang, L.Y., Nevin, K.P., Holmes, D.E., 2017. Stimulation of the anaerobic digestion of the dry organic fraction of municipal solid waste (OFMSW) with carbon-based conductive materials. Bioresour. Technol. 238, 30–38.

De Bok, F.A.M., Plugge, C.M., Stams, A.J.M., 2004. Interspecies electron transfer in methanogenic propionate degrading consortia. Water Res. 38, 1368–1375.

Eaktasang, N., Kang, C.S., Lim, H., Kwean, O.S., Cho, S., Kim, Y., Kim, H.S., 2016. Production of electrically-conductive nanoscale filaments by sulfate-reducing bacteria in the microbial fuel cell. Bioresour. Technol. 210, 61–67.

Ferguson, R.M.W., Coulon, F., Villa, R., 2016. Organic loading rate: A promising microbial management tool in anaerobic digestion. Water Res. 100, 348–356.

Hu, Q., Sun, D., Ma, Y., Qiu, B., Guo, Z., 2017. Conductive polyaniline nanorods enhanced methane production from anaerobic wastewater treatment. Polymer (Guildf). 120, 236–243.

Kato, S., Hashimoto, K., Watanabe, K., 2012. Methanogenesis facilitated by electric syntrophy via (semi) conductive iron-oxide minerals. Environ. Microbiol. 14, 1646–1654.

Lee, H.-S., Dhar, B.R., An, J., Rittmann, B.E., Ryu, H., Santo Domingo, J.W., Ren, H., Chae, J., 2016a. The roles of biofilm conductivity and donor substrate kinetics in a mixed-culture biofilm anode. Environ. Sci. Technol. 50, 12799–12807.

Lee, J.-Y., Lee, S.-H., Park, H.-D., 2016b. Enrichment of specific electro-active microorganisms and enhancement of methane production by adding granular activated carbon in anaerobic reactors. Bioresour. Technol. 205, 205–212.

Li, L.-L., Tong, Z.-H., Fang, C.-Y., Chu, J., Yu, H.-Q., 2015. Response of anaerobic granular sludge to single-wall carbon nanotube exposure. Water Res. 70, 1–8.

Li, Y., Zhang, Y., Yang, Y., Quan, X., Zhao, Z., 2017. Potentially direct interspecies electron transfer of methanogenesis for syntrophic metabolism under sulfate reducing conditions with stainless steel. Bioresour. Technol. 234, 303–309.

Lin, R., Cheng, J., Zhang, J., Zhou, J., Cen, K., Murphy, J.D., 2017. Boosting biomethane yield and production rate with graphene: the potential of direct interspecies electron transfer in anaerobic digestion. Bioresour. Technol. 239, 345–352.

Liu, F., Rotaru, A.-E., Shrestha, P.M., Malvankar, N.S., Nevin, K.P., Lovley, D.R., 2012. Promoting direct interspecies electron transfer with activated carbon. Energy Environ. Sci. 5, 8982–8989.

Logan, B.E., 2009. Exoelectrogenic bacteria that power microbial fuel cells. Nat. Rev. Microbiol. 7, 375.

Lovley, D.R., 2011. Live wires: direct extracellular electron exchange for bioenergy and the bioremediation of energy-related contamination. Energy Environ. Sci. 4, 4896–4906.

Malvankar, N.S., Lau, J., Nevin, K.P., Franks, A.E., Tuominen, M.T., Lovley, D.R., 2012. Electrical conductivity in a mixed-species biofilm. Appl. Environ. Microbiol. 78, 5967–5971.

Morita, M., Malvankar, N.S., Franks, A.E., Summers, Z.M., Giloteaux, L., Rotaru, A.E., Rotaru, C., Lovley, D.R., 2011. Potential for direct interspecies electron transfer in methanogenic wastewater digester aggregates. MBio. 2, e00159–11.

Panigrahi, S., Dubey, B.K., 2019a. Electrochemical pretreatment of yard waste to improve biogas production: Understanding the mechanism of delignification, and energy balance. Bioresour. Technol. 292, 121958.

Panigrahi, S., Dubey, B.K., 2019b. A critical review on operating parameters and strategies to improve the biogas yield from anaerobic digestion of organic fraction of municipal solid waste. Renew. Energy. 143, 779–797.

Panigrahi, S., Sharma, H.B., Dubey, B.K., 2019. Overcoming yard waste recalcitrance through four different liquid hot water pretreatment techniques – Structural evolution, biogas production and energy balance. Biomass Bioenerg. 127, 105268.

Panigrahi, S., Sharma, H.B., Dubey, B.K., 2020. Anaerobic co-digestion of food waste with pretreated yard waste: A comparative study of methane production, kinetic modeling and energy balance. J. Clean. Prod. 243, 118480.

Prakash, O.M., Gihring, T.M., Dalton, D.D., Chin, K.-J., Green, S.J., Akob, D.M., Wanger, G., Kostka, J.E., 2010. *Geobacter daltonii* sp. nov., an Fe (III)-and uranium (VI)-reducing bacterium isolated from a shallow subsurface exposed to mixed heavy metal and hydrocarbon contamination. Int. J. Syst. Evol. Microbiol. 60, 546–553.

Rotaru, A.-E., Shrestha, P.M., Liu, F., Markovaite, B., Chen, S., Nevin, K.P., Lovley, D.R., 2014a. Direct interspecies electron transfer between Geobacter metallireducens and Methanosarcina barkeri. Appl. Environ. Microbiol. 80, 4599–4605.

Rotaru, A.-E., Shrestha, P.M., Liu, F., Shrestha, M., Shrestha, D., Embree, M., Zengler, K., Wardman, C., Nevin, K.P., Lovley, D.R., 2014b. A new model for electron flow during anaerobic digestion: direct interspecies electron transfer to *Methanosaeta* for the reduction of carbon dioxide to methane. Energy Environ. Sci. 7, 408–415.

Sarathi, V.G.S., Nahm, K.S., 2013. Recent advances and challenges in the anode architecture and their modifications for the applications of microbial fuel cells. Biosens. Bioelectron. 43, 461–475.

Shin, S.G., Han, G., Lim, J., Lee, C., Hwang, S., 2010. A comprehensive microbial insight into two-stage anaerobic digestion of food waste-recycling wastewater. Water Res. 44, 4838–4849.

Shrestha, P.M., Rotaru, A.-E., 2014. Plugging in or going wireless: strategies for interspecies electron transfer. Front. Microbiol. 5, 237.

Singh, S.P., Prerna, P., 2009. Review of recent advances in anaerobic packed-bed biogas reactors. Renew. Sustain. Energy Rev. 13, 1569–1575.

Storck, T., Virdis, B., Batstone, D.J., 2016. Modelling extracellular limitations for mediated versus direct interspecies electron transfer. ISME J. 10, 621.

Tao, J., Qin, L., Liu, X., Li, B., Chen, J., You, J., Shen, Y., Chen, X., 2017. Effect of granular activated carbon on the aerobic granulation of sludge and its mechanism. Bioresour. Technol. 236, 60–67.

Thauer, R.K., 1998. Biochemistry of methanogenesis: a tribute to Marjory Stephenson: 1998 Marjory Stephenson prize lecture. Microbiology. 144, 2377–2406.

Tian, T., Qiao, S., Li, X., Zhang, M., Zhou, J., 2017. Nano-graphene induced positive effects on methanogenesis in anaerobic digestion. Bioresour. Technol. 224, 41–47.

Wang, L.-Y., Nevin, K.P., Woodard, T.L., Mu, B.-Z., Lovley, D.R., 2016. Expanding the diet for DIET: electron donors supporting direct interspecies electron transfer (DIET) in defined co-cultures. Front. Microbiol. 7, 236.

Wei, J., Liang, P., Huang, X., 2011. Recent progress in electrodes for microbial fuel cells. Bioresour. Technol. 102, 9335–9344.

Wu, W., Wu, Z., Yu, T., Jiang, C., Kim, W.-S., 2015. Recent progress on magnetic iron oxide nanoparticles: synthesis, surface functional strategies and biomedical applications. Sci. Technol. Adv. Mater. 16, 23501.

Yong, Y.-C., Dong, X.-C., Chan-Park, M.B., Song, H., Chen, P., 2012. Macroporous and monolithic anode based on polyaniline hybridized three-dimensional graphene for high-performance microbial fuel cells. ACS Nano. 6, 2394–2400.

Zhang, J., Lu, Y., 2016. Conductive Fe_3O_4 nanoparticles accelerate syntrophic methane production from butyrate oxidation in two different lake sediments. Front. Microbiol. 7, 1316.

Zhang, Q., Huang, J., Qian, W., Zhang, Y., Wei, F., 2013. The road for nanomaterials industry: A review of carbon nanotube production, post-treatment, and bulk applications for composites and energy storage. Small. 9, 1237–1265.

Zhao, Z., Zhang, Y., Holmes, D.E., Dang, Y., Woodard, T.L., Nevin, K.P., Lovley, D.R., 2016a. Potential enhancement of direct interspecies electron transfer for syntrophic metabolism of propionate and butyrate with biochar in up-flow anaerobic sludge blanket reactors. Bioresour. Technol. 209, 148–156.

Zhao, Z., Zhang, Y., Woodard, T.L., Nevin, K.P., Lovley, D.R., 2015. Enhancing syntrophic metabolism in up-flow anaerobic sludge blanket reactors with conductive carbon materials. Bioresour. Technol. 191, 140–145.

Zhao, Z., Zhang, Y., Yu, Q., Dang, Y., Li, Y., Quan, X., 2016b. Communities stimulated with ethanol to perform direct interspecies electron transfer for syntrophic metabolism of propionate and butyrate. Water Res. 102, 475–484.

7 Biogas Recovery from Poultry and Piggery Waste

A Review

David O. Olukanni and Chukwuebuka N. Ojukwu

7.1 INTRODUCTION

Energy demands have been increasing due to the rise in population, various technological developments, and modern industrialisation. In order for countries to be sustainably developed, economic growth and constant energy production are key factors (Akyürek, 2018). As a key component in development of countries, various methods have been adopted to create cleaner energy using low-cost alternatives. Solid waste disposal, on the other hand, is presently a persistent problem facing most economies because of the rising output of waste. The speed at which waste is currently generated is greatly influenced by population, economy, and a rise in standards of living. Towns assigned with the purpose of managing waste have the challenge of making available an effective and sufficient system to carry out this function (Olukanni & Olatunji, 2018; Abarca et al., 2013; Olukanni et al., 2018; Olukanni et al., 2020).

Agricultural waste, a type of solid waste which includes crops straw, husks, and livestock manure, and domestic waste have not been properly disposed of over the years, which has led to significant side effects. When these wastes are not properly disposed of, large numbers of chemical substances such as nitrous oxide (N_2O), sulphur dioxide (SO_2), methane (CH_4), and carbon dioxide (CO_2) are produced, and this seriously affects the environment on a large scale. Especially in most developing nations, common household burning of crops or livestock waste can lead to air pollution and a significant increase in the dangers to human health (He et al., 2019; Akyürek, 2018).

Anaerobic digestion is a safe method of eliminating solid waste, such as food and agricultural waste. It is a naturally occurring process in an oxygen deficient climate, with a series of reactions simultaneously occurring to produce biogas through microbial activities (Liu et al., 2018; Silva et al., 2019). Conversion of livestock waste to biogas using anaerobic digestion is advantageous as problems such as unpleasant odours, improper waste disposal, and harmful microbial pathogens are taken care of. Biogas consists of CO_2 mixture and combustible CH_4 with traces of H_2S, nitrogen (N_2), hydrogen (H_2) and carbon monoxide (CO). It is a renewable energy source that has the bonus of being environmentally friendly. The energy value of biogas is around 400–700 BTU/ft^3 as compared to 1000 BTU/ft^3 for natural gas (Akyürek, 2018; Olugasa et al., 2014; Ogundare & Olukanni, 2020).

Fossil fuel is gaining attention as a source of energy generation in Nigeria. Their combustion is a main source of environmental pollution, and due to this, alternative solutions to clean energy are sought after (Afazeli et al., 2014). One of the renewable sources that are sustainable and environmentally friendly is the use of waste matter, also known as biomass, which can be gotten from slaughterhouses, markets, and animal husbandries. Anaerobic digestion of waste and conversion to biogas is the best way to utilise the waste accumulated, as this will be used to generate energy at its full potential. The waste would include various parts of the livestock, wastewater produced, and blood content. All these would be digested anaerobically to give off methane that would in turn be used for energy production.

Therefore, this study was aimed at exploring the biogas recovery potential from poultry and piggery waste. Its objectives are to: a) extract the improvements made through co-digestion with

DOI: 10.1201/9781003204435-9

various substrates, b) explore the different anaerobic digestion methods and possible results in the literature, and c) examine the different models adopted in the literature and the factors that affect the model sensitivity (Table 7.1).

7.2 RESULTS OBTAINED FROM REVIEW

There are different factors in anaerobic digestion that affect the production of biogas from waste (i.e. substrates used, pH, temperature, organic loading rate, total solids used, etc.). The results were extracted based on the substrate that was used in co-digestion with either poultry or piggery waste. Some of the literature works made use of inorganic substrates and organic substrates in co-digestion with the waste in question and obtained improvements.

7.3 INORGANIC SUBSTRATES

Magnetite was made use of in co-digestion with swine manure and chicken litter, respectively. This led to an increase in biogas that was produced due to enhanced methanogenic activity (Zhang et al., 2019; Guadalupe Stefanny Aguilar-Moreno et al., 2020). Ferric oxide was used in the anaerobic digestion of swine manure at 8% total solids. It was found that ferric oxide increased the production of methane by 11.06% (Lu et al., 2019). Thermophilic anaerobic digestion was carried out on swine manure with copper salts where it was observed that methane production increased by 28.78% (Wu et al., 2020). Antibiotics (tetracycline and sulfamethoxydiazine) were used on pig manure using large-scale laboratory digesters to analyse its effects on methane production. The antibiotics were seen to delay the production of biogas (Shi et al., 2011).

7.4 ORGANIC SUBSTRATES

Organic substrates like buffalo dung, sheep waste, and energy crop residues were used in some of the literature. In this section, the organic substrates are in two categories: plant waste and livestock waste.

7.4.1 Plant Waste

Digested swine manure was digested with rice straw at various total solids concentrations of 2%, 3%, and 4%. The addition of rice straw was found to result in better output and more production of biogas (Darwin et al., 2014). Biogas output from the co-digestion of chicken manure and cereal residues (rice straw, wheat straw, corn stalks) was analysed. This resulted in significant increase from each cereal substrate with corn stalks having the highest potential (Zhang et al., 2014). Apple waste when co-digested with swine manure at mesophilic and thermophilic temperatures was seen to lead to a slight increase in methane production (Kafle & Kim, 2013). Together with swine manure, corn stover was digested and planned biogas output increased by about 14 times (Gontupil et al., 2012). In another work, swine manure was digested with maize, sunflower residues, and rapeseeds. Maize was seen to give rise to a higher yield than sunflower residues and rapeseeds. Experiments were also carried out with poultry waste using wheat straw and meadow grass as co-substrates, which led to improvements in biogas production as problems concerning carbon to nitrogen ratios (C:N) were solved (Cuetos et al., 2011; Rahman et al., 2017).

Poultry litter was combined with thin stillage under thermophilic temperatures in a completely stirred tank reactor (CSTR) stabilised by just poultry litter to see its performance. This resulted in enhanced biogas production and an increase in digester performance, as it could accommodate a substrate ratio of 60% thin stillage and 40% poultry litter (Sharma et al., 2013). The effects of loading and temperature were tested using the co-digestion of poultry waste and straw in another study. The digester was observed to be feasible at 34°C with a loading rate of kg VS/m^3 (Roshani et al., 2012).

TABLE 7.1
Summary of Motivations and Findings from Different Studies

S/N	Author/Year	Aim	Research Methodology	Findings
1	Darwin et al. (2014)	Assess the production of biomethane from co-digested rice straw with digested swine manure.	The authors used six reactors of 500 mL. Three were filled with rice straw and digested swine manure, while the others were without rice addition (control reactors). The experiment was carried out at total concentrations of 2%, 3%, and 4% for each reactor.	It was observed that the most methane generation was at 3%. Highest chemical oxygen demand removal and volatile solid reduction was also noticed with a value of (52.97% ± 1.46%) and (61.81% ± 1.04%), respectively.
2	Yasin & Wasim (2011)	Demonstrate the production of methane gas through anaerobic digestion of buffalo dung, plus poultry litter, and buffalo dung plus sheep waste.	Buffalo dung (BD), sheep waste (SW), and poultry litter (PL) were used as the substrates. BD, BD + PL, and BD + SW were tested out in different plants. BD was mixed with PL and SW at a ratio of 1:1. Then the individual mixtures were diluted with water at a ratio of 1:1.5 before they were put into separate plants.	It was observed that the various organic loading rates had influence over the temperature and pH value of the slurry. Gas production was shown to significantly increase for BD, BD + PL, BD + SW at 40 kg treatment level (0.82 m^3, 0.92 m^3, 0.79 m^3 per day, respectively). pH values were found to be 6.6, 6.5, and 6.8 for BD, BD + PL, BD + SW, respectively.
3	Yu et al. (2020)	Investigate the effect of aqueous pyrolysis liquid (APL) on the anaerobic digestion of swine manure at different concentrations.	Swine manure and APL were used. Inoculum was mixed with the swine manure at a ratio of 9:3:1. APL was then diluted in different concentrations of 5 times (A5), 50 times (A50) and 100 times (A100) and then added to separate mixtures.	High APL concentrations, as in A5, were shown to inhibit methane production, whereas A50 and A100, with low concentrations, increased methane yields. It was also found via various analyses that APL included components that could relieve ammonia and acid inhibition during the anaerobic digestion process.
4	Zhang et al. (2014)	Analysis of biogas output from digestion of anaerobic chicken manure and residues of cereals as co-substrates.	Chicken manure (CM) and cereal residues (CR) made up of rice straws, wheat straw, and corn stalks were used. A 1-L conical flask was used to accommodate the co-digestion of chicken manure and the three variations of cereal residues. Total solid content of 8% was used in each setup. Each of the cereal residues were mixed with chicken manure in the following ratios: CM:CR = 83.3:16.7, 75:25, 50:50, 25:75, 16.7:83.3.	It was noticed in the experiment that the most amount of biogas was obtained at the ratio 50:50 over the sixty-day period for the hydraulic retention time.

(Continued)

TABLE 7.1 *(Continued)*
Summary of Motivations and Findings from Different Studies

S/N	Author/Year	Aim	Research Methodology	Findings
5	Kafle & Kim (2013)	Determine the production rate of biogas for apple waste (AW) and swine manure (SM). Determine the anaerobic digester's efficiency. Examine the efficiency of continuous digesters with various AW:SM ratios.	The research made use of SM and crushed AW. Both were stored at a temperature of 4°C. The biogas production rate was tested in experiment one and in experiment two, the test was carried out with an AW:SM feed mix ratio of 0:100 and 33:67 under thermophilic and mesophilic conditions. The experiment was conducted in a 5.5 L totally stirred tank reactor with a working volume of 4.5 L for continuous setup and temperature regulated chamber. Temperatures were maintained at 36–38°C. The continuous anaerobic digestion test was carried out at various mixing ratios of SM and AW. For 30 days, different mixing ratios of SM and AW were fed. The apple concentration gradually increased from 25% to 50%.	Biogas production was shown to increase with the use of a combination of AW and SM (33:67) when compared to swine manure alone. At thermophilic temperature, it was seen that biogas production was higher for mixed feeds.
6	Zhang et al. (2019)	Analyse the effects of magnetite on anaerobic digestion of swine manure.	A 3:1 ratio (total solids, TS) of swine manure and inoculum sludge was mixed and transferred to 0.4 L working volume bottles. After that, magnetite was applied to the bottles at concentrations of 0 mmol (control), 5 mmol (M5), 75 mmol (M75), 150 mmol (M150) and 350 mmol (M350).	Methane production was shown to increase by 16.1% due to the addition of magnetite ($p<0.05$) to the anaerobic digestion of swine manure.
7	Guadalupe Stefanny Aguilar-Moreno et al. (2020)	Analyse the effect of different concentrations of Fe_3O_4 nanoparticles, synthesised by coprecipitation, in the anaerobic digestion of chicken litter on methane production.	Fe_3O_4 nanoparticles were created using the coprecipitation method. The best nanoparticle of hydrodynamic diameter 79.37 nm was selected. Chicken litter containing chicken waste, feed waste, and chips of wood were handled with different quantities of nanoparticles as follows: 20 mg/L, 40 mg/L, 60 mg/L of both Fe_3O_4 nanoparticles and chicken litter.	It was noticed that Fe_3O_4 nanoparticles led to a considerable rise in production of methane. This took place at nanoparticle concentration of 20 mg/L which resulted in 2.55 mL CH_4/g_{vsf} being produced per day.
8	Shi et al. (2011)	Examine the effects of tetracycline (TC) and sulfamethoxydiazine (SMD) on the anaerobic digestion of pig manure using laboratory-scale reactors.	A mixture of 100 g of dried pig manure and 100 mL of inoculum was put in laboratory scaled digesters with the addition of water to make 1 L of feed. Treatment levels of 25 mg and 50 mg were used for both TC and SMD. This treatment was carried out in darkness for a digestion cycle of 20 days.	It was observed that concentration of TC and SMD led to inhibition of methane production. Over time, it was also seen that the antibiotics were readily absorbed within the first 12 hours of digestion.

(Continued)

TABLE 7.1 *(Continued)*
Summary of Motivations and Findings from Different Studies

S/N	Author/Year	Aim	Research Methodology	Findings
9	Wu et al. (2020)	Investigate the effects of three kinds of copper (Cu) salts (cupric sulfate, cupric glycinate, and the 1:1 mixture of the two) on anaerobic digestion performance and antibiotic resistance gene (ARG) changes during thermophilic anaerobic digestion of swine manure.	Swine manure was co-digested in this experiment with various Cu salts. The experiment was carried out in batches using anaerobic flasks of working volume 800 ml and incubated at 50°C. Three Cu salts were used with a concentration of 800 mg/kg dry weight. Four anaerobic digestion groups were made: Control set without Cu, inorganic set without cupric sulfate, organic set with cupric glycinate and mixed set with cupric sulfate and cupric glycinate spiked mixture (1:1 ratio).	Methane production was noticed to be at its highest on the 7th day in the mixed Cu set (38.24 ± 1.91 mL/g VS_{added}). The production rose by 28.78%. In the mixed Cu sample, the overall ARG abundance was reduced by 26.94%.
10	Gontupil et al. (2012)	Investigate the performance of anaerobic co-digestion of swine manure and corn stover in completely mixed and semi-continuously fed reactors.	Wastewater and sludge of swine manure were obtained and preserved at 5°C while corn stover was dried. 1 L of sludge and 13 L of wastewater were added to each digester of 14 L working volume. This digester was equipped with a stirring system and temperature control system. Each reactor was operated for a 25-day hydraulic retention period where 560 mL of wastewater was added per day. The process continued at a temperature of 35°C.	It was noted that when corn stover was put in the digester, biogas output multiplied by about 14 times, but methane yield in Reactor 2 was 50.7% less than Reactor 1 (67.8%). It was also noticed that corn stover can help reduce ammonia inhibition.
11	Gopalan et al. (2012)	Characterisation profile of effluent originating from different production stages of swine.	Samples were obtained from the various pig types: dry sow, farrowing, weaner, grower, and finisher, and their diet base was noted. The biochemical processing of methane was performed on the waste after mixing with inoculum obtained from an anaerobic digester in a 1:1 ratio. Precision gas-tight syringe and a water-filled manometer were used to measure the gas volume.	Effluents were found to break down to a higher degree from the finisher, grower, and weaner sheds and have nearly twice the methane capacity obtained from the other waste. This variance in methane capacity was related to the industry's feeding strategies and waste handling methods.
12	Zhu et al. (2018)	Investigate mixing of poultry litter and wheat straw ratios dependent on various total solids (TS) and added volatile solids of wheat straw to produce biogas through co-digestion.	Poultry litter was gathered here and sieved through a screen of 2.38 mm then placed in a refrigerator. The wheat straw was processed to a size of 20 mesh (0.85 mm). Substrate was prepared at 0%, 25%, and 50%. The volatile solids were further described, with poultry litter and wheat straw mixed in water to form three total solid concentrations, that is 2%, 5%, and 10% for each TS level.	Maintaining a proper C:N ratio was found not to guarantee a maximum release of biogas. To achieve maximum yield of poultry litter, TS content should be limited to 6.8% and 4.15% if combined with wheat straw.

(Continued)

TABLE 7.1 *(Continued)*
Summary of Motivations and Findings from Different Studies

S/N	Author/Year	Aim	Research Methodology	Findings
13	Chaump et al. (2019)	Compare the relative digestibility of poultry litter leachates, insoluble materials, and whole litter.	Serum bottles (160 mL) with working volume of 60 mL were operated in batch modes and experiments were carried out in triplicates. The first analysis involved only the digestion of solid litter samples; 10 g/L of total volatile solids were filled for each reactor. The volatile solid was made up of both the litter and inoculum. Litter to inoculum ranged from 0 to 80%.	Soluble leachate digestion has been found to contain nearly twice the methane potential. This is due to the low C:N ratios found. Initially, waste litter had a fast production of biogas, but lower production occurred at the final stages relative to fresh litter. This contributes to the likelihood of storing litter to the detriment of the biogas production.
14	Lu et al. (2019)	Ascertain the effects of Fe_2O_3 addition on methane production during anaerobic digestion (AD) of swine manure while also evaluating optimal Fe_2O_3 addition levels.	A batch reactor system comprising three parts was used. The first was an AD unit consisting of 15 500 mL glass blue-cap bottles each. The second part consists of 15 100 mL glass bottles to extract CO_2 and H_2S filled with 90 mL of 3 M NaOH. The third portion was composed of a gas volume and data storage unit.	There was a 31.9% rise in the overall methane production rate observed by adding 25 mmol of Fe_2O_3. Methane production increased by 11.04%. Instead of acidogenesis and hydrolysis, methanogenesis was improved.
15	Borowski & Doman (2014)	Measure the efficacy of the anaerobic digestion process for sewage treatment with swine and poultry manure sludge.	Six experiments were carried out, with each having a digester of working volume of 1 dm^3. The batch experiment contained the following trials: 100% Sewage sludge (SS), 90% SS + 10% swine manure (SM), 80% SS + 20% SM, 70% SS + 30% SM, 60% SS + 40% SM, 50% SS + 50% SM. This was measured by weight. There were no replicates for the experiment. Two reactors with working volume of 3 dm^3 were later carried out in semi-batch conditions at a temperature of $35 \pm 1°C$. A weight-based mixture of sewage sludge and swine manure, SS:SM = 70:30, was analysed using one of the reactors.	It was observed that with the addition of 30% of swine manure to sewage sludge, biogas production rose by almost 40% relative to swine manure alone. For the second reactor, this was not the case, because the addition of 10% poultry resulted in a decrease in the release of biogas because of ammonia. The effluent digestate still contained pathogens, making it unsuitable for agricultural purposes.
16	Indren et al. (2020)	Identify the impacts of wood-pellet biochar on methane production from poultry litter as a function of the digester total solids (TS) content.	The experiment made use of 500 mL glass bottles as digesters. Biochar was added at TS values of 5%, 10%, 20% to the mixture in various digesters. This was done with an equal dose through the three digesters at an equal mass to that of poultry litter.	There was a decrease in lag time before biogas production and this increased with the increasing total content of solids. The highest daily methane output and highest daily production occur earlier due to biochar.

(Continued)

TABLE 7.1 *(Continued)*
Summary of Motivations and Findings from Different Studies

S/N	Author/Year	Aim	Research Methodology	Findings
17	Rahman et al. (2017)	Optimise the biodegradation of volatile solids and the biomethane quantity and generation patterns of poultry droppings and lignocellulosic-rich substrates using substrate composition and C:N ratio.	Experiments on poultry waste with wheat straw and poultry waste with meadow grass were conducted. Different mixtures were performed separately for lignocellulose substrates and poultry droppings 90:10, 70:30, 50:50, 30:70, 10:90.	Anaerobic digestion of poultry droppings and lignocellulosic substrates was found to solve the C:N ratio problem, leading to a rise in biochemical methane production (BMP). The highest BMP was gotten from 70:30 mixing ratio for poultry dropping and wheat straw and 50:50 ratio for poultry droppings and meadow grass.
18	Indren et al. (2018)	Develop a relationship between methane production rate and changes in the chemical conditions in the digester due to biochar addition.	Poultry litter was obtained from a chicken at the end of their growth cycle. Biochar of 10–20 mm in length and 4 mm in diameter was used. This was obtained through reaction of pellets inside a top-lit updraft gasifier. Glass bottles of 250 mL were used as anaerobic digesters. At a volatile solid-based feedstock/inoculum ratio of 0.5, 1, 2, feedstock and inoculant were added. The added biochar was of the same weight as the poultry litter.	Interactions between microorganisms and biochar, to improve anaerobic digestion efficiency was achieved by decreasing the time required before development of methane begins. However, using biochar, ammonia inhibition was not completely overcome.
19	Markou (2015)	Investigate the anaerobic digestion performance of poultry litter (PL) with low nitrogen content derived by a two-phase treatment.	PL was first stored under anaerobic conditions for a 60-day hydraulic retention period at a temperature of 17–22°C to increase the ammonia content. To eliminate the ammonia content that was stored up, treated PL was put in an oven and heated at 80°C, then aerated for a day. The digestion was performed using both PL and treated PL under total solids of 5%, 10%, 15% and 20%. This was run in duplicates.	It was observed that lowering the nitrogen content present in poultry waste improves biomethane yield significantly.
20	Roman et al. (2015)	Analyse poultry litter bio-methanation with co-substrates of cow dung and poultry droppings.	Four different reactors were prepared. Reactor 1 made use of poultry litter alone, Reactor 2 used 75% poultry litter and 25% cow dung, Reactor 3 used 50% poultry litter and 50% cow dung; Reactor 3 also used 70% poultry litter. Reactor 4 used 30% poultry droppings.	It was shown that the substrate mixture of poultry litter and cow dung had a high potential for producing biogas.

(Continued)

TABLE 7.1 (Continued)
Summary of Motivations and Findings from Different Studies

S/N	Author/Year	Aim	Research Methodology	Findings
21	Sharma et al. (2013)	Test whether a thermophilic, continuously stirred tank reactor (CSTR) digester previously stabilised on poultry litter could accommodate thin stillage during co-digestion, and to test the limits of thin stillage as co-substrate.	Bench scale digesters were used. This experiment was performed within 15 days of the hydraulic retention date. It started in all three digesters with 100% poultry litter. Thin stillage was then applied to poultry litter in the following ratios after the successful conclusion of two hydraulic retention times (HRTs) using 100% poultry litter: 20:80, 40:60, 60:40, 80:20. Each treatment had a period of 2 HRTs.	Enhanced methane production was observed by applying thin stillage to the poultry litter at 60% stillage (40% poultry litter), however it was observed that the efficiency of the digesters dropped.
22	Cuetos et al. (2011)	Investigate the performance of reactors co-digesting swine manure with three types of energy crop residues (ECRs) from: maize, rapeseed, and sunflowers.	Different ratios of swine manure (SM) to ECR were used. At 25%, 50%, and 75% volatile solids (VS) ratios. SM and ECR were prepared (Percentages indicate ECR VS as a percentage of total). For the continuous test, four continuously stirred tank reactors made of methacrylate were used to digest swine manure and mixtures with ECRs.	Swine manure co-digestion with maize gave the highest methane yield in batch digestion. Due to presence of lignin content within sunflower residues and rapeseeds, lower amounts of methane were obtained.
23	Fotidis et al. (2013)	Investigate the impact on the mesophilic effect of various natural zeolite dosages on anaerobic digestion of inoculated poultry waste with a non-acclimatised inoculum of ammonia.	Experiments were carried out in triplicates a) without zeolite addition b) 5 g zeolite/L addition c) 10 g zeolite/L addition in three different experimental sets. There was a working volume and total volume of 900 mL and 1000 mL, respectively, for each reactor.	Zeolite dosages (5 g/L and 10 g/L) were seen to improve methane production yield during the digestion. At high concentrations of ammonia, zeolite (> 6 g/L) had more benefits in the biomethane production process than demerits.
24	Kougias et al. (2013)	Investigate the effect of different natural zeolite concentrations on the anaerobic digestion of poultry waste.	Four batch trials were conducted in duplicates. Zeolite and untreated poultry manure were the products used. The four experiments included: two batch reactors containing only poultry manure and inoculum, two poultry reactors containing inoculum and 5 g of zeolite/ dm^3 of waste (Z5), two poultry reactors containing inoculum and 10 g of zeolite (Z10), two inoculum and water reactors.	There was increased biogas efficiency during mesophilic anaerobic digestion with zeolite addition. Zeolite absorption caused a reduction in ammonium ions present in treatment Z5 and Z10.

(Continued)

TABLE 7.1 *(Continued)*
Summary of Motivations and Findings from Different Studies

S/N	Author/Year	Aim	Research Methodology	Findings
25	Lazor et al. (2010)	Monitor biogas production and its composition from agro-food wastes.	A semi-continuous reactor with a working volume of 15 L was used with the vacant space acting as an improvised gas holder. The organic fraction ratio of kitchen waste oil and poultry manure used was 9:1. Temperature was maintained at 37°C and pH value at 7.8.	It was observed that storing of waste for two weeks or more led to increased amounts of ammonia. Free ammonia in high concentrations negatively influenced the anaerobic process.
26	Roshani et al. (2012)	Investigate the effect of loading and temperature on the anaerobic co-digestion of poultry manure and wheat straw.	A cylindrical CSTR reactor having a total volume of 60 L was used to carry out this experiment. Poultry manure and straw were stirred together in ratio 20:80 based on weight.	It was observed that temperature fluctuations have influence with an increase in biogas production.
27	Bayrakdar et al. (2017)	Investigate the performance of the laboratory-scale anaerobic co-digester operated by feeding with a mix of chicken manure and spent poppy straws.	Chicken manure and poppy straws in a ratio of 4.3:1 were used as substrates to carry out experiments in a laboratory scaled anaerobic digester.	When total ammonia nitrogen and free ammonia nitrogen were held below 4000 and 3000 mg/L respectively, chicken manure was found to be digestible with spent poppy straws and methane produced as high as 0.36 $lgVS^{-1}$.
28	Astals et al. (2012)	Investigate the effect of anaerobic co-digestion of pig manure and glycerol in biogas production and the resulting digestate.	Two digesters were used; digester D1, which was fed with only pig manure and digester D2, which contained many glycerol and pig manure mixtures. Two batches of manure were used within day 1–99 and 100–196, while glycerol content remained same throughout.	It was found that co-digestion of pig manure and glycerol contributed to an increase in organic loading volume, a balance in the C:N ratio, and a decrease in free ammonia concentration.
29	Borowski & Weatherley (2013)	Anaerobic comparison of mixed raw sewage sludge digestion with this sludge co-digestion but in the presence of poultry.	Experiments were conducted using two separate reactors. Sewage sludge alone was fed in one of the reactors while a mixture of sludge and poultry manure with a 70:30 ratio was in the other.	When sewage sludge was digested with 30% of poultry waste, gas production was observed to increase by about 50%.
30	Deepanraj et al. (2017)	Investigate the feasibility of co-digesting food waste with poultry manure (10–40%) as an external source of nitrogen in the development of biogas.	2 L lab digesters with a working volume of 1.6 L were used. In percentages of 10%, 20%, 30%, and 40%, poultry manure was digested along with food waste. This experiment was conducted and calculated using the water displacement method at a thermophilic state of 50°C.	Compared to other substrates, digestion of poultry waste with food waste resulted in higher biogas release and high degradation efficiency.

Chicken waste was also co-digested with spent poppy straws using laboratory-scale digesters at total ammonia concentrations and organic loading rate up to 6500 mg/l and 3.56 g VS/ld. Methane yields obtained from the experiment were high (Bayrakdar et al., 2017).

7.4.2 Animal Waste

Buffalo dung was digested with poultry litter and sheep waste separately at an organic loading rate of 15, 30, and 40 kg per day. It was seen that the substrate and its loading rate greatly affected the pH, temperature, and gas production. Highest gas production was obtained from an OLR of 40 kg per day (Yasin & Wasim, 2011). Swine effluents from different swine types (dry sow, farrowing, grower, weaner, and finisher) were anaerobically digested and this resulted in grower, weaner, and finisher sheds having more methane potentials than the others (Gopalan et al., 2012). In relation to the digestion of poultry litter leachate and poultry litter, biogas production and nutrient formation were examined. It was noticed that digestion of soluble leachates led to almost twice the biogas production and storage of waste led to losses in biogas production (Chaump et al., 2019).

Sewage sludge was anaerobically digested with swine waste and this led to an almost 40% increase in biogas production compared to digestion of just sewage sludge alone. The same experiment was also carried out with poultry waste, and total gas production showed a 50% increase (Borowski & Doman, 2014; Borowski & Weatherley, 2013). In another instance, poultry waste was previously treated to reduce its nitrogen percentage and improve the C:N ratio of the waste before actual anaerobic procedure was carried out. Results obtained from the experiment indicated that methane yield significantly improved (Markou, 2015). Poultry waste was also co-digested with cow dung and it was observed that there was a high potential for biogas present (Roman et al., 2015).

7.5 MODELS USED

Models used in this literature are mathematical representations used to explain and observe the behaviour of real time objects used in anaerobic digestion (Britannica, 2011). The models used in the literature review are of four types: Best fit regression model, first order kinetic model, logistic model, and modified Gompertz model.

In studying the anaerobic digestion of poultry droppings and lignocellulosic-rich substrates, the best fit regression model was used and it was found that there was consistency between the data obtained and expected (Rahman et al., 2017). In the study of data obtained from the co-digestion of chicken manure and cereal residues, this model was also used and the estimated result was also matched with the result obtained (Zhang et al., 2014).

To carry out a kinetic analysis on the anaerobic digestion of swine manure and apple waste, a first order kinetic model and updated Gompertz model were used. In the first order kinetic model, methane yield was found to be greater than the modified Gompertz model, so the latter was used (Kafle & Kim, 2013). In the anaerobic co-digestion of magnetite and swine manure, a modified model of Gompertz was used. The overall methane production rate was increased by 8% (Zhang et al., 2019). The modified Gompertz model showed less root mean square error when used in the kinetic analysis of anaerobic co-digestion of poultry and food waste, relative to the first order kinetic model and logistic model (Deepanraj et al., 2017).

7.6 CONCLUSION AND RECOMMENDATIONS

7.6.1 Conclusion

i. Substrates digested alongside poultry and piggery waste affect the amount of biogas produced.
ii. Storing of waste (substrate) to be used for two weeks or more can lead to increase in ammonia concentration, leading to decrease in biogas production.

iii. Model used in an anaerobic experiment influences biogas produced.
iv. In the digestion of organic waste, temperature is also a consideration that should not be ignored, as it can impact the biogas that is processed. Thermophilic temperatures contribute to more biogas production than mesophilic temperatures.
v. The high free ammonia content of substrates leads to a substantial drop in the efficacy of the anaerobic digestion process.
vi. Zeolite addition led to a decrease in the ammonium content as a substrate, leading to improved anaerobic digestion process performance.

7.6.2 Recommendations

i. Further work should be carried out on the anaerobic co-digestion of various livestock waste.
ii. Models should be used in anaerobic digestion as they help to affirm or validate predicted values.
iii. Research work should be carried out on the anaerobic digestion of poultry waste at various growth stages and how this affects the biogas produced.

REFERENCES

Abarca, L., Maas, G., & Hogland, W. (2013). Solid waste management challenges for cities in developing countries. *Waste Management, 33*(1), 220–232. https://doi.org/10.1016/j.wasman.2012.09.008

Afazeli, H., Jafari, A., Ra, S., & Nosrati, M. (2014). An investigation of biogas production potential from livestock and slaughterhouse wastes. *Renewable and Sustainable Energy Reviews 34*, 380–386. https://doi.org/10.1016/j.rser.2014.03.016

Akyürek, Z. (2018). Potential of biogas energy from animal waste in the Mediterranean Region of Turkey, *Journal of Energy Systems* 2(4), 160–167. https://doi.org/10.30521/jes.455325

Astals, S., Nolla-Ardèvol, V., & Mata-Alvarez, J. (2012). Anaerobic co-digestion of pig manure and crude glycerol at mesophilic conditions: Biogas and digestate. *Bioresource Technology, 110*, 63–70. https://doi.org/10.1016/j.biortech.2012.01.080

Bayrakdar, A., Molaey, R., Onder, R., & Sahinkaya, E. (2017). International Biodeterioration & Biodegradation Biogas production from chicken manure: Co-digestion with spent poppy straw. *International Biodeterioration & Biodegradation 119*. https://doi.org/10.1016/j.ibiod.2016.10.058

Borowski, S. & Doman, J. (2014). Anaerobic co-digestion of swine and poultry manure with municipal sewage sludge. *Waste Management 34*, 513–521. https://doi.org/10.1016/j.wasman.2013.10.022

Borowski, S. & Weatherley, L. (2013). Co-digestion of solid poultry manure with municipal sewage sludge. *Bioresource Technology, 142*, 345–352. https://doi.org/10.1016/j.biortech.2013.05.047

Britannica. (2011). *Scientific modeling*. https://www.britannica.com/science/scientific-modeling

Chaump, K., Preisser, M., Shanmugam, S. R., Prasad, R., Adhikari, S., & Higgins, B. T. (2019). Leaching and anaerobic digestion of poultry litter for biogas production and nutrient transformation. *Waste Management, 84*, 413–422. https://doi.org/10.1016/j.wasman.2018.11.024

Cuetos, M. J., Fernández, C., Gómez, X., & Morán, A. (2011). Anaerobic co-digestion of swine manure with energy crop residues. *Biotechnology and Bioprocess Engineering 1052*, 1044–1052. https://doi.org/10.1007/s12257-011-0117-4

Darwin, Cheng, J. J., Liu, Z. M., Gontupil, J., & Kwon, O. S. (2014). Anaerobic co-digestion of rice straw and digested swine manure with different total solid concentration for methane production. *International Journal of Agricultural and Biological Engineering, 7*(6), 79–90. https://doi.org/10.3965/j.ijabe.20140706.010

Deepanraj, B., Sivasubramanian, V., & Jayaraj, S. (2017). Experimental and kinetic study on anaerobic co-digestion of poultry manure and food waste. *Desalination and Water Treatment 59*, 72–76. https://doi.org/10.5004/dwt.2016.0162

Fotidis, I. A., Kougias, P. G., Zaganas, I. D., Kotsopoulos, T. A., & Martzopoulos, G. G. (2013). Inoculum and zeolite synergistic effect on anaerobic digestion of poultry manure. *Environmental Technology* 37–41. https://doi.org/10.1080/09593330.2013.865083

Gontupil, J., Darwin, M., Liu, Z., Cheng, J. J., & Chen, H. (2012). Anaerobic co-digestion of swine manure and corn stover for biogas production. *Environmental Technology 7004*(12), 12–17.

Gopalan, P., Jensen, P. D., & Batstone, D. J. (2012). Anaerobic digestion of swine effluent: Impact of production stages. *Biomass and Bioenergy*, *48*, 121–129. https://doi.org/10.1016/j.biombioe.2012.11.012

Guadalupe Stefanny Aguilar-Moreno, E. N.-C., Velázquez-Hernández, A., Hernández-Eugenio, G., & Aguilar-Méndez, Miguel Ángel, Teodoro, E.-S. (2020). Enhancing methane yield of chicken litter in anaerobic digestion using magnetite nanoparticles. *Renewable Energy 147*, 204–213. https://doi.org/10.1016/j.renene.2019.08.111

He, K., Zhang, J., & Zeng, Y. (2019). Science of the Total Environment – Knowledge domain and emerging trends of agricultural waste management in the field of social science: A scientometric review. *Science of the Total Environment 670*, 236–244. https://doi.org/10.1016/j.scitotenv.2019.03.184

Indren, M., Birzer, C. H., Kidd, S. P., & Medwell, P. R. (2020). Effect of total solids content on anaerobic digestion of poultry litter with biochar. *Journal of Environmental Management*, *255*(August 2019), 109744. https://doi.org/10.1016/j.jenvman.2019.109744

Indren, M., Birzer, C., Medwell, P., & Kidd, S. (2018). Biochar addition in high-solids anaerobic digestion of poultry litter. *2018 IEEE Global Humanitarian Technology Conference (GHTC)*, 1–8.

Kafle, G. K. & Kim, S. H. (2013). Anaerobic treatment of apple waste with swine manure for biogas production: Batch and continuous operation. *Applied Energy*, *103*, 61–72. https://doi.org/10.1016/j.apenergy.2012.10.018

Kougias, P. G., Fotidis, I. A., Zaganas, I. D., Kotsopoulos, T. A., & Martzopoulos, G. G. (2013). Zeolite and swine inoculum effect on poultry manure biomethanation. *International Agrophysics* 169–173. https://doi.org/10.2478/v10247-012-0082-y

Lazor, M., Hutňan, M., Sedláček, S., Kolesárová, N., & Špalková, V. (2010). Anaerobic co-digestion of poultry manure and waste kitchen oil. *Biotechnology and Bioprocess Engineering*, *16*, 1399–1406.

Liu, C., Chufo, A., Tong, H., Shi, S., & Zhang, L. (2018). Biogas production and microbial community properties during anaerobic digestion of corn stover at different temperatures. *Bioresource Technology*, *261*(October 2017), 93–103. https://doi.org/10.1016/j.biortech.2017.12.076

Lu, T., Zhang, J., Wei, Y., & Shen, P. (2019). Bioresource technology effects of ferric oxide on the microbial community and functioning during anaerobic digestion of swine manure. *Bioresource Technology*, *287*(March), 121393. https://doi.org/10.1016/j.biortech.2019.121393

Markou, G. (2015). Improved anaerobic digestion performance and biogas production from poultry litter after lowering its nitrogen content. *Bioresource Technology*, *196*, 726–730. https://doi.org/10.1016/j.biortech.2015.07.067

Ogundare, O. J. & Olukanni, D. O. (2020). Potential recovery of biogas from lime waste after juice extraction using solid–liquid extraction process. *Recycling 5*(2), 14. https://doi.org/10.3390/recycling5020014

Olugasa, T. T., Odesola, I. F., & Oyewola, M. O. (2014). Energy production from biogas: A conceptual review for use in Nigeria. *Renewable and Sustainable Energy Reviews*, *32*, 770–776. https://doi.org/10.1016/j.rser.2013.12.013

Olukanni, D. O., Aipoh, A. O., & Kalabo, I. H. (2018). Recycling and reuse technology: Waste to wealth initiative in a private tertiary institution, Nigeria. *Recycling*, *3*(3). https://doi.org/10.3390/recycling3030044

Olukanni, D. O. & Olatunji, T. O. (2018). Cassava waste management and biogas generation potential in selected local government areas in Ogun State, Nigeria. *Recycling*. https://doi.org/10.3390/recycling3040058

Olukanni, D. O., Pius-Imue, F. B., & Joseph, S. O. (2020). Public perception of solid waste management practices in Nigeria: Ogun State experience. *Recycling*, *5*(2), 8. https://doi.org/10.3390/recycling5020008

Rahman, A., Møller, H. B., Kumer, C., Alam, M., Wahid, R., & Feng, L. (2017). Optimal ratio for anaerobic co-digestion of poultry droppings and lignocellulosic-rich substrates for enhanced biogas production. *Energy for Sustainable Development*, *39*, 59–66. https://doi.org/10.1016/j.esd.2017.04.004

Roman, M., Kalam, A., Rahman, L., Rajibul, M., Pulak, A., & Rouf, A. (2015). Production of biogas from poultry litter mixed with the co-substrate cow dung. *Integrative Medicine Research*. https://doi.org/10.1016/j.jtusci.2015.07.007

Roshani, A., Shayegan, J., & Babaee, A. (2012). Methane production from anaerobic co-digestion of poultry manure. *Journal of Environmental Studies 38*(62), 22–24.

Sharma, D., Espinosa-solares, T., & Huber, D. H. (2013). Thermophilic anaerobic co-digestion of poultry litter and thin stillage. *Bioresource Technology*, *136*, 251–256. https://doi.org/10.1016/j.biortech.2013.03.005

Shi, J. C., Liao, X. D., Wu, Y. B., & Liang, J. B. (2011). Effect of antibiotics on methane arising from anaerobic digestion of pig manure, *Animal Feed Science and Technology 167*, 457–463. https://doi.org/10.1016/j.anifeedsci.2011.04.033

Silva, C. E. D. F., Gois, G. N. S. B., Abud, A. K. S., Amorim, N. C. S., Girotto, F., Markou, G., Carvalho, C. M., Tonholo, J., & Amorim, E. L. (2019). Anaerobic Digestion : Biogas Production from Agro-industrial Wastewater, Food Waste, and Biomass. Springer International Publishing. https://doi.org/10.1007/978-3-030-14463-0

Wu, X., Tian, Z., Lv, Z., Chen, Z., Liu, Y., & Yong, X. (2020). Effects of copper salts on performance, antibiotic resistance genes, and microbial community during thermophilic anaerobic digestion of swine manure. *Bioresource Technology, 300*(30), 122728. https://doi.org/10.1016/j.biortech.2019.122728

Yasin, M., & Wasim, M. (2011). Anaerobic digestion of buffalo dung, sheep waste and poultry litter for biogas production. *Journal of Agricultural Research 49*(1), 73–82.

Yu, X., Zhang, C., Qiu, L., Yao, Y., & Sun, G. (2020). Anaerobic digestion of swine manure using aqueous pyrolysis liquid as an additive. *Renewable Energy, 147*, 2484–2493. https://doi.org/10.1016/j.renene.2019.10.096

Zhang, J., Lu, T., Wang, Z., Wang, Y., Zhong, H., Shen, P., & Wei, Y. (2019). Effects of magnetite on anaerobic digestion of swine manure: Attention to methane production and fate of antibiotic resistance genes. *Bioresource Technology, 291*(July), 121847. https://doi.org/10.1016/j.biortech.2019.121847

Zhang, T., Yang, Y., Liu, L., Han, Y., Ren, G., & Yang, G. (2014). Improved biogas production from chicken manure anaerobic digestion using cereal residues as co-substrates. *Energy Fuels, 28*(4), 2490–2495.

Zhu, J., Wu, S., & Shen, J. (2018). Toxic/hazardous substances and environmental engineering: Anaerobic co-digestion of poultry litter and wheat straw affected by solids composition, free ammonia and carbon/nitrogen ratio. *Journal of Environmental Science and Health, Part A*, 1–7. https://doi.org/10.1080/10934529.2018.1546494

8 Mini Review
Solid-State Stratified-Bed Reactor Design for Leaf Litter Digestion

Aastha Paliwal, Ashritha J., and H. N. Chanakya

8.1 INTRODUCTION

Anaerobic digestion (AD) of organic fractions of municipal solid waste (OFMSW) for biogas production and compost generation is rapidly becoming one of the appropriate, economic, sustainable, and popular routes for its treatment and recycling. Although technologies for this have been tried since the 1980s, its utility is being harnessed only recently. As source segregation is made compulsory and is becoming successful, segregated household wastes have become predominantly wet and organic [Shwetmala and Ramachandra, 2011]. Decentralised anaerobic digesters not only provide effective and immediate containment to such rapidly fermenting OFMSW, avoiding fly, insect, and odour nuisance, it also obviates need to transport over 85% of the daily domestic waste [Shwetmala and Ramachandra, 2011, Chanakya and Moletta, 2004, 2005]. It escapes the usual problems of dripping of leachate and attendant health, odour, environment, and aesthetic issues, while also significantly reducing handling costs. More recently, it has been shown that many value-added products can also be accrued from such AD systems, depending upon the feedstocks used [Chanakya and Malayil, 2012].

Despite the increased packing material content in domestic municipal solid waste (MSW), such as plastics, the organic fraction of MSW still remains the major fraction of fresh domestic MSW, with an average of approximately 45%, sometimes reaching up to more than 72% of the total MSW collected and processed in some places of India [Joshi and Ahmed, 2016; Chanakya and Moletta 2005; Shwetmala and Ramachandra, 2011; Swachh Bharat Mission, Municipal Solid Waste Management Manual, Part 2, 2016]. When household or domestic MSW is mixed with predominantly organic components of street sweepings, such as leaf litter, garden cuttings, and discarded paper, these too are attractive feedstock for biomethanation. Lignocellulosic biomass constitutes about 35% of MSW [Kumar et al., 2019] and consists of both fresh biomass (such as grass cuttings, land weeds such as *Parthenium, Synedrella* sp.,) and dry leaf litter (such as species of *Swietenia, Polyalthia, Delonix, Ficus,* etc.). Leafy biomass thus becomes an important potential feedstock for AD to biogas.

AD of leaf litter, agro-residues, and other herbaceous biomass acquire even more significance in India where primary productivity of about 1.13 billion tonnes [Jagadish et al., 1998] and a surplus of 300–500 million tonnes is annually available for its sustainable conversion to biogas, compost, and various value-added by-products [Chanakya and Malayil, 2012]. Biogas plants deployed extensively in rural India will improve working conditions for women and remove various forms of household drudgery, improve quality of life in rural areas, and divert human endeavour into other productive areas, factors that are rarely quantified or measured in cost or development price [Sasse, 1988]. Yet it is only recently that serious work to digesters converting them to biogas has begun.

Digestion of lignocellulosic biomass, be it leaf litter, MSW components, or agro-residues, is, however, not straight forward, and the older, slurry-based completely stirred tank reactors (CSTRs) function poorly with this feedstock, even when they are powdered and rendered into an aqueous slurry similar to cattle dung. They pose unique difficulties for biomethanation, especially because composition is highly variable and they have a significant lignin content. Lignin of leaf litter and agro-residues render them difficult for access to biodegrading enzymes, and they need long

DOI: 10.1201/9781003204435-10

residence times for effective conversion to biogas compared to the wastes of fruit and vegetable origin emerging as segregated wet wastes from households. While wet wastes (in MSW, predominantly fruits, vegetables, and discarded food) are ground into a slurry and pumped to typical CSTR-type digesters, their pectin-held cellular structures permit rapid conversion to intermediate volatile fatty acids (VFA), leaf litter and agro-residues are not amenable to slurrification. Even if powdered, these particles tend to float in the digesters and pose problems for operation and digested material removal [Chanakya et al., 1992]. There has been a need to find alternative fermentation options that either avoid the large water content in a slurry (solid-state fermentation) or escape this phenomenon by clever design (rotary digesters, plug-flow reactors, etc).

While various reactor designs for biomethanation of leafy biomass such as two-phase, single-phase, and plug-flow reactors have been developed, solid-state stratified-bed reactors (SSBRs) with an overall loading rate of 2 kg TS/m^3/d [Chanakya and Moletta, 2004] achieve gas yields similar to or better than floating drum–type CSTR reactors (floating drum cattle dung digesters are in large numbers and have been used for benchmarking plant performance). SSBRs are also reported to require low maintenance and are easy to operate [Chanakya et al., 1997, 1999]. Such digesters provide a potential reactor design for handling non-slurry modes of fermentation while also escaping several attendant problems when using feedstocks such as MSW, garden cuttings, leafy biomass, agro-residues, and biomass-based packing materials. SSBR fermenters function on the principle of high-rate stratified leach bed biomass reactors, wherein the leach bed stratifies into an upper zone of freshly fed leaf biomass that predominantly undergoes the acidogenesis (VFA production) stage, where up to a third of total solids (TS) are broken down (Figure 8.1). When digester liquid is recycled on the top of the above bed, the accumulated VFA intermediates leach down to lower layers of older and partially decomposed feedstock richly colonised by methanogens and is converted to biogas. Here, just as in the case of other forms of solid-state reactors, there is poor transfer of VFA

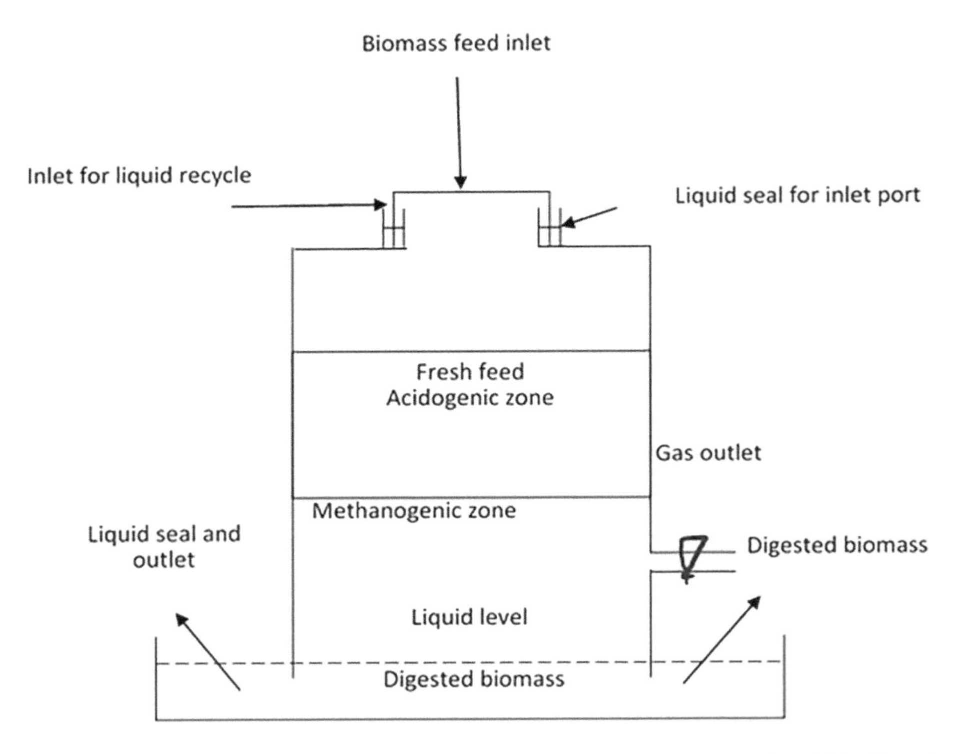

FIGURE 8.1 Single-stage solid-state stratified-bed reactor. (Chanakya et al., 1997, 1999; Chanakya and Moletta, 2005).

to methanobacteria, which can become rate limiting. Similarly, there are obstacles for interspecies hydrogen (H_2) transfer on one side slowing down hydrogenotrophy, and on the other creating bottlenecks for the production of the preferred VFA intermediate, acetate. And yet high feed rates are possible. This requires a closer review of the SSBR system. In this chapter, the reactor design is presented with regard to the advantages it presents, the challenges it overcomes, and the limitations it faces in the digestion of leafy biomass.

8.2 LIGNOCELLULOSIC BIOMASS AS FEEDSTOCK FOR AD AND ASSOCIATED CHALLENGES

There are approximately 12,500 genera of angiosperms and 70 of gymnosperms [Gunaseelan, 1997]. These lignocellulosic biomass types are composed of different concentrations of cellulose, hemi-cellulose, lignin, and pectin. Further, the varying extent of cellulose crystallinity, pore volume, surface area, particle size, and the structural characteristics of lignin makes these a very heterogenous feedstock [Liew et al., 2012]. This also makes predicting the biogas potential of the lignocellulose very difficult because biodegradability of the lignocellulose depends not only on the relative ratios of these components, but also on the interaction between them.

During AD, biomass feedstocks exhibit two clear stages of decomposition. In the first phase, the easily accessible, digestible fraction undergoes rapid conversion to VFA via hydrolytic and acidogenic bacteria generating high VFA flux [Chanakya et al., 1999]. As a result, fed biomass undergoes 30–40% volatile solids (VS) loss in 10 days [Chanakya et al., 1999]. Biomass feedstock where such easily digestible fraction is accessible for digestion are said to undergo disintegrative/VFA-flush fermentation. On the other hand, biomass feedstocks where lignin or other recalcitrant constituents block access to rapid initial decomposition, initial decomposition and VFA releases are low and AD is termed as access-deficient fermentation [Chanakya et al., 1999]. The second phase of digestion is slow because of the limited access to celluloses by cellulolytic and xylanolytic enzymes. This difference in rate and extent of lignocellulosic biomass degradation complicates the process engineering of AD with varying mix of feedstocks [Liew et al., 2012].

Biomass feedstocks fed to conventional (slurry) digesters tend to float and come out of the liquid layer. This limits the access to decomposing bacteria and their enzymes (low initial bulk density of approximately 20–130 kg/m^3). This necessitates the continuous mixing of the reactor to keep the floating biomass in constant contact with digester liquid. In solid-state fermenters, the low bulk density of the biomass feedstock imposes space limitations and reduces the maximum achievable organic loading rate.

Many species of leafy biomass also undergo disintegrative fermentation. In such cases, the rapid digestion of pectin-bound tissues creates small pieces of fermenting biomass devoid of their original rigid structure and rapidly become compacted and pulpy during digestion. Such a compaction renders the feedstock quite impervious to digester liquid and in turn to leaching of intermediate VFA to methanogens. Such a property creates significant mass-transfer limitations [Lissens et al., 2001] and its removal from the digester is also rendered difficult. For biomass feedstocks that exhibit a significant level of access-limited fermentation behaviour, one way to overcome low digestion rates is to carry out pretreatment such as pulverising, delignification, and chemical softening.

8.2.1 PRETREATMENT

Mechanical, chemical, thermal, and biological pretreatment have been used for enhancing digestibility of lignocellulosic biomass. Mechanical means for particle size reduction (mechanical pretreatment) attempts to increase surface area and render feedstock into a slurry to achieve working digester concentrations of approximately 5–8% TS (slurry fermentation) and 15–25% TS (solid-state fermentation). Increased surface area and exposed cellulosics in turn increase the rate and efficiency of hydrolysis [Kim et al., 2003], albeit only to a small extent. Chemical pretreatment methods such as acid pretreatment or alkali hydrolysis aim to remove lignin and its obstructing linkages from the cellulosic complex [Boe, 2006]. It also helps in softening the bast fibres. Alkali treatment results in the 2.3-fold increase of methane (CH_4) yield while acid treatment increases CH_4 yields 0.2–2 times [Zheng et al., 2014], depending upon the biomass species used. When combining both thermal and chemical pretreatments, gas yields of >690 L/kg of VS, an 86% increase in cumulative biogas production in comparison to untreated sample (at 70°C), have also been reported [Rafique et al., 2010]. Aerobic biological pretreatment such as pre-composting also softens biomass feedstock and enables higher feed rates. However, this approach, while it makes many feedstocks such as a water hyacinth amenable for efficient digestion, sacrifices about 30–40% VS that would have otherwise been converted to biogas [Chanakya et al, 1999]. Breaking of certain links in the hemicellulose-lignin polymeric system increases diffusivity of hydrolytic enzymes [Zheng et al., 2014]. The addition of commercial enzymes or pure cultures of lignin-removing bacteria can be used for the purpose of aerobic treatment [Chanakya and Malayil, 2012]. Enzymatic pretreatment has been reported to increase gas yields up to 34% in a few cases [Zheng et al., 2014], but is yet to be used commercially.

In order to balance the technological potential of various approaches and basic tenets of sustainability in the technologies used, early research in CSTR identified the following "guiding principles" or path for sustainable conversion of biomass to biogas as a source of sustainable energy and nutrient recycling options for the tropical world. Biomass feedstocks in the tropics would be numerous, and a single property or profile would not be appropriate for designing the AD process or reactors. Biomass feedstocks would float at all stages of fermentation, therefore there is a need to accept this property and not fight it. Biomass feedstocks will always be available as a mix of feedstocks and no single or unique property would be available for determining a fermentation technique. Biomass feedstocks cannot be made into a stable slurry and will always segregate into liquid and solid layers. Even after a 40-day fermentation, the digested biomass would remain predominantly afloat. Sustainable AD processes and reactor designs need to incorporate and accept these properties, or else they would become less efficient [Chanakya et al, 1997; Chanakya and Moletta, 2005]. The SSBR design exploits the natural tendencies of the biomass feedstocks exhibiting a predominantly access-deficient type of fermentation properties. The process is created around these properties instead of fighting them in a CSTR system. It achieves biogas production rates comparable to cow-dung fed biogas plants, without adopting a pretreatment step. This combined with low reactor volume and passive reactor operation (no mixing parts), gives unmixed SSBRs an energy and resource economy advantage over the wet digesters for digestion of lignocellulose.

8.3 SOLID-STATE FERMENTATION

AD processes may be classified into wet or slurry type processes (<10% TS), semi-solid (10–20% TS) and dry or solid-state (>20% TS) fermentation on the basis of TS concentration fed to the reactor [García-Bernet et al., 2011]. TS concentrations below approximately 23% results in density-based separation of fed solids, with low-density particles floating at top of liquid slurry and denser particles settling at the reactor bottom [Elsharkawy et al., 2019]. Only the dissolved particles remain accessible to the bulk liquid. Slurry-based digesters, therefore, necessitate constant stirring at great energy costs [Jagadish et al., 1998]. Typical biomass feedstocks possess bulk densities of approximately 20–130 kg/m^3 and float on the top of the digester liquid. Further, bubbles rising from the produced biogas carry adhered biomass to the surface and increases the mass of floating biomass [Chanakya and Malayil.2012]. Pretreatment by powdering and slurrification is of little or no help. In another trial, it is interesting to note that despite draining away all the digester liquid, the solid residue continued to produce appreciable biogas. Biomass feedstock was forcibly held under digester liquid for 5 days and partially digested biomass was drained free of digester liquid and placed in a reactor. The biogas production continued over the 30-day trial period. From this the authors concluded that after the initial rapid degradation phase, the digester liquid had a very little role to play [Chanakya and Moletta, 2005]. This suggested that an adequate number of methanogens had colonised the biomass, that there was a balance between acidogenesis and methanogenesis at this stage, and that the digester liquid could be near completely dispensed for such a state of feed.

Dry/solid-state fermenters may be classified into mixed and unmixed reactors: VALORGA, where mixing is achieved by recycling biogas, and KOMPOGAS, with lateral mixing of fed biomass, constitute the mixed solid-state reactors [Edelmann and Engeli, 2005]. DRANCO, BEKON, plug-flow, two-phase, and single-phase SSBRs constitute the unmixed reactors, however, in DRANCO process, fresh feed is externally mixed with digestate in 1:6–1:8 ratio [Six and De Baere, 1992]. Unmixed reactors give an edge in operational costs, which are significantly reduced owing to absence of any active mixing parts. Further, solid-state reactors operated at thermophilic temperatures, spent >45% of input energy to warming substrate and >21% to heat loss [Li et al., 2018]. A DRANCO-like solid-state reactor expends 30–50% of obtained energy in process operation [Six and De Baere, 1992].

8.4 INFLUENCE OF PROCESS PARAMETERS ON PERFORMANCE OF UNMIXED SSBR

8.4.1 Effect of Digester Liquid Recycling

Recycling of biogas digester liquid (BDL) has proven to enhance biogas production in unmixed solid-state fermenters [Chugh et al., 1998]. BDL recycling redistributes VFA flux from the site of production to the site of consumption, the methanogen-rich bed below in the SSBR. Earlier studies of CSTRs have clearly implicated that biomass after a certain extent of acidogenic fermentation is rapidly colonised by methanogens and is responsible for over 20 times the required specific methanogenic activity [Chanakya et al., 1998]. This capacity of partially digested biomass below then converts the excess of VFA leached down into the lower redox regions of the reactor to methanogenic products, namely CH_4 and carbon dioxide (CO_2). Such a leaching/diffusion-dominant redistribution of VFA also prevents formation of acid pockets, which would have otherwise retarded the dissociation of long-chain fatty acids to acetate, H_2 +CO_2 [Chanakya and Malayil, 2012]. Recirculation of leached BDL also helps to a smaller extent in reintroduction of washed-off microbial population, rapid initiation of acidogenesis, and increasing mean cell residence time. This is especially important for increasing methanogenic population in the reactor. Further, in the absence of any active mixing (solid-solid, solid-liquid), recirculated BDL is the only bacterial and wetting source.

8.4.2 Temperature

AD has been shown to be functional at psychrophilic temperatures (<20°C), mesophilic (25–40°C), thermophilic (45–65°C), and extreme thermophilic conditions (>60°C). However, mesophilic (25–40°C) and thermophilic (45–65°C) are two zones showing high conversion rates [Boe, 2006]. The effect of temperature on AD is two-fold. Temperature positively impacts the thermodynamic and kinetic properties of the reactor. Increased kinetic properties of the substrate and the enzyme and increased solubility increases the probability of substrate-enzyme interaction, contributing to more efficient conversion [Sreekrishnan et al., 2004, Boe, 2006]. The positive correlation of temperature to reaction thermodynamics further increases the reaction rates. These, along with the enhanced solid-liquid and liquid-gas mass transfer rates, reduce the required retention time to achieve VS to biogas conversion, allowing higher specific gas production rates [(L gas/L reactor/d), Sreekrishnan et al., 2004]. Faster disintegration and enhanced rates of hydrolysis in thermophilic digestors can yield approximately 36% higher rates of CH_4 production than in mesophilic digestion [Zhou and Wen, 2019]. Rising temperature increases the diffusion coefficient and makes the diffusion gradient steeper, and the oxidation rates of organic acid increases. Production of LCFA, VFA, and other intermediates is favoured and H_2 concentration in the reactor increases [De Bok et al., 2004].

Thermodynamics favours syntrophic acetate oxidation (ΔG = 104 kJ/mol) with increase in temperature contributing up to 14% of CH_4 from acetate [Petersen and Ahring, 1991]. However, until a temperature of 60–63°C (temperature optima of thermophilic aceticlastic methanogen: *Methanosarcina thermophila*) the aceticlastic route of CH_4 production still dominates. At temperatures of more than 65°C, syntrophic acetate oxidation overtakes and more than 95% of acetate is directed to CH_4 through acetate oxidation route. At psychrophilic temperatures, especially below 15°C, homo-acetogenesis is the dominant route for oxidation of free H_2, and CH_4 production through homo-acetogenesis can be as high as 95% of total CH_4. At psychrophilic temperatures, the dominant methanogen for converting acetate to CH_4 is *Methanosaeta* [Boe, 2006]. At mesophilic temperatures, hydrogenotrophy (ΔG = –135 kJ/mol) and aceticlasty (ΔG = –31 kJ/mol) are favoured over homo-acetogenesis (ΔG = –104 kJ/mol) and acetate oxidation (ΔG = 104 kJ/mol), respectively. *Methanosarcina* dominates methanogen population, followed by *Methanobacteriales*, *Methanomicrobiales,* and *Methanococcales* (Hydrogenotrophic methanogens) [Karakashev et al., 2005]. However, *Methanosarcina* are sensitive to ammonia concentration and dominate the methanogen population only at non-inhibitory concentrations of ammonia [Schink, 1997].

The downside of high temperature is decreased pKa of ammonia and increased pKa of VFA [Boe, 2006]. This results in dominance of free NH_3 and undissociated forms of VFA, both of which impede methanogenesis. Enzyme denaturation at high temperature is another factor that puts an upper limit to increased biogas production rates. Therefore, despite positive correlation of temperature to reaction rates and efficiency, mesophilic range (25–40°C) is the preferred operation range for anaerobic fermentation. Reduced operating cost by preventing the energy required to maintain constant high temperature, approximately 45–60°C (>45% of process energy), favours the economics of the process and is a major contributor in the preference of a mesophilic range [Chanakya and Malayil, 2012, Li et al., 2018].

8.4.3 pH

Effect of pH on reactor health is two-fold. The pH affects the acid-base equilibrium of various compounds in the digester wherein at low pH, free VFA causes weak acidic forms of the compounds to dominate absorption and intake (acetic acid or propionic acid instead of acetate or propionate) by microbial cells. However, at a neutral and higher pH, free ammonia is the dominant form of cation. While buffering ammonia also inhibits production of both CH_4 and H_2 [Boe, 2006], pH level also has a significant effect on enzyme activity. As each enzyme is active only at a specific pH range and has an activity maxima at its optimal pH, the role of pH then becomes crucial in maintaining the structural and functional integrity of enzymes of participating microorganisms [Boe, 2006]. The pH range of fermentative bacteria is 4–8.5, while methanogens are active in 5.5–8.5, with an optimal pH range of 6.5–8 [Nielsen, 2006]. A mixed culture anaerobic digester has optimum pH of 6.6–7.7 [Moosbrugger et al., 1993].

A reactor's pH is governed by the feed provided and feed's buffering capacity. A reactor with low buffering capacity experiences rapid pH changes when intermediate VFA is produced as a flux and is poorly buffered. This affects the digestion rate and ultimately the overall methane production rates. In such cases, increasing the leachate recycling rate and addition of small doses of lime [buffering cations] has proven to be of help [Chanakya et al., 1995]. Generally, herbaceous biomass has been reported to possess high buffering capacity, owing to the large mineral content in the leaf (ash). This resists large pH changes over high OLR [organic loading rate] and high VFA accumulation [Chanakya et al., 1993a]. High partial pressure of H_2 instead affects the performance in such cases.

8.4.4 Feed Inlet Design

Continuous reactor systems require that biomass feedstocks are fed daily. However, the anaerobic nature of the process requires the feed to be introduced without simultaneous entry of air into the reactor. In a slurry-based reactor system, only the slurry is introduced into the reactor and there is very little chance of O_2 entering the reactor system. However, in SSBRs, substrate is fed as a solid mass. These leaf litter solid feedstocks have significant voids and makes entry of air along with the feed inevitable. Chanakya et al. [1993a] conducted experiments with and without cross flow of N_2 to understand the extent of reduction in CH_4 and lower biogas production rates due to air exposure while feeding in SSBRs. No significant reduction in gas production rates and high methane percentage suggested possible dominance of acidogenic microbes in the layers closer to feed inlet. It was hypothesised that rapid consumption of the introduced oxygen prevented the methanogens in lower zones of the SSBR from exposure to oxygen and concomitant inhibition of methanogen activity. Studies have also shown SSBRs to be tolerant to a day-long exposure to air (open feed inlet) and the AD resumes unabated upon shutting of the air exposure [Chanakya et al., 1993a].

This understanding of the underlying process formed the basis of the method evolved to feed SSBRs. Partially digested biomass (retention time [RT] > 7 days) gradually builds up the required methanogenic populations. This decomposing biomass bed therefore undergoes a stratification with regard to predominant decomposition activity and underlying microbial species along the reactor height. Upper layers are constantly under acidogenesis and are only marginally affected by the air introduced while feeding [Chanakya et al., 1993a]. Methanogens are dominant in the lower layers accompanied by lower redox regimes (≤–150 mV). This in turn protects methanogens from the oxygen introduced at the top of the bed during feeding [Chanakya et al., 1993b]. From an operational outlook, this led to the conclusion that if headspace of the reactor is kept very low to enable the lowest possible volume for entry of oxygen, stratification established in SSBRs is able to tolerate the exposure to air while feeding [Chanakya et al., 1993a], while also simplifying the reactor design and operation.

8.4.5 Start-Up Options

The SSBR design employs a water seal at the bottom which covers the outlet and prevents the biogas from escaping through it. This BDL is also recycled daily to retain its recycled leach-bed mode of operation. However, there is a need to develop a methanogenic bed at the lower part of the SSBR prior to start-up, for which an external source of digested (or spent) biomass from another SSBR is preferred, so as to obtain a rapid start-up [Chanakya et al., 2007, Chanakya and Kumar, 2002]. Early experiments suggest approximately 25% of reactor volume to be occupied by spent biomass to ensure enough starting microbial population [Chanakya et al., 2007]. Another, however less efficient and less common, approach for obtaining a methanogen-rich biomass bed is to load a pit with alternating thick layers of cow dung and fresh leaves with abundant water. The soggy mixture thus obtained can then serve as a source of methanogens [Chanakya and Kumar, 2002].

8.5 SINGLE-STAGE SOLID-STATE STRATIFIED-BED REACTOR

The use of a single reactor with an upper zone predominantly carrying out acidogenesis and a lower zone carrying out predominantly methanogenesis combined in a single reactor reduces the complexities of a two-phase system, especially when targeted for rural areas or complex and varying mix of biomass in the feedstock typical for leaf litter in MSW or agro-residues [Chanakya et al., 1993a, 1993b]. During continuous operation, the SSBR forms a stratified bed with an acidogenic zone above the methanogenic zone in the same reactor (Figure 8.1), where the position and transition of these zones is determined by the fermentation properties of the feedstock mix and, to a small extent, on the method of operation. This stratification of microbes is a natural process observed in nature (e.g. bioreactor landfills) formed as a result of the microenvironment surrounding the microbial fauna [Chanakya et al., 2004, Chanakya and Moletta, 2004, Anand et al., 1991]. However, while the overall reaction occurs over a lifetime of about 5,000 days in the bioreactor landfill, this process is expected to happen in about 50 days [Lissens et al., 2001] in the SSBR.

8.5.1 Reactor Design, Rationale, and Functioning

SSBRs consist of a packed bed of biomass feedstock fed generally in a form as found or as collected without powdering and often in a dry state. As discussed before, this mode of fermentation requires a mix of biomass species, which on the whole represents the access-deficient mode of solid-state digestion, wherein the digested mass still retains a certain degree of the original tissue structure to remain porous enough to allow it to be leached, as well as encourage a methanogen-rich microbial colonisation. Irrespective of the initial bulk density, the fermenting mass within should not exceed a bulk density of 600 kg/m^3. Above this density and with an increased level of disintegrative biomass types, the bed becomes impervious and retards the rate of VFA rich BDL percolating through it and its conversion to biogas. Most leaf biomass species undergo a rapid initial decomposition where approximately 40–50% of the total VS removed occurs during the initial–20 day solids retention time [Chanakya et al., 1999]. Higher growth rates of hydrolytic and acidogenic bacteria result in VFA fluxes that are 8–10 times faster than turnover rates of methanogens [Chanakya et al., 1998]. In SSBR, this digestion behaviour of lignocellulosic biomass leads to pockets of high VFA that are formed in and around the fresh feedstock at a solids retention time of <15 days [Chanakya et al., 1997]. In this case when VFA levels are not controlled or recycling is poor, they witness increased H_2 partial pressure (>10^{-5} atm) and a tendency to accumulate long-chain VFA such as propionic acid, butyric acid, and valeric acid [Boe, 2006]. Often even longer-chain VFA of C_5–C_{10} species are recorded, bringing the overall decomposition nearly to a halt [Chanakya et al., 1998]. When such a situation spreads across the bed, it can severely impede reactor health. For a healthy reactor, concentration of long-chain fatty acids to propionic and butyric acid needs to be present at <0.3 g/L [Chanakya et al., 1993a] and pH between 6.8–7.2 [Boe, 2006; Sreekrishnan et al., 2004].

A packed bed of biomass, with a bare minimum of water, poses another operational challenge of maintaining an adequate moisture level to sustain microbial activity across the digester volume. This control of transporting and regulating the concentrations of VFA produced from fresh biomass and adequate moisture levels on digesting biomass to sustain microbial activity is attained via liquid recycling. Liquid for recycling is generally obtained as starting inoculum from any functioning biogas reactors (Gobar gas plant is the most common choice). It is also initiated using sewage sludge or anaerobic MSW digestate or with anoxic water when start-up is carried out as indicated earlier.

Recycled BDL acts as an inoculum to initiate degradation of fresh biomass added on top of the existing pile. It also transports VFA intermediates that are produced from the biomass bed [Chanakya et al., 1993a] to regions that are rich with methanogens. Earlier studies [Chanakya et al., 1998] show that the methanogenic population colonised and adhered very strongly to partially digested biomass. This methanogen-rich biofilm has 10–30 times the CH_4 production capability (often described as specific methanogenic assay) of the extent of biogas actually produced or expected from a biogas reactor. Although a total specific methanogenic assay (SMA) (hydrogenotrophy and aceticlasty) 1.5 mL biogas/g biomass/d is adequate to yield 0.5 L biogas/L/d reactor, H-SMA of 20 mL biogas/g biomass/d and A-SMA of 8–12 mL biogas/g biomass/d has been recorded (CST-unpublished data; Chanakya et al., 1998). This methanogen-rich biofilm on partially digested biomass at the lower zone in the SSBR and its proper functioning is important for rapid conversion of VFA leached with BDL to biogas [Chanakya et al., 1995, 1998, 1999, Chanakya and Moletta, 2004]. BDL recycling therefore lies at the centre of the reactor operation control. Some research pertaining to understanding the reaction dynamics related to BDL recycling volume and frequency as a spray or sprinkle and its effect on biogas production from fed biomass was attempted in the 1990s. The objective was to optimise BDL recycling volume and frequency [Chanakya et al., 1993a, 1993b, 1995, 1997, 1999, Anand et al., 1991]. Table 8.1 presents the reactor performance of single-stage SSBR with few-lignocellulosic biomass. SSBRs can achieve biogas yields of 0.4–0.6 m^3/m^3/d for a feed rate of 12 kg (fresh weight)/m^3/day [Chanakya and Moletta, 2004]. More research

TABLE 8.1
Biogas Production from SSB with Different Lignocellulosic Biomass

Reactor Type	Biomass	Reactor Volume (L)	Biogas Production (L/kg TS)	Study Period (d)	Feed (TS)	Mode of Operation	References
SSB	Fresh water hyacinth	2	291	300	1.175 (gTS/L/d)	Fed Batch	Chanakya et al. (1993b)
SSB	Leaf litter (Paper mulberry)	10.6	250	70	2 (gTS/L/d)	Fed Batch	Chanakya et al. (2007)
SSB	Fresh paper mulberry leaves	6.2	~412	78	343.9 (gTS)	Fed Batch	Chanakya et al. (1999)
SSB	Dry leaves	6.2	351	78	211.1 (gTS)	Fed Batch	Chanakya et al. (1999)
SSB	*Acacia auriculiformis*	6.2	189	78	441.3 (gTS)	Fed Batch	Chanakya et al. (1999)
SSB	*Synedrella nodiflora*	6.2	439	78	235.5 (gTS)	Fed Batch	Chanakya et al. (1999)

pertaining to engineering BDL recycling rates specific to biomass is required to enable successful translation of SSBR to field.

8.6 CONCLUSION

The leaf litter component of MSW poses sustainability challenges of transporting organic matter and nutrients away from site of production and causing environmental nuisance at the site of disposal. Decentralised biogas plants are a good solution. For easy digestion of leafy biomass, SSBRs shows promise as the appropriate reactor design, with low maintenance, passive design, and small reactor size in comparison to slurry-based reactors. The biogas production rates (approximately 0.3–0.6 L/L/d) are also comparable to that obtained from cow dung–based reactors and show promise of small-scale economic feasibility. This makes SSBRs an attractive choice for decentralised leaf litter digestion and conversion to biogas. However, despite the potential of generating approximately 1 L/L/d of biogas, only 0.3–0.6 L/L/d of biogas has been achieved in most of the documented studies. Research on scalability of this design is now required, as this approach is likely to solve a third of the MSW problem across India and the tropical world, where leaf litter forms a significant component of OFMSW collection.

ABBREVIATIONS

AD = anaerobic digestion
BDL = biodigester liquid
CSTR = continuously stirred tank reactor
OFMSW = organic fraction of municipal solid waste
RT = retention time
SMA = specific methanogenic assay
SSBR = solid-state stratified-bed reactor
SSR = solid-state reactor
TS= total solids
VFA = volatile fatty acids
VS = volatile solids

REFERENCES

Anand, V., Chanakya, H.N. and Rajan, M.G.C., 1991. Solid-phase fermentation of leaf biomass to biogas. *Resources, Conservation and Recycling*, *6*(1), pp. 23–33.

Boe, K., 2006. Online monitoring and control of the biogas process. Technical University of Denmark, Copenhagen.

Chanakya, H. and Kumar, S., 2002. Bioreactors for clean coffee effluents – User Manual (5th ed.). Bangalore.

Chanakya, H.N. and Malayil, S., 2012. Anaerobic digestion for bioenergy from agro-residues and other solid wastes—An overview of science, technology and sustainability. *Journal of the Indian Institute of Science*, *92*(1), pp. 111–144.

Chanakya, H.N. and Moletta, R., 2005, August. Performance and functioning of USW plug-flow reactors in a 3-zone fermentation model. In *Proc. 4th Intl. Symp. on Anaerobic Digestion of Solid Wastes* (Vol. 1, pp. 277–284).

Chanakya, H.N. and Moletta, R., 2004. Emerging trends in small scale biogas plants for agro residues and biomass feedstocks–A case study from India. Session 5C: Agricultural feed Stock. In *10th World Congress-Anaerobic Digestion*, pp. 550–555.

Chanakya, H.N., Borgaonkar, S., Meena, G. and Jagadish, K.S., 1993a. Solid-phase fermentation of untreated leaf biomass to biogas. *Biomass and Bioenergy*, *5*(5), pp. 369–377.

Chanakya, H.N., Borgaonkar, S., Meena, G. and Jagadish, K.S., 1993b. Solid-phase biogas production with garbage or water hyacinth. *Bioresource Technology*, *46*(3), pp. 227–231.

Chanakya, H.N., Borgaonkar, S., Rajan, M.G.C. and Wahi, M., 1992. Two-phase anaerobic digestion of water hyacinth or urban garbage. *Bioresource Technology*, *42*(2), pp. 123–131.

Chanakya, H.N., Ganguli, N.K., Anand, V. and Jagadish, K.S., 1995. Performance characteristics of a solid-phase biogas fermentor. *Energy for Sustainable Development*, *6*(1), pp. 43–46.

Chanakya, H.N., Rajabapaiah, P. and Modak, J.M., 2004. Evolving biomass-based biogas plants: The ASTRA experience. *Current Science*, pp. 917–925.

Chanakya, H.N., Ramachandra, T.V. and Vijayachamundeeswari, M., 2007. Resource recovery potential from secondary components of segregated municipal solid wastes. *Environmental Monitoring and Assessment*, *135*(1-3), pp. 119–127.

Chanakya, H.N., Srikumar, K.G., Anand, V., Modak, J. and Jagadish, K.S., 1999. Fermentation properties of agro-residues, leaf biomass and urban market garbage in a solid-phase biogas fermenter. *Biomass and Bioenergy*, *16*(6), pp. 417–429.

Chanakya, H.N., Srivastav, G.P. and Abraham, A.A., 1998. High rate biomethanation using spent biomass as bacterial support. *Current Science*, pp. 1054–1059.

Chanakya, H.N., Venkatsubramaniyam, R. and Modak, J., 1997. Fermentation and methanogenic characteristics of leafy biomass feedstocks in a solid-phase biogas fermentor. *Bioresource Technology*, *62*(3), pp. 71–78.

Chugh, S., Clarke, W., Pullammanappallil, P. and Rudolph, V., 1998. Effect of recirculated leachate volume on MSW degradation. *Waste Management & Research*, *16*(6), pp. 564–573.

De Bok, F.A.M., Plugge, C.M. and Stams, A.J.M., 2004. Interspecies electron transfer in methanogenic propionate degradaing consortia. *Water Research*, *38*(6), pp. 1368–1375.

Edelmann, W. and Engeli, H., 2005. More than 12 years of experience with commercial anaerobic digestion of the organic fraction of municipal solid wastes in Switzerland. *In ADSW 2005 Conference Proceedings*, 1, pp. 27–33.

Elsharkawy, K., Elsamadony, M. and Afify, H., 2019. Comparative analysis of common full scale reactors for dry anaerobic digestion process. *In E3S Web of Conferences*, 83, p. 01011. EDP-Sciences.

García-Bernet, D., Buffière, P., Latrille, E., Steyer, J.P. and Escudié, R., 2011. Water distribution in biowastes and digestates of dry anaerobic digestion technology. *Chemical Engineering Journal*, *172*(2-3), pp. 924–928.

Gunaseelan, V.N., 1997. Anaerobic digestion of biomass for methane production: A review. *Biomass and Bioenergy*, *13*(1-2), pp. 83–114.

Jagadish, K.S., Chanakya, H.N., Rajabapaiah, P. and Anand, V., 1998. Plug-flow digestors for biogas generation from leaf biomass. *Biomass and Bioenergy*, *14*(5-6), pp. 415–423.

Joshi, R. and Ahmed, S., 2016. Status and challenges of municipal solid waste management in India: A review. *Cogent Environmental Science*, *2*(1), p. 1139434

Karakashev, D., Batstone, D.J. and Angeljdaki, I., 2005. Influence of environmental conditions on methanogenic compositions in anaerobic biogas reactors. *Applied and Environmental Microbiology*, *71*(1), pp. 331–338.

Kim, J., Park, C., Kim, T.H., Lee, M., Kim, S., Kim, S.W. and Lee, J., 2003. Effects of various pretreatments for enhanced anaerobic digestion with waste activated sludge. *Journal of Bioscience and Bioengineering*, *95*(3), pp. 271–275.

Kumar, D.R., Chanakya, H.N. and Dasappa, S., 2019. Anaerobic digestion potential of leaf litter: Degradability and gas production relationships. In *Waste Valorisation and Recycling* (pp. 553–556). Springer, Singapore.

Li, Y., Xu, F., Li, Y., Lu, J., Li, S., Shah, A., Zhang, X., Zhang, H., Gong, X. and Li, G., 2018. Reactor performance and energy analysis of solid-state anaerobic co-digestion of dairy manure with corn stover and tomato residues. *Waste Management*, *73*, pp. 130–139.

Liew, L.N., Shi, J. and Li, Y., 2012. Methane production from solid-state anaerobic digestion of lignocellulosic biomass. *Biomass and Bioenergy*, *46*, pp. 125–132.

Lissens, G., Vandevivere, P., De Baere, L., Biey, E.M. and Verstraete, W., 2001. Solid waste digestors: Process performance and practice for municipal solid waste digestion. *Water Science and Technology*, *44*(8), pp. 91–102.

Moosbrugger, R.E., Wentzel, M.C., Ekama, G.A. and Marais, G.V.R., 1993. A 5 pH point titration method for determining the carbonate and SCFA weak acid/bases in anaerobic systems. *Water Science and Technology*, *28*(2), pp. 237–245.

Nielsen, H.B., 2006. Control Parameters for Understanding and Preventing Process Imbalances in Biogas Plants. Emphas on VFA Dynamics.

Petersen, S.P. and Ahring, B.K., 1991. Acetate oxidation in a thermophilic anaerobic sewage sludge digester: the importance of non-aceticlastic methanogenesis from acetate *FEMS Microbiology Letters*, *86*(2), pp. 149–152.

Rafique, R., Poulsen, T.G., Nizami, A.S., Murphy, J.D. and Kiely, G., 2010. Effect of thermal, chemical and thermo-chemical pretreatments to enhance methane production. *Energy*, *35*(12), pp. 4556–4561.

Sasse, L., 1988. Biogas plants. Vieweg & Sohn, Wiesbaden.

Schink, B., 1997. Energetics of syntrophic cooperation in methanogenic degradation. *Microbiology and Molecular Biology Review*, *61*(2), pp. 262–280.

Shwetmala, C.H. and Ramachandra, T.V., Decentralised energy potential of Bangalore solid waste. In *Kerala Environment Congress* 2011 (p. 228).

Six, W. and De Baere, L., 1992. Dry anaerobic conversion of municipal solid waste by means of the DRANCO process. *Water Science and Technology*, *25*(7), pp. 295–300.

Sreekrishnan, T.R., Kohli, S. and Rana, V., 2004. Enhancement of biogas production from solid substrates using different techniques—A review. *Bioresource Technology*, *95*(1), pp. 1–10.

Zheng, Y., Zhao, J., Xu, F. and Li, Y., 2014. Pretreatment of lignocellulosic biomass for enhanced biogas production. *Progress in Energy and Combustion Science*, *42*, pp. 35–53.

Zhou, H. and Wen, Z., 2019. Solid-state anaerobic digestion for waste management and biogas production. *Solid State Fermentation: Research and Industrial Applications*, pp. 147–168.

9 Anaerobic Co-Digestion of Drain Sludge with Fermentescibles Municipal Waste of Sokodé (Togo)

Nitale M'Balikine Krou, Gnon Baba, and Ogouvidé Akpaki

9.1 INTRODUCTION

One of the consequences of the improvement of the living conditions of the populations is the high demand for thermal energy for households. In developing countries, the cost of living is an obstacle to all social layers, due to the fact that they cannot reach supply of butane gas for their energy needs. Generally, in these countries, charcoal is the most used source of household energy. The excessive use of firewood and charcoal leads to deforestation, which in turn has a negative effect on the environment. According to the 2010 World Bank report on the energy balance in Togo, biomass remains the main source of energy consumed. It represents on average 70–80% of total final consumption and is used to supply the domestic and craft sectors. According to the same report, in Togo, 61% of households use only firewood, 24% charcoal and firewood, and 15% only charcoal; 75% of rural households use firewood with a consumption of 347 kg/year/person, while 72% of urban households use charcoal of about 59 kg/year/person (World Bank, 2010). With this alarming consumption of firewood, which is likely to increase in the coming years, and as no concrete measures have been taken to balance the consumption of firewood energy with the production capacity of forests, it can be estimated that Togo is trending towards a chronic inability to meet its wood energy needs and to the disappearance of the forests that constitute the sources of supply.

Moreover, the problem of management of faecal sludge (FS) and urban solid waste has not yet found an adequate solution in Togo. In the city of Sokodé, this waste is often dumped in and around rivers, leading to risks of pollution of surface water. Optimising their management is therefore more than necessary. There is also an increasingly growing need for firewood or charcoal for cooking in households. With this alarming consumption of firewood, which is likely to increase in the years to come and could lead to the disappearance of forests, the search for alternatives is essential. This ever-increasing rate of consumption of firewood and the strong degradation of forests in Togo is due to the cost of fossil fuels and the difficulty of serving landlocked regions. At a time when renewable energies are increasingly explored to supplement or gradually replace fossil fuels, anaerobic digestion is an opportunity for the diversification of energy resources and to sustainably manage the environment. The objective of this work is to evaluate the interest of co-digestion of FS with fermentable solid waste through their maximum production of biogas and methane (CH_4). Anaerobic digestion has significant potential through its double capacity for recovering energy from organic waste and reducing greenhouse gas emissions. It could be seen as an economic, decentralised, and ecological solution to these problems through energy autonomy and sustainable agricultural development in rural areas.

This study takes place in the context of energy, environmental, and sanitation challenges in the city of Sokodé. Indeed, we are witnessing growing needs for butane gas, firewood, or charcoal for cooking and lighting in the city of Sokodé, as in the other cities of the same size. Thus, by the

DOI: 10.1201/9781003204435-11

production of energy in situ, the condition of the populations will be able to improve significantly thanks to the recovery of sludge and fermentable fractions of solid waste into biogas. The wood and charcoal that constitute the traditional biomass in households in Sokodé generate smoke, which is a threat to public health. Co-fermentation of FS with fermentable waste of domestic or industrial origin to produce methane can be a good alternative for domestic cooking.

Small-scale biogas fermentation systems offer the possibility of using waste substrates, and therefore saving public health, instead of using wood. The CH_4 or biogas recovered can be used as fuel in kitchens and households. The resulting digestate serves as an organic amendment to agricultural soils with good fertilising value. In fact, all the nitrogen contained in the methanised material is conserved during anaerobic digestion. However, there is a change from the original organic form to an ammoniacal NH_4^+ form, which is more assimilable by plants (Almansour, 2011). The present study specifically aims at evaluating the interest of co-digestion of FS with fermentable waste through their maximum production of biogas and methane.

As no study has ever been reported on the co-digestion of sewage sludge with fermentable municipal waste in Togo, this study could provide a baseline database in Togo using inexpensive and available raw materials that complement each other. Tests on a laboratory scale have been carried out in order to validate the value of the co-digestion of the sludge with the fermentable solid waste. The results obtained will enable the process to be validated and implemented on a large scale at the sewage sludge treatment station in Sokodé.

9.2 METHODOLOGY

Organic waste used in this research comes from Sokodé, the main town of the Centrale Region and the district of Tchaoudjo. It is located between 8° and 9° N, and 1° and 3° E and, 346 km north of Lomé. It has long been the regional capital of the northern zone and especially the second center of settlement of the country. With a population estimated at 111,258, Sokodé enjoys a tropical climate of transition between Guinean (wet) and Sudano-Sahelian (dry).

The sludge used comes from public toilets. Preliminary modelling studies of biogas production by sludge from dry latrines (DL), public septic tanks (PST), and domestic septic tanks (DST) from the city of Sokodé have shown that DST and DL sludge have a low production of biogas compared to that of PST sludge. Thus, only fresh FS (from public toilets) is suitable for the production of biogas, with a dryness of at least 3% (Klingel, 2002). Fermentable fractions are obtained after sorting solid waste from hotels, markets, and households. The samples used for the experiments were taken in accordance with good sampling and conservation practices. The test includes four scenarios:

- Scenario 1: filling the digester with the FS alone
- Scenario 2: filling the digester with the fermentable fractions of household waste (FFHW) only
- Scenario 3: filling the digester with the fermentable fractions of assimilated waste (FFAW) alone
- Scenario 4: filling the sludge drain digester mixed with the fermentable fractions of solid waste (households, hotels, and markets)

Tests carried out under a mesophilic regime (37°C) consisted in monitoring the performance of biogas production by these residues.

The solid/liquid ratio (S:L) equal to 1:5 was used (Deressa et al., 2015) because it makes it possible to avoid inhibitions and reach the maximum production of biogas for most substrates (Mata-Alvarez, 2014; Walker, 2010). Thus, to prevent inhibitions, the amount of substrate used was not too large compared to the amount of distilled water. This optimal S:L ratio of 1:5 fixed also allows bacteria to ensure the fermentation of organic matter, limiting the concentration of nitrogen and/

or volatile fatty acids (Eskicioglu and Ghorbani, 2011). In the laboratory, before the launch of the experiment, the solid matter (SM) and volatile solid matter (VSM) contents were determined. These results made it possible to calculate the quantities of co-substrate to be introduced into each bottle. The VSM sludge to VSM waste ratio set in this study is equal to 0.3 because it is considered to be optimal for good management of waste of various kinds (Nsavyimana et al., 2012). In addition to the dry matter and volatile dry matter contents, it was necessary to assess the carbon to nitrogen ratio (C:N), the humidity, and the hydrogen potential.

The pH was determined using a laboratory pH meter (accuracy of ± 0.01 pH unit).

SM is obtained by drying in an oven and VSM by the loss-on-fire method (Kelly, 2002). Total nitrogen, the sum of ammonia nitrogen and organic nitrogen, was measured using the Kjeldahl method. The technique used for the determination of total organic carbon was that of Walkley and Black, modified according to the MA analysis method (Center of Expertise in Environmental Analysis of Quebec, MA. 405 – C 1.1, 2014).

After mixing the substrates and distilled water, the vials were closed tightly. The vials were then incubated at 37°C to perform digestion under mesophilic conditions. The experiments were stopped after 42 days of production when the daily production of biogas was very low.

Reducing the particle size of the substrate increases its homogeneity. This reduction in size is necessary because the specific surface that can be attacked by bacteria and enzymes increases with the reduction of particles (Hu et al., 2005). In addition, large particles can be introduced into small laboratory bottles by grinding. Studies on the effect of grinding food scraps have also been carried out; the results showed that a reduction in particle size to 2 mm allows an increase in methane production of at least 20% (Huyard, 2012). Grinding would release easily degradable particles, which are not always available to bacteria without reducing the particle size (Izumi et al., 2010; Bruni et al., 2010). In this study, the grinding was done using a laboratory mortar to reduce the particle size. Photos 1 and 2 show the samples of fresh FS and of FFAW.

The protocol is based on preliminary studies in laboratory pilot. This protocol, once implemented, made it possible to obtain the total biogas production of the effluent representative of its level of anaerobic biodegradability. The experimental device was made up of digesters (1500 mL bottles), rubber pipes, graduated cylinders for collecting biogas, a solution of sodium chloride (NaCl) contained in a beaker, and supports. In this assembly, the pilot digesters were connected through gas supply pipes to inverted test tubes (allowing the visual measurement of the volume

PHOTO 9.1 Sample of fresh sewage sludge.

PHOTO 9.2 Sample of fermentable fraction of assimilated waste.

of biogas produced daily) immersed in a beaker containing a trap solution consisting of saturated water in NaCl and in acid (hydrochloric acid 5%, NaCl 20%, and pH = 2) in order to minimise the dissolution of carbon dioxide (CO_2) from the biogas (Pouech and Marcato, 2005; Akpaki, 2015). This aqueous solution uses the displacement of the liquid to properly quantify the daily production of biogas (Photo 9.3).

PHOTO 9.3 Biogas production system.

9.3 RESULTS AND DISCUSSION

In order to validate the value of the co-digestion of sewage sludge with the fermentable fractions of municipal solid waste, a comparative study of the production of biogas from these different fermentable substrates was carried out. To do this, it was necessary to determine some physicochemical characteristics of the substrates.

The evaluation of the physicochemical parameters such as the pH, the level of SM and VSM as well as the C:N ratio is essential in order to validate their interest in anaerobic digestion. Table 9.1 presents the results of these global parameters.

- *FFHW: Fermentable Fractions of Household Waste (Corn paste, prepared rice, prepared pasta and mango).*
- *FFAW: Fermentable Fractions of Assimilated Waste (Mango, banana, bread, and avocado).*
- *PST: Public Sceptic Tank.*

From this chart, it is obvious that the FS from PST has higher water content (83.38%) and the FFHW the lowest (57.04%). Overall, the water contents are high, which is an asset for anaerobic digestion. The dry matter content of the substrates is between 16.82% (PST) and 42.96% (FFHW). The VSM contents are between 70.01% (FFHW) and 81.13% (PST). Drainage sludge has the lowest percent SM content. The association of this sludge with the fermentable fractions of solid waste (with a higher content) can increase this content and promote good anaerobic digestion.

The C:N ratio of FS from PST was 8.98. This result indicates high nitrogen content. Indeed, it has shown in studies of biodegradability of sludge that the NH_4^+ to NTK ratio of sludge from PST, which was 14.05, is high and indicates that the mineralisation phase of organic matter is less advanced in these pits because of their emptying frequencies. The effluent from FFHW and FFAW had a C:N ratio of between 20 and 30. The optimal ratio for anaerobic digestion must be between 20 and 30, extreme values to ensure the biological stability of the system (Gunaseelan, 2004; Chulalaksananukul, 2012). Indeed, if the C:N ratio is very high, methanogenic bacteria tend to consume nitrogen quickly to meet their protein needs. Thus, the carbon will not be degraded. Consequently, the production rate of biogas would be very low (Chandra et al., 2012). Conversely, if the C:N ratio is very low, the nitrogen will be released and will accumulate in the form of ammonium (NH_4^+). The excessive formation of ammonium increases the pH of the medium through the formation of the NH_3 form. Above a pH of 8.5, toxic effects on methanogenic bacteria appear (Chandra et al., 2012; Akindele et al., 2017) and consequently, the production of biogas decreases. The effluents studied had a suitable ratio for anaerobic digestion.

Acid pHs are not favourable for biological digestion. The pH of the mixtures was 6.44–7.8. This is comparable to the optimum pH (6.5 and 7.5) for the production of biogas (Chua et al., 2008; Deressa et al., 2015). This result shows that the microorganisms in the digester will not be affected by the pH of the co-substrates. Thus, there will be no inhibition for the production of biogas.

In view of the results on the characterisation of the substrates, a hypothesis on the investigation in relation to the modelling of biogas production from these substrates was formulated and executed.

TABLE 9.1
Characteristics of the Substrates Used

Substrates	%H	%SM	%VSM	pH	NTK (%MS)	Corg (% SM)	C:N
Scenario 1: PST	83.38	16.82	81.13	7.8	3	26.93	8.98
Scenario 2: FFHW	57.04	42.96	70.01	6.82	1.44	35.01	24.3
Scenario 3: FFAW	61.21	38.79	78.72	6.44	1.56	39.36	55.2

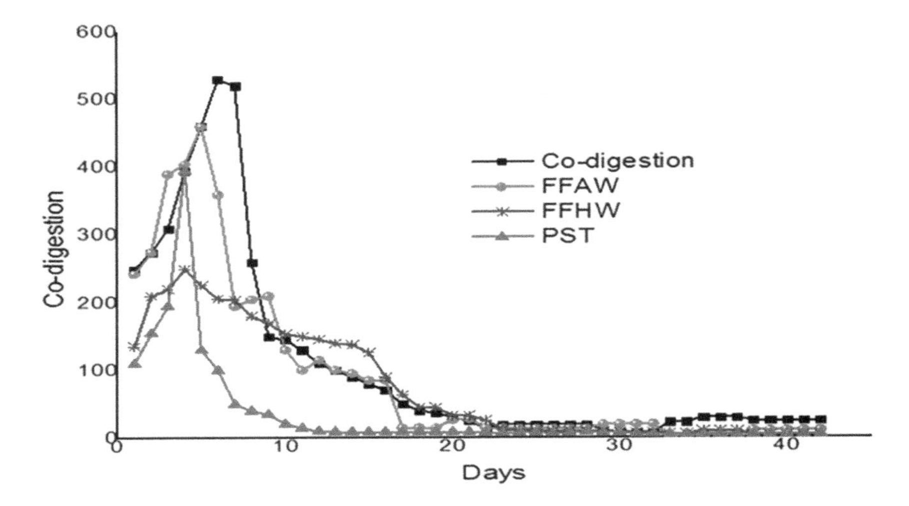

FIGURE 9.1 Daily biogas production.

The production measured represents the maximum amount of biogas that can be produced by a given fermentable substrate. This measurement is specific to each fermentable substrate and representative of its level of biodegradability.

The comparative study of daily production shows a higher production of biogas from the first five days by the sludge from PST compared to that of the FFHW. However, it remained lower than the daily production of the FFAW and co-digestion. After five days of production, there was a drop in production in the sludge from PST compared to the production of biogas in the FFHW. The maximum daily production was reached from the sixth day (Figure 9.1). This result is explained by the variation in the organic matter content of each substrate and the accessibility to this organic matter by the microorganisms in these different substrates.

Figure 9.2 shows the cumulative biogas production for an anaerobic digestion of 42 days. Depending on the type of substrate, biogas production varies depending on whether it is rich in easily biodegradable organic matter.

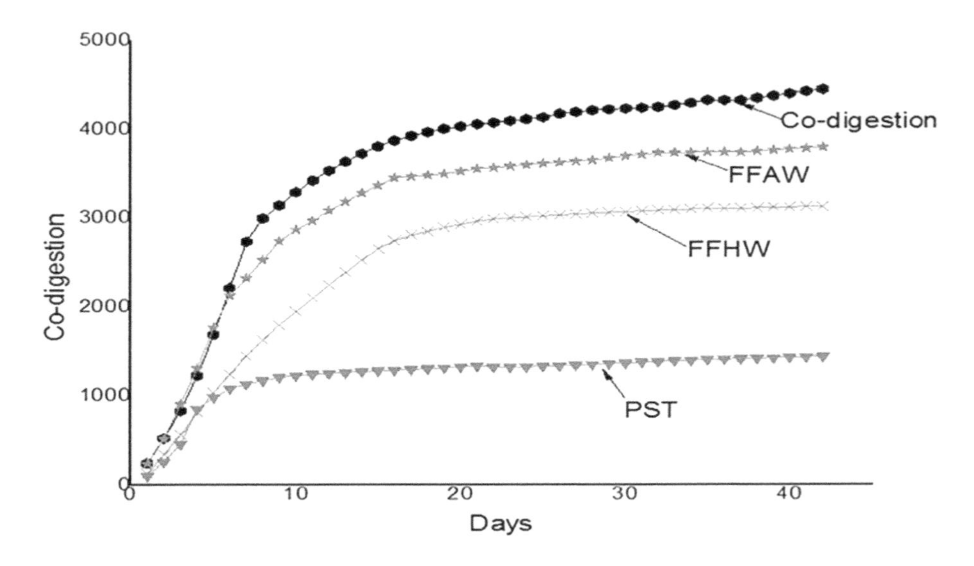

FIGURE 9.2 Cumulative biogas production.

This figure shows that, from the same quantity of substrate, the biogas production potential of FFAW is high compared to that of FFHW, which is also higher than that of FS from PSTs. This result is explained by the fact that the biologically degradable part is higher at the fermentable fractions of solid waste than at the level of the FS; the FS having already undergone decomposition in the pits, thus reducing their digestible-matter content. The high biogas production of FFHW (leftover corn dough, prepared rice, prepared pasta) compared to that of sewage sludge is attributed to the presence of residual food starch in these effluents (Pouech et al., 2005). Waste materials have a low level of biodegradability compared to other substrates. It may be a problem of accessibility of the organic matter, the composition of the organic matter, or else the limitation of the fermentation by ammonia in more or less significant concentration in these sludge (their ratio C:N being weak) (Pouech et al., 2005). The best performance of FFAW effluents is attributed to their high organic matter content in the waste because the amount of gas produced by a substrate is positively correlated with the amount of organic matter present. Indeed, most of the fruits used in this scenario are rich either in lipids (avocado), starch (bread), or carbohydrates (banana). Studies have also shown that the main constituents of mango are water, carbohydrates, proteins, fats (carotene), minerals, pigments, tannins, and vitamins (Fréhaut, 2001). However, lipid products have a high potential for biogas production, as well as the methane content (Pouech et al., 2005). In addition, it has been shown that fruit residues often have high methanogenic potential and are easily assimilated by bacteria. This assimilation is due to the high contents of pectin at the level of their cell walls and of easily biodegradable organic acids, giving them a relatively rapid kinetics of anaerobic degradation (Chanakya et al., 2009). The organic matter provided by FFAW is much more easily degradable than that from FFHW, which results in their much higher gasification potential.

The results presented indicate that the association of these fermentable fractions of various origins will improve individual performance, which allowed us to formulate another hypothesis on the investigation compared to the anaerobic co-digestion of these sludge with the fermentable fractions of municipal solid waste.

9.3.1 Interest of Co-digestion

The results of co-digestion (Figure 9.2) show that the production yield of biogas improved significantly compared to the digestion of these substrates alone. The biogas production potential increased by almost 35% compared to the phase without the addition of co-substrates. The contribution of co-substrates of lipidic nature allowed an increase in the quantity of biogas produced daily as well as the content of methane (Pouech and Marcato, 2005). The addition of fermentable fractions of municipal solid waste also lead to an increase in the rate of organic matter compared to the phase of FS alone; as a result, there was a greater production of biogas. The present study shows that a ratio equal to 0.3 gives better biogas production. This result is comparable to that obtained by Nsavyimana, where a sludge volatile solid matter to waste volatile organic matter ratio of 0.3 was considered optimal for good digestion of waste of a different nature (Nsavyimana et al., 2012).

The biogas produced from the co-digestion of sewage sludge with fermentable market waste contains on average 65.6% of CH_4 against 34.4% of CO_2 (Nsavyimana et al., 2012). On this basis, we estimated volumes of CH_4 and CO_2 in the biogas produced. Indeed, from 1 L of co-substrate in 42 days of co-digestion, an average production is estimated to be approximately 4 L/g SM; that is to say, a volume of approximately 3 L of CH_4 and approximately 1 L of CO_2.

From these results, an extrapolation of the quantities of biogas, CH_4, and CO_2 by 2035 was established (Table 9.2). This extrapolation is made from the equation of the curve obtained during the co-digestion tests. These tests assisted in establishing the production of biogas by varying the quantity of co-substrates.

Table 9.2 shows that by 2025, the fermentable fractions of solid waste and FS would produce approximately 36,337 m^3 of CH_4. In 2030, they would produce around 44,392 m^3 and in 2035, the value would be around 53,747 m^3. These quantities are quite large and could be used in households and solve the problem of gas stock in the city of Sokodé and limit deforestation.

TABLE 9.2
Quantities of Biogas, Methane, and Carbon Dioxide Available by 2035

Years	Quantity of FFSW (t.MS)	Quantity of FS (t.MS)	Total Quantity (t.MS)	Volume of Biogas (m³)	Volume of CH_4 (m³)	Volume of CO_2 (m³)
2018	1,654.91	198.32	1,853.22	36508	23,949.25	12,558.75
2019	2,247.8	201.76	2,449.54	48255	31,655.28	16,599.72
2020	2,265.15	205.20	2,470.34	48665	31,924.24	16,740.76
2025	2,588.42	223.34	2,811.76	55391	36,336.50	19,054.50
2030	3,182.89	252.20	3,435.08	67671	44,392.18	23,278.82
2035	3,878.37	280.58	4,158.96	81931	53,746.74	28,184.26

Abbreviations: FFSW, fermentable fractions of solid waste; FS, faecal sludge.

9.4 CONCLUSION

The general objective of this study was to produce and recover methane from fermentable waste produced in the city of Sokodé. It is particularly a question of evaluating the interest of co-digestion of drain sludge with fermentable waste through their maximum production of biogas and methane. The results of the pilot tests on a laboratory scale demonstrate the significant variability that may exist between fermentable residues of different origins. From the same quantity of substrate, the biogas production potential of the FFAW (from markets and hotels) is high compared to that of the FFHW, which is also higher than that of emptying from PST.

FS alone has a low production of biogas compared to the fermentable fractions of municipal solid waste. The addition to fermentable sludge under controlled conditions of the fermentable fractions of municipal solid waste with a higher dry matter content has made it possible to increase the production performance of biogas compared to FS alone. The biogas production potential increased by almost 35% compared to the phase without the addition of co-substrates (on average 4 L/g SM compared to 1 L/g SM) for the sludge. From 1 L of co-substrate in 42 days of co-digestion, an average production is estimated to be approximately 4 L/g SM, a volume of approximately 3 L of CH_4 and approximately 1 L of CO_2. The results thus obtained clearly show the interest of the co-digestion of this waste and constitute basic data for the large-scale implementation of this co-digestion on the sewage sludge treatment station of the city of Sokodé. The gas produced will reduce the gas stock problem and supply households with butane gas. The use of this green technology will avoid the emission of harmful greenhouse gases and contribute positively to environmental objectives. The anaerobic digestion of Sokodé's fermentable substrates would produce clean renewable energy available on site without being dependent on imports. The quantities estimated by 2035 are enormous and therefore the anaerobic co-digestion of the fermentable fractions of the city's waste could contribute to reducing deforestation in the environment.

9.5 CONFLICT OF INTEREST

All authors have no conflict of interest concerning the work reported in this paper.

ACKNOWLEDGMENTS

I am grateful to Professor Gnon Baba, the Director of the Laboratory of Sanitation, Water Science and Environment and supervisor of this work, for his financial support.

REFERENCES

Akindele, Sartaj, 2017, "The toxicity effects of ammonia on anaerobic digestion of organic fraction of municipal solid waste," Waste Management.

Akpaki, 2015, Physico-chemical characterization of the sludge from Attidjin (district of Golf-Togo), Lomé, Thesis, University of Lomé, Water and Environment Chemistry, p. 167.

Almansour, 2011, Energy and environmental assessments of biogas sectors: Approach by standard sector., University of Bordeaux, 1, p. 147.

Bruni E., Jensen AP., Pedersenc ES., Angelidaki I., 2010, "Anaerobic digestion of maize focusing on variety, harvest time and pretreatment," Applied Energy, 7(187), pp. 2212–2217. https://doi.org/10.1016/j.apenergy.2010.01.004

Centre of Expertise in Environmental Analysis of Quebec, MA. 405 – C 1.1, 2014, Determination of total organic carbon in solids: Assay by titration, Quebec, p. 9.

Chanakya HN., Sharma I., Ramachandra TV., 2009, "Micro-scale anaerobic digestion of point source components of organic fraction of municipal solid waste," Waste Management, 4(129), pp. 1306–1312. https://doi.org/10.1016/j.wasman.2008.09.014.

Chandra, Vijay VK., Subbarao PMV., Khura TK, 2012, "Production of methane from anaerobic digestion of jatropha and pongamia oil cakes," Applied Energy, 193, pp. 148–159. DOI: 10.1016/j.apenergy.2010.10.049.

Chua, Yip, 2008, "A case study on the anaerobic treatment of food waste and gas formation."

Chulalaksananukul, Sinbuathong N., and Chulalaksananukul W., 2012, "Bioconversion of Pineapple solid Waste under Anaerobic Condition through Biogas production," KKU Research Journal, 5(117), pp. 734–742.

Deressa, Libsu S., Chavan, RB., Manaye D., Dabassa, A., 2015, "Production of biogas from fruit and vegetable wastes mixed with different wastes," Environment and Ecology Research, 3(13), pp. 65–71.

Eskicioglu C., Ghorbani M., 2011, "Effect on inoculum/substrate ratio on mesophilic anaerobic digestion of bioethanol plant whole stillage in batch mode," Process Biochemistry, 146, pp. 1682–1687. https://doi.org/10.1016/j.procbio.2011.04.013

Fréhaut, 2001, "Study of the biochemical composition of mango (*Mangifera Indica L.*) according to its stage of maturity," Biological Engineering, p. 65.

Gunaseelan VN., 2004, "Biochemical methane of fruits and vegetable solid waste feedstocks," Biomass and Bioenergy, 4(126), pp. 389–399. https://doi.org/10.1016/j.biombioe.2003.08.006

Hu ZH., Yu HQ., Zhu RF., 2005, "Influence of particle size and pH on anaerobic degradation of cellulose by ruminal microbes," International Biodeterioration & Biodegradation, 3(155), pp. 233–238. https://doi.org/10.1016/j.ibiod.2005.02.002

Huyard, 2012, "Anaerobic co-digestion of wastewater sludge and food waste pulp from a unpacking process Enhancement of biogas production and co-metabolism," IWA.

Izumi K., Okishio YK., Nagao N., Niwa C., Yamamoto S., Toda T., 2010, "Effects of particle size on anaerobic digestion of food waste," International Biodeterioration & Biodegradation, 7(164), pp. 601–608. https://doi.org/10.1016/j.ibiod.2010.06.013

Kelly, 2002, Solid Waste Biodegradation Enhancements and the Evaluation of Analytical Methods Used to Predict Waste Stability, Virginia, Environmental Science and Engineering, p. 72.

Klingel, 2002, Faecal Sludge Management, Eawag, Sandec (Water and Sanitation in Developing Countries), First ed. p. 63.

Mata-Alvareza J., Dosta J., Romero-Güiza MS., Fonoll X., Peces M., Astalsa S., 2014, "A critical review on anaerobic co-digestion achievements between 2010 and 2013," Renewable and Sustainable Energy Review, 136, pp. 412–427. https://doi.org/10.1016/j.rser.2014.04.039

Nsavyimana, Bigumandondera P., Ndikumana T., Vasel J.L., 2012, "Anaerobic co-digestion of septic tank emptying sludge and fermentable market waste with a view to their recovery: Case of Burundi," Poster Gaston-UB Conference (uliege.be), p. 1.

Pouech, Coudur R.E., Marcato CE., 2005, «Interest of co-digestion for the development of slurry and the treatment of fermentable waste at the scale of a territory.,» Pig Research Days., 137, pp. 39–44.

Walker, 2010, "Residual biogas potential test for digestates, OFW004-005," Waste and Resources Action Programme, p. 53.

World Bank, 2010, "Environmental Analysis of Togo with Emphasis on the Agriculture, Energy and Mining Sectors," Lomé.

10 Effect of Inoculum Concentration and Particle Size of the Substrate on Anaerobic Digestion of Yard Waste

Ravi Kumar D., H. N. Chanakya, Swati Bhatia, and Dasappa S.

10.1 INTRODUCTION

Tropical countries such as India are endowed with a large extent of year-round leaf litter or yard waste, which can be used for energy production through anaerobic digestion (Chanakya et al., 1997). Large campuses such as the Indian Institute of Science have a huge problem with yard waste disposal. Yard waste of such campuses is usually composted with significant effort, whereas alternative technologies for sustainable usage are few (Chanakya et al., 2009). The decomposable organic fraction may be used at source for anaerobic digestion for methane generation (Chanakya et al., 2007). The yard waste is a mix of various species of leaves, with a varying extent of fractions or subcomponents such as pectin, hemi-cellulose, cellulose, and lignin. Within each of the species, this composition may also vary depending upon many parameters, like season, recycling practices, and culture (Gunaseelan, 2016). In the United States, yard waste is the third most abundant component of the total municipal solid waste at 13.5%. Anaerobic digestion of such leaves produced gas at around 123 mL of methane/gVS (Chynoweth et al., 1993).

In this study, unsegregated yard waste (leaves) was chosen to assess the biogas generation potential and pattern. It has often been reported that the size reduction of biomass feedstocks increases the surface area and in turn increases the gas production rates and extent. This hypothesis was examined to determine whether and to what extent the particle sizes played a role in methane production, in addition to the effects of inoculum concentration.

10.2 MATERIALS AND METHODS

The sample of unsegregated leaves was collected from a designated place where the yard waste would be stacked upon for further processing. The collected sample was dried at 70°C until it reached a constant weight indicating a moisture content of less than 5%.

Total solids and volatile solids (VS) were measured through American Public Health Association (APHA, 1995) standard methods.

The extent of subcomponents was determined in various feedstocks by Chesson's method of sequential digestion (Chesson, 1978).

10.2.1 Biochemical Methane Potential (BMP) Assay

The total biogas production and methane content were estimated through the downward displacement of water in a burette and gas chromatography (Angelidaki et al., 2009; Owens and Chynoweth,

DOI: 10.1201/9781003204435-12

1993; Raposo et al., 2011). The substrate and the inocula were mixed at various proportions and placed in 135 mL serum vials. The gas evolved was measured at intervals such that appreciable volumes were captured and errors minimised. Several factors, such as temperature, the origin of inocula, substrate components and their interlinkages, and other physicochemical properties, affect the assay (Angelidaki et al., 2009; Raposo et al., 2011).

In this particular study, the Biochemical Methane Potential (BMP)) was carried out in 135 mL serum bottles with a working volume of 70 ml. The substrate to inoculum ratios of 2, 1, 0.5, and 0.25 were used to understand the effect of inoculum concentration, and various particle sizes were also chosen as part of the study. The serum bottles were sealed with butyl rubber stoppers and crimped with aluminium crimps. The bottles were first flushed with a mixture of methane and carbon dioxide to remove all traces of oxygen and subsequently flushed with pure nitrogen to remove all the traces of methane. These bottles were incubated upside down under laboratory conditions (25 ± 2°C) and the gas production was measured at intervals through the downward displacement of acidified water. The composition of the gas was analysed using a portable gas chromatograph (Mayura analytics) fitted with a Hayesep A column. A thermal conductivity detector was used with hydrogen as a carrier gas. The carrier gas flow rate was maintained at 30 mL/min.

10.3 RESULTS AND DISCUSSIONS

10.3.1 Feedstock Composition

The feedstock composition was obtained through sequential extraction by Chesson's method (Chesson, 1978). From Table 10.1 we see that the yard waste has a high extent of water extractives (predominantly pectin 1 and less of pectin 2). The lignin content is equally high at 24%, which is comparable with Cherosky (2012).

10.3.2 Gas Production

The gas production was studied for the different particle sizes at an abundant level of inoculum, making sure that there was no problem with acidification, or any other process-related problems. Figure 10.1 clearly shows a ten-day lag phase; this may be due to the abundance of inocula and lack of food. The average gas production was 145 mL/gVS for all the particle sizes, which is substantially higher than obtained by Cherosky (2012).It goes on to show that the particle size does not play any significant role in the final process of gas production. This also proves that any pulverisation and segregation of leaves would be a wasteful process leading to no significant gas production.

The methane content at the same substrate to inoculum ratio and different particle size shows that the particle size plays no significant role in methane production. There is an average of 40% methane content at a steady state. From Figure 10.2, it can be noted that there is no difference in methane content at different particle sizes, showing that the gas production was consistent and there was no change in the process with a change in particle size. Decreasing particle size in the

TABLE 10.1
Composition of the Yard Waste

Component	% of VS
Pectin 1	26.5
Pectin 2	4.3
Hemi-cellulose	17.6
Cellulose	19.2
Lignin	24.4
Ash	7.8

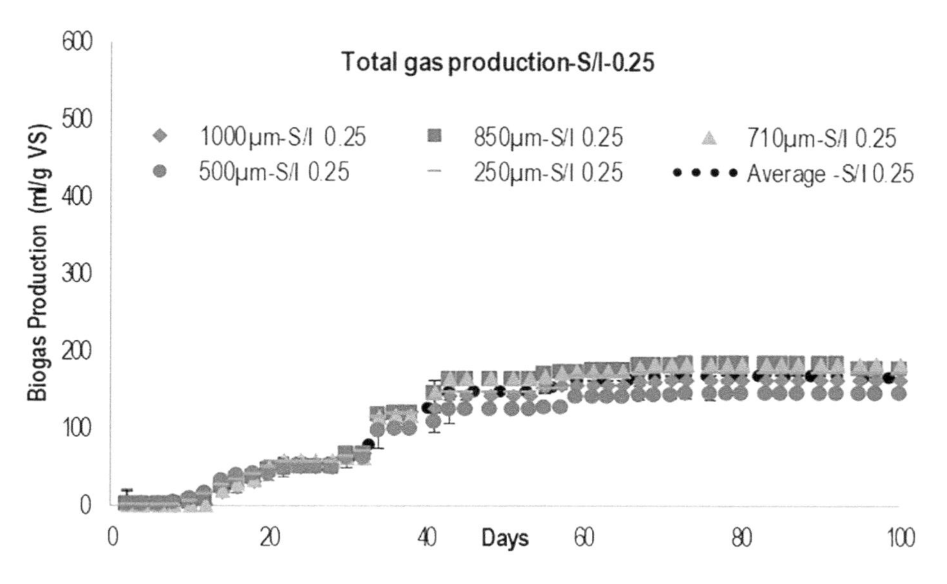

FIGURE 10.1 Cumulative biogas production at different particle size at a substrate to inoculum ratio of 0.25.

range studied did not increase the acidogenic rates, as expected, and neither was this translated into a high gas production level. All of these suggest that even at the lowest particle sizes studied, the acidogenic phase was rate-limiting. And, as hypothesised earlier, these represent an access-deficient mode of biomass decomposition.

Yard waste is a nonsegregated mix of different species of leaves. From Table 10.1, it was clear that there could be gas production from these mixed leaves, as they had water extractives, which could produce some amount of gas (Cherosky 2012.; Li et al., 2014). Lignin would be a hindrance that would limit the gas production, owing to its recalcitrant nature. From Figure 10.3, it can be observed that the yard waste with a particle size of 1 mm produced on an average of 100 mL/gVS of biogas (Li et al., 2014) (<30 mL/gVS in 30 days). There was little effect of the substrate to

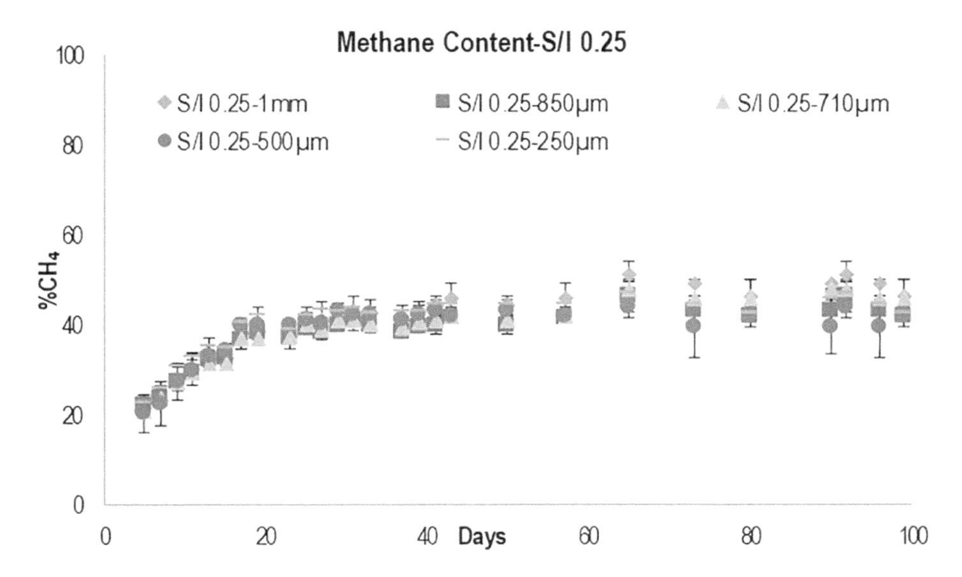

FIGURE 10.2 Methane concentration at different stages of anaerobic digestion of yard waste at different particle sizes.

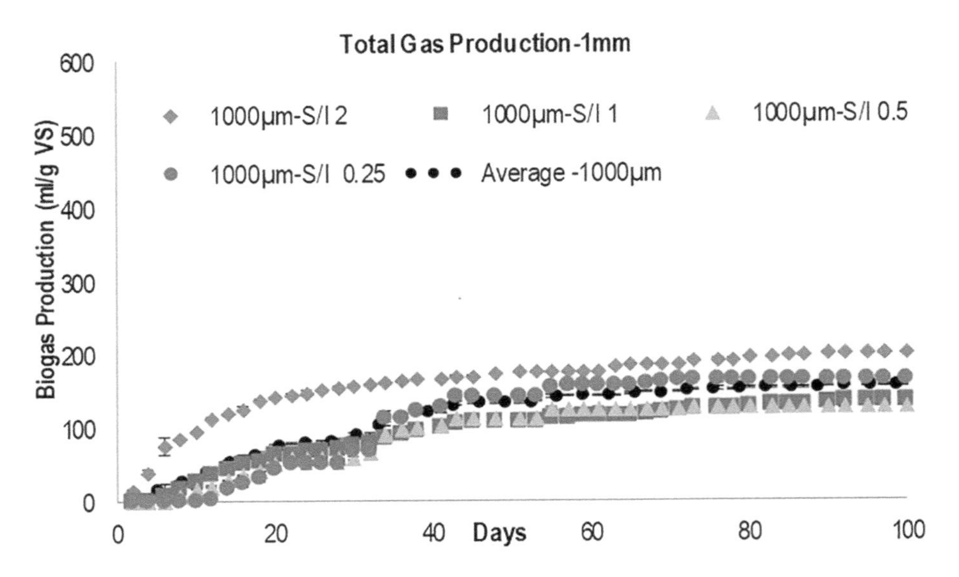

FIGURE 10.3 Cumulative biogas production at a different substrate to inoculum ratio.

inoculum ratio on the gas production from the yard waste. Any noticeable effect of the substrate to inoculum ratio would indicate the ease or difficulty of degradation of the yard waste. A substrate to inoculum ratio of 2 has a marginally higher gas production. There is also an initial rapid production over 15 days, and it reaches saturation by day 20. At a substrate to inoculum ration of 1–0.25, the increase in gas production is slower.

The methane content in the biogas from the yard waste at the different substrate to inoculum ratio and a particle size of 1 mm seems to be in a narrow range of 21–26% during the initial days and reaches a steady state in about 15 days, with a methane concentration of an average of 40% and a maximum of 52%. The methane content shows that the inoculum content has only a minor effect on biogas production, as seen in Figure 10.4. The low methane content at the early stages of

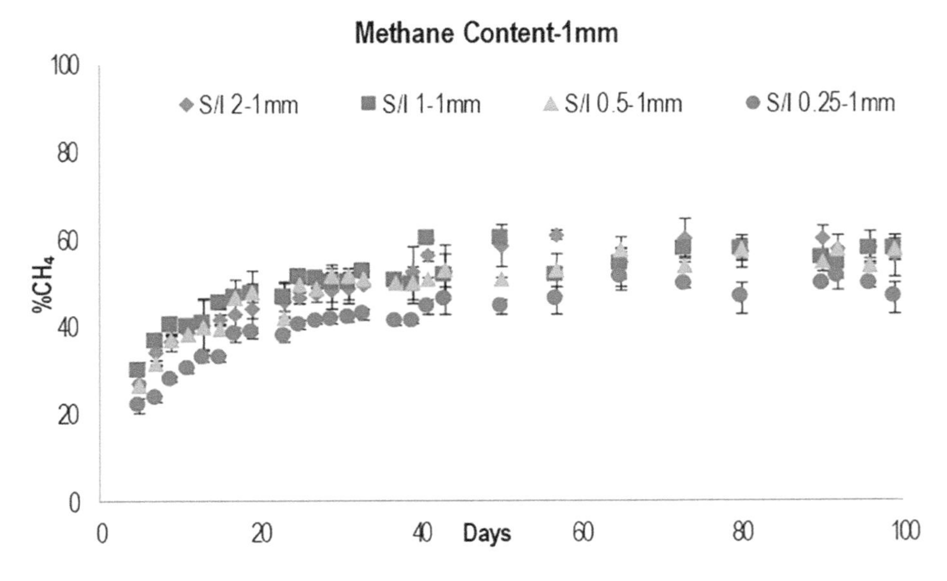

FIGURE 10.4 Methane concentration change over 100 days at a different substrate to inoculum ratio.

decomposition indicates the predominance of acidogenesis, suggesting the rapid release of extractables as the key source of biogas at this stage of fermentation.

Owing to the small size of the fermentation samples, it was not possible to determine the composition of the residual biomass. The disappearance of the extractables and the presence of more recalcitrant ligno-cellulosic complex would confirm the existence of access-deficient decomposition, as hypothesised (Chanakya et al, 2009). This is, therefore, much-needed proof and an area for further research.

10.4 CONCLUSION

Yard waste, depending upon the mixture of leaves, has an adequate amount of water extractives and is simultaneously rich in lignin. While there is rapid early decomposition due to the removal and fermentation of the extractables, the presence of lignin makes the waste degrade at a slower rate and degree in the subsequent phase. The biogas production levels obtained in the study are comparable or higher than some studies, which bears out the relevance. In this study, there was no effect of the particle size on methane production and methane concentrations, and the gas production levels remained the same. There was also no effect of the inocula concentration on the production of biogas.

REFERENCES

American Public Health Association (APHA), American Water Works Association, Water Environment Federation, Federation WE. Standard Methods for the Examination of Water and Wastewater. Stand Methods 1999:541. doi:10.2105/AJPH.51.6.940-a.

Angelidaki, I., Alves, M., Bolzonella, D., Borzacconi, L. Campos, J.L., Guwy, A.J., Kalyuzhnyi, S., Jenicek, P., and van Lier, J.B. 2009. Defining the biomethane potential (BMP) of solid organic wastes and energy crops: a proposed protocol for batch assays. Water Science Technology, *59*(5), p. 927.

Chanakya, H.N., Ramachandra, T.V. and Vijayachamundeeswari, M., 2007. Resource recovery potential from secondary components of segregated municipal solid wastes. Environmental Monitoring and Assessment, *135*(1-3), pp. 119–127.

Chanakya, H.N., Sharma, I. and Ramachandra, T.V., 2009. Micro-scale anaerobic digestion of point source components of organic fraction of municipal solid waste. Waste Management, *29*(4), pp. 1306–1312.

Chanakya, H.N., Venkatsubramaniyam, R. and Modak, J., 1997. Fermentation and methanogenic characteristics of leafy biomass feedstocks in a solid phase biogas fermentor. Bioresource Technology, *62*(3), pp. 71–78.

Cherosky, P.B., 2012. *Anaerobic digestion of yard waste and biogas purification by removal of hydrogen sulfide* (Doctoral dissertation, The Ohio State University).

Chesson, A., 1978. The maceration of linen flax under anaerobic conditions. *Journal of Applied Bacteriology*, *45*(2), pp. 219–230.

Chynoweth, D.P., Turick, C.E., Owens, J.M., Jerger, D.E. and Peck, M.W., 1993. Biochemical methane potential of biomass and waste feedstocks. *Biomass and Bioenergy*, *5*(1), pp. 95–111.

Gunaseelan, V.N., 2016. Biochemical methane potential, biodegradability, alkali treatment and influence of chemical composition on methane yield of yard wastes. *Waste Management & Research*, *34*(3), pp. 195–204.

Li, W., Zhang, G., Zhang, Z. and Xu, G., 2014. Anaerobic digestion of yard waste with hydrothermal pretreatment. *Applied Biochemistry and Biotechnology*, *172*(5), pp. 2670–2681.

Owens, J.M. and Chynoweth, D.P., 1993. Biochemical methane potential of municipal solid waste (MSW) components. *Water Science and Technology*, *27*(2), pp. 1–14.

Raposo, F., Fernández-Cegrí, V., De la Rubia, M.A., Borja, R., Béline, F., Cavinato, C., Demirer, G.Ö.K.S.E.L., Fernández, B., Fernández-Polanco, M., Frigon, J.C. and Ganesh, R., 2011. Biochemical methane potential (BMP) of solid organic substrates: evaluation of anaerobic biodegradability using data from an international interlaboratory study. *Journal of Chemical Technology & Biotechnology*, *86*(8), pp. 1088–1098.

11 Mathematical Modelling of Aerobic Digestion in the Activated Sludge Process

Alomoy Banerjee and S. Raghuraman

11.1 INTRODUCTION

The activated sludge process shown in Figure 11.1 is the most frequently used suspended growth process for municipal wastewater treatment because of its technical efficiency and economic viability. The principal water treatment steps of an activated sludge plant include:

1. Wastewater aeration in presence of microbial suspension
2. Solid and liquid separation in clarifier
3. Discharge of treated effluent
4. Wastage of surplus biomass
5. Recycling of remaining biomass through aerator tank

Sewage water primarily consists of carbonaceous organic compounds and nitrogenous waste products in the form of ammonia [6]. Both of these components contribute to the biological oxygen demand of sewage water. Hence, for aerobic digestion of the organic and nitrogenous substrate, aeration becomes an indispensable part of the activated sludge process. The efficiency of the aeration process, termed as oxygen transfer efficiency (OTE), fluctuates, owing to the variation in quality of influent wastewater.

It is imperative for the aeration tank to be supplied with adequate oxygen in order to maintain the population of the aerobic bacteria necessary to digest the substrate in sewage water. However, the aeration of sewage water comes at the cost of power consumption from the aerators. Hence, keeping the sludge water in the aeration tank for longer periods will actually reduce the efficiency of the continuous stirred tank reactor, as the aerators will consume more power to treat the same volume of sewage water. Therefore, it is important to predict the minimum number of days for which the sewage water needs to be kept in the aerator tank for complete aerobic digestion of the substrate.

Aerobic digestion of the substrate primarily occurs in two phases. The carbonaceous organic compounds are first consumed by heterotrophic bacteria. The reaction rate is rapid and reaction conditions like pH, temperature, and dissolved oxygen concentration are lenient. The next phase involves the consumption of nitrogenous waste products by Nitrosomonas and Nitrobacter in the presence of oxygen. Ammonia is initially converted into nitrite by Nitrosomonas, which in turn gets converted into nitrate by Nitrobacter to complete the aerobic digestion of nitrogenous waste product. The conversion of ammonia to nitrate in the aerator tank takes the longest time and demands strict reaction conditions that include strictly neutral pH, adequate dissolved oxygen, accurate sludge water temperature, and high sludge retention time.

11.2 LITERATURE REVIEW

Henze Mogens et al. [1] illustrates a summary of ASM1 [2], ASM2, ASM2d [7], and ASM3, thus providing a comparative analysis between them. The book provides an exhaustive list of wastewater characteristic and kinetic and stoichiometric constants that give accurate results while mathematically

DOI: 10.1201/9781003204435-13

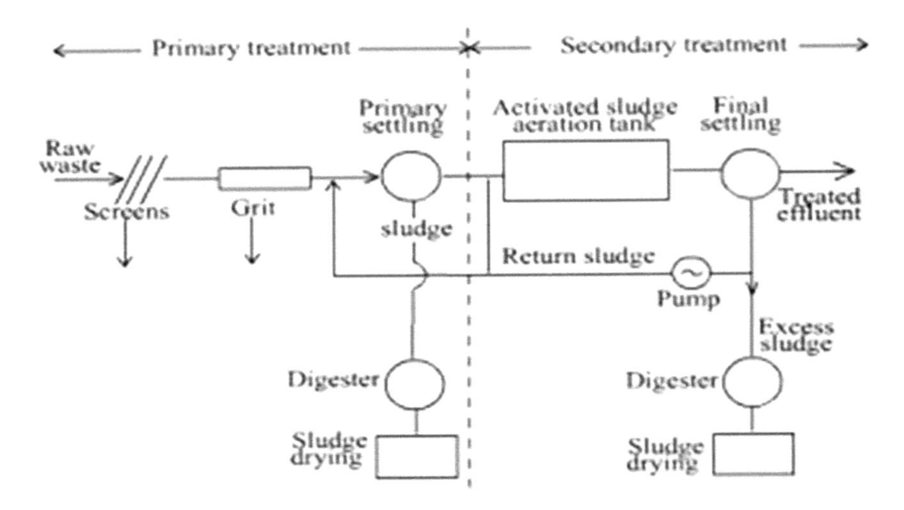

FIGURE 11.1 Activated sludge process in municipal wastewater treatment.

modelling the activated sludge system. Gernaey et al. [3] carried out an extensive review of white box modelling techniques used in wastewater treatment plants. The work combined AI methodologies with modular agent-based systems to deduce a calibrated mathematical model for sewage water treatment plants. Miata Jun et al. [8] simulated a process model for oxygen ditch facilities using IWA-approved mathematical model ASM2d. Shao-Yuan Leu [5] developed a dynamic model to predict offgas mole fraction and backed up the simulation data with experimental evidence to establish the authenticity of the model. Researchers [7, 9, 10] developed a simplified version of ASM1, based on benchmark simulation model. The model was verified by comparing the COD values of simulation results and practical testing.

11.3 CHEMICAL REACTIONS IN AERATION TANK

There are four chemical reactions that occur in the aerator tank to oxidise the organic and nitrogenous substrate of sewage water. The reactions are represented in the form of chemical equations below:

i. The organic substrate reduction in sludge water due to the growth of heterotrophic bacteria is represented by the following chemical equation:

$$C_aH_bO_c + YNH_4^+ + \left(a + \frac{b}{4} - \frac{c}{4}\right)O_2$$

$$\rightarrow YC_5H_7NO_2 + (a-5Y)CO_2 + \left(\frac{b}{2} - 2Y\right)H_2O + YH^+$$

ii. Oxidation of ammonia to nitrite by Nitrosomonas is described by the following equation:

$$NH_4^+ + 5Y_{NS}CO_2 + \left(\frac{3}{2} - \frac{13}{2}Y_{NS}\right)O_2$$

$$\rightarrow Y_{NS}C_5H_7NO_2 + (1-Y_{NS})NO_2^- + (1-3Y_{NS})H_2O + (2-Y_{NS})H^+$$

iii. Oxidation of nitrite to nitrate by Nitrobacter is depicted by the following equation:

$$NO_2^- + 5Y_{NB}CO_2 + 3Y_{NB}H_2O\left(\frac{1}{2} - 7Y_{NB}\right)O_2 + Y_{NB}H^+$$

$$\rightarrow Y_{NB}C_5H_7NO_2 + (1 - Y_{NB})NO_3^-$$

iv. Decay of bacterial biomass to production of carbon dioxide and ammonia at long sludge retention time is illustrated by the following equation:

$$C_5H_7NO_2 + 5O_2 + H^+ \rightarrow 5CO_2 + 2H_2O + NH_4^+$$

11.4 MONOD FUNCTION

French scientist Jacques Monod derived a function to represent the microbial growth kinetics in continuous stirred tank reactors. The function takes into consideration microbial growth & decay rates, as well as substrate concentration. The function is represented by

$$\mu_1 = \frac{\mu_s S}{K_s + S} - K_d X$$

In order to incorporate the influence of dissolved oxygen in microbial growth kinetics, double Monod-type growth rate kinetic function was developed by Produska. The double Monod function is represented by

$$\mu_2 = \frac{\mu_s S}{K_s + S}\frac{DO}{DO + K_{SDO}}$$

μ_s = Maximum biomass growth rate
S = Substrate concentration
K_s = Half velocity coefficient
K_d = Decay rate
DO = Dissolved oxygen concentration
K_{SDO} = Half Saturation Coefficient

11.5 SLUDGE RECYCLE RATE

After aeration in the tank, as illustrated in Figure 11.1, the effluent is separated in the clarifier and the waste sludge is eliminated. The rest of the sludge water is recycled back to the aeration tank for aeration along with fresh sewage water. The recycle rate of the sludge may be mathematically denoted by

$$P = -\frac{Q_W}{V} * \frac{Q + Q_R}{Q_R}$$

Q = Sludge flow rate from sewage tank Q_w = Waste sludge flow rate
Q_R = Recycle sludge flow rate

11.6 GOVERNING DIFFERENTIAL EQUATIONS

i. Substrate concentration (S):

$$\frac{ds}{dt} = \frac{Q}{V}(S_0 - S) - \frac{\mu_2}{Y_{mass}} X$$

ii. Biomass concentration (X):

$$\frac{dX}{dt} = (P + \mu_2 - K_D)X$$

iii. Nitrosomonas concentration (X_{NS}):

$$\frac{dX_{NS}}{dt} = (P + \mu_2 - K_{DNS})X_{NS}$$

iv. Nitrobacter concentration (X_{NB}):

$$\frac{dX_{NB}}{dt} = (P + \mu_2 - K_{DNB})X_{NB}$$

v. Dissolved oxygen concentration (DO):

$$\frac{dDO}{dt} = \frac{Q}{V}(DO_0 - DO) + \gamma_{DOTR} - \gamma_{DO1} - \gamma_{DO2} - \gamma_{DO3} - \gamma_{DO4} - \gamma_{DO5} - \gamma_{DO6}$$

vi. Carbon dioxide stripping (DCD):

$$\frac{dDCD}{dt} = \frac{Q}{V}(DCD_0 - DCD) + \gamma_{CDSTRP} - \gamma_{DCD1} + \gamma_{DCD2} - \gamma_{DCD3} + \gamma_{DCD4} - \gamma_{DCD5} + \gamma_{DCD6}$$

vii. Ammonia stripping (N_{NH3}):

$$\frac{dN_{NH3}}{dt} = \frac{Q}{V}(N_{NH30} - N_{NH3}) - \gamma_{NH_3 1} + \gamma_{NH_3 2} - \gamma_{NH_3 3} + \gamma_{NH_3 4} + \gamma_{NH_3 5}$$

viii. Nitrite oxidation (N_{NO2}):

$$\frac{dN_{NO2}}{dt} = \frac{Q}{V}(N_{NO20} - N_{NO2}) + \gamma_{NO_2 1} - \gamma_{NO_2 2}$$

ix. Nitrate formation (N_{NO3}):

$$\frac{dN_{NO3}}{dt} = \frac{Q}{V}(N_{NO30} - N_{NO3}) + \gamma_{NO_3 1}$$

x. Alkalinity (Z):

$$\frac{dZ}{dt} = \frac{Q}{V}(Z_0 - Z) + \frac{\gamma_{Z1} - \gamma_{Z2} + \gamma_{Z3}}{M_W - N}\gamma_{CDSTRP}$$

11.7 SOLUTION METHODOLOGY OF GOVERNING DIFFERENTIAL EQUATIONS

11.7.1 Assumptions

1. The fundamental theorem of activated sludge model is based on mass conservation equation.

2. The reactor is assumed to be a continuous flow stirred tank reactor.
3. The clarifier completely separates liquids and solids.
4. All the chemical reactions occur only in the aeration tank.
5. The clarifier is assumed to be a zero-volume container.

11.7.2 Solver

The governing differential equations are solved in MATLAB using ode45. The numerical technique used to solve the equations is an explicit Runge–Kutta Dormand–Prince method, with an adaptive-step size-integration algorithm. Linear stability analysis is done to ensure the stability and convergence of the solution obtained.

11.8 RESULTS AND DISCUSSION

Figure 11.2 demonstrates the decrease in substrate concentration over time. The rate of reaction depends on the substrate concentration, according to Monod's function. Hence, when the substrate concentration was high initially, the rate of consumption by heterotrophic bacteria was high as well. As the substrate concentration reduced, the reaction rate stabilised.

Figure 11.3 illustrates three plots that depict the progress of the nitrification process. The concentration of initial nitrogenous waste in sewage water, ammonia, reduces in two distinct phases. In phase 1, the rate of decrease in concentration is steep, as initially the heterotrophic bacteria consume ammonia to increase its population. In phase 2, the growth rate dips to a gradual curve, as the ammonia is gradually converted to nitrite by Nitrosomonas. Nitrite is an intermediate product in the nitrification process. The initial rise in Nitrosomonas concentration occurs because of the low

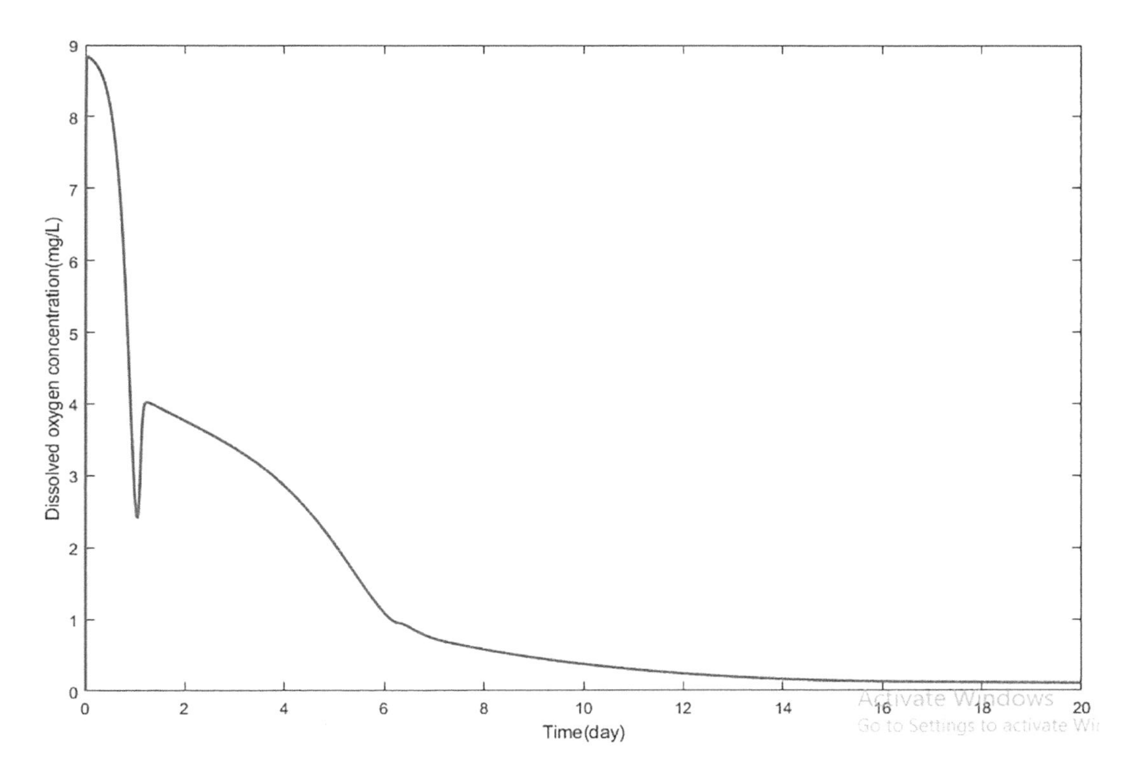

FIGURE 11.2 Simulation results of consumption of organic substrate.

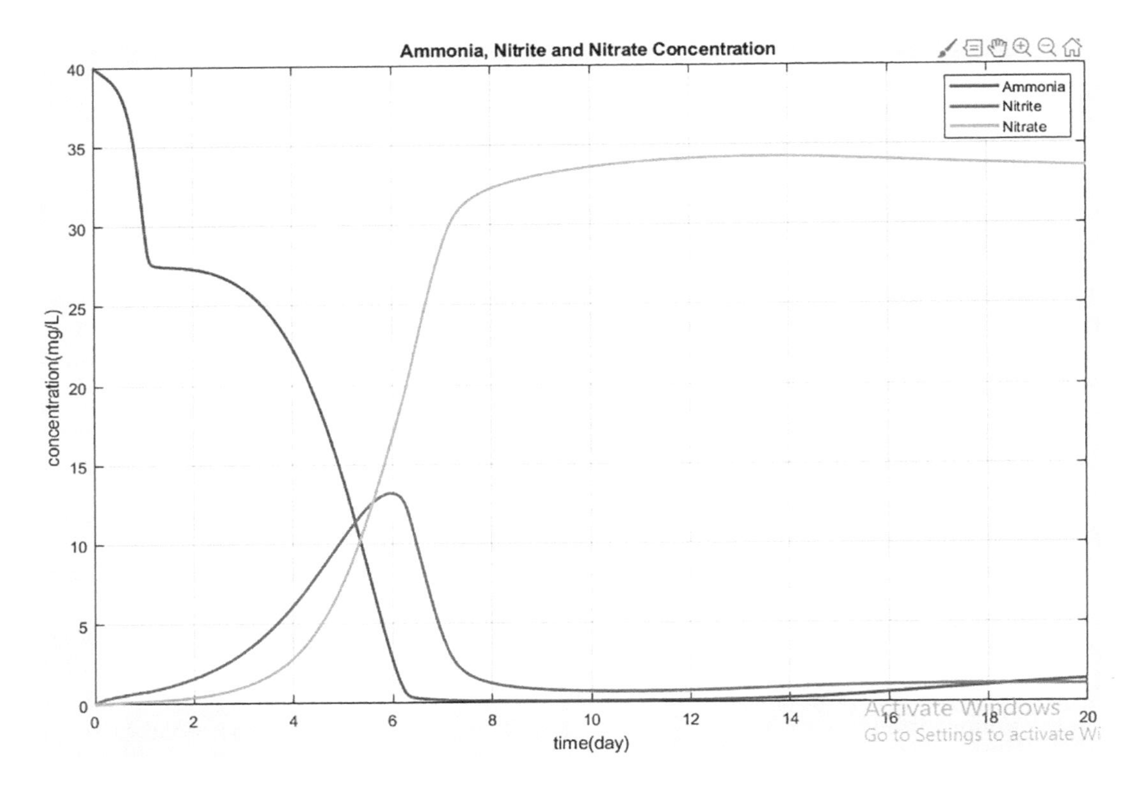

FIGURE 11.3 Simulation results of nitrification.

concentration of Nitrobacter to begin with. Then as the Nitrobacter population increases, the nitrite concentration drops from its peak, along with the simultaneous increase in concentration of nitrate.

Figure 11.4 illustrates the change in dissolved oxygen and carbon dioxide concentration over time. The steep rise in carbon dioxide and the drop in dissolved oxygen in the initial phase of the reaction is because of the rapid consumption of organic substrate by heterotrophic bacteria. This phase consumes oxygen and generates carbon dioxide. The increase in carbon dioxide concentration then drops down to a gradual slope, as a portion of it starts getting consumed by Nitrosomonas and Nitrobacter. Finally, the generation and usage of both carbon dioxide and oxygen reach a steady state, which gets maintained through the rest of the reaction.

In Figure 11.5, the pH varies between 7 and 8, the ideal pH that is required for aeration of sewage water. The drop in pH is due to the generation of H^+ ions during oxidation of organic substrate. The rise in pH after one to two days is primarily because of the consumption of H^+ ions in the formation of nitrate. Finally, the pH stabilises and maintains a strict pH closest to 7, which aids the nitrification process.

11.9 CONCLUSION

A mathematical model simulating the several components in an activated sludge wastewater treatment process was developed. The target components or properties include carbonaceous pollutants (substrate), nitrogenous pollutants (ammonia, nitrite, and nitrate), heterotrophic bacteria concentration, nitrifying bacteria (Nitrosomonas and Nitrobacter) concentrations, gas and liquid phase oxygen concentrations, gas and liquid phase carbon dioxide concentrations, alkalinity, and pH.

From the provided figures, it can be inferred that the minimum number of days for which the sewage water needs to be stored in the aerator tank for complete consumption of organic and inorganic

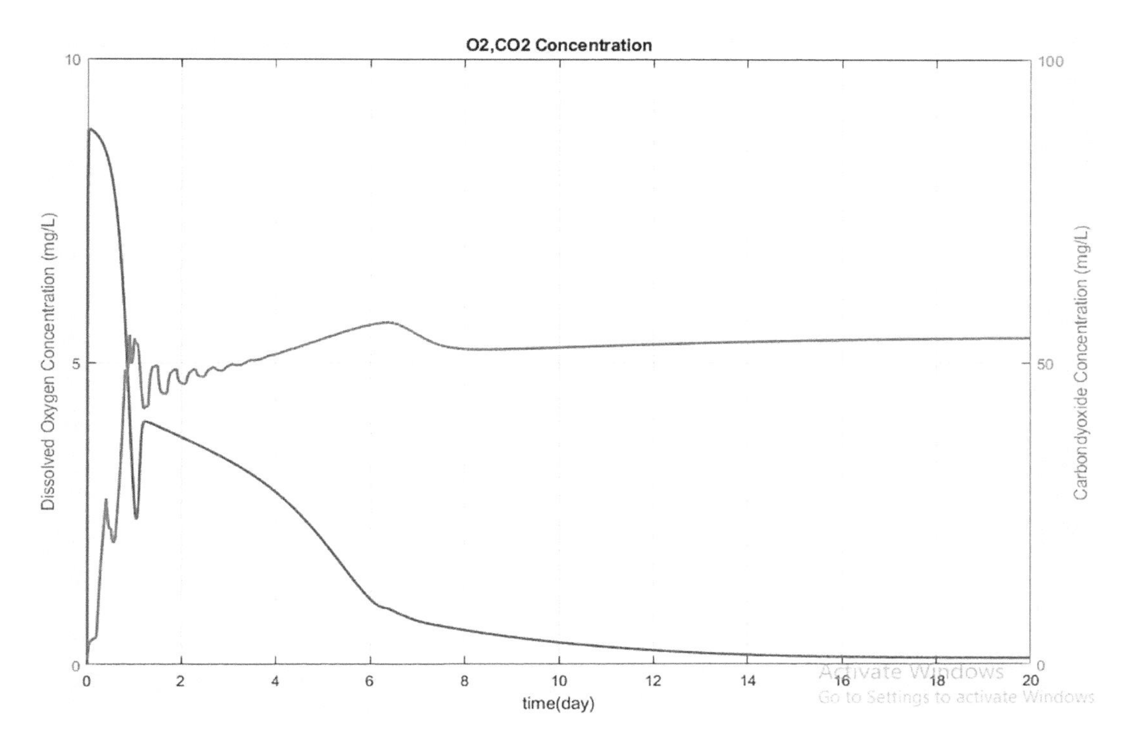

FIGURE 11.4 Simulation results of DO and DCD.

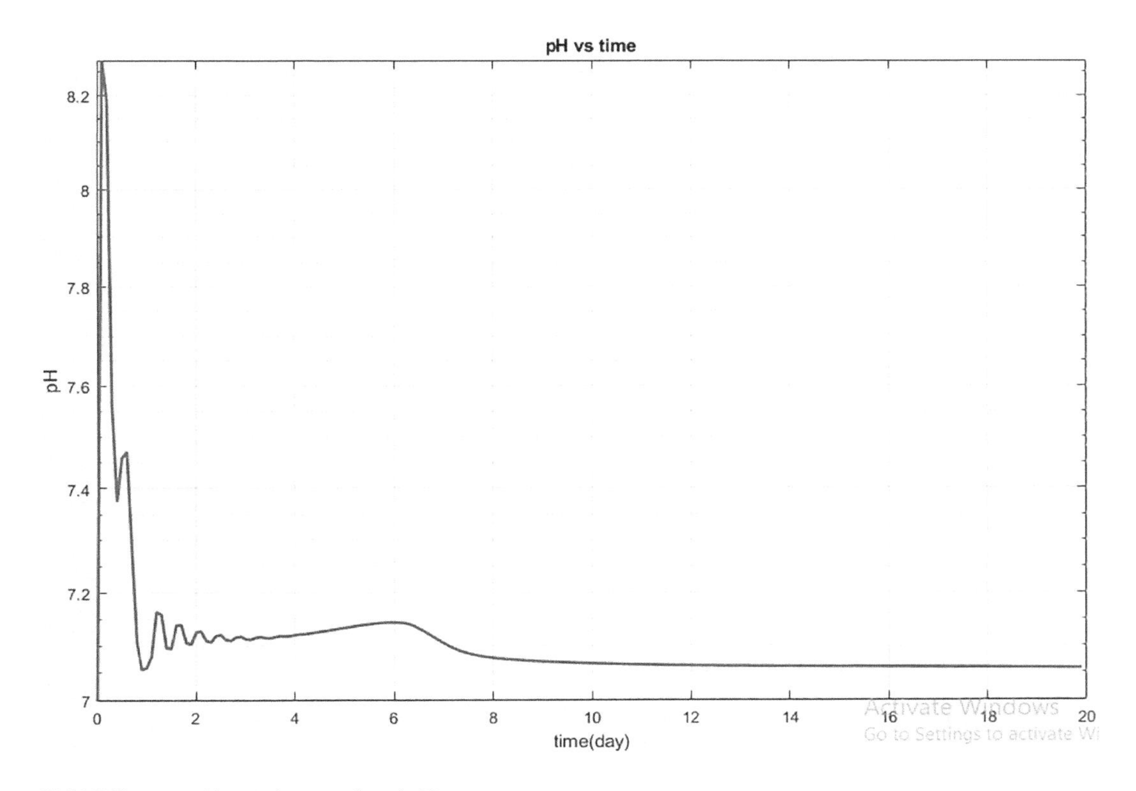

FIGURE 11.5 Simulation results of pH.

substrate is eight days. After eight days, all the chemical processes, as inferred from the graphs, stabilise, thus allowing the water to move to the clarifier.

The efficiency of oxygen transfer through the aerators is calculated to be 14.66%, which is within the stipulated range of OTE (10–15%) of standard aerators.

REFERENCES

1. Henze, M. (2011). Biological wastewater treatment: Principles, modelling and design. London: IWA Publishing.
2. Tillman, G. M. (2017). Activated sludge. Wastewater Treatment, 69–78. doi: 10.1201/9780203734216-10.
3. Krist V Gernaey et al. (2004). Activated Sludge Wastewater Treatment Plant Modelling and Simulation: State of the Art. Environmental Modelling and Software, 19(9), 763–783. doi: 10.1016/j.envsoft.2003.03.005.
4. Orhon, D. (2009). Industrial wastewater treatment by activated sludge. Water Intelligence Online, 8. doi: 10.2166/9781780401836.
5. Leu, S. (2004). A dynamic model for predicting off-gas mole fraction from the nitrifying activated sludge process.
6. Trzcinski, A. P. (2018). Conventional waste activated sludge process. Advanced Biological, Physical, and Chemical Treatment of Waste Activated Sludge, 1–20. doi: 10.1201/9780429437960-1.
7. Jia Zhou (2015). Research advance of the model of activated sludge process to wastewater treatment. Information Technology and Mechatronics Engineering Conference. DOI: 10.2991/itoec-15.2015.48.
8. Miata Jun, Sudo Taka, TSUBONE Toshiaki (2004). Wastewater Treatment Processing Simulation Technology Using "Activated Sludge Model" Content. *JFE Technical Report 27–34.*
9. Wei, Yao, Wu, Li, Ojao, Junfei (2010). ICICTA'10: Proceedings of the 2010 International Conference on Intelligent Computation Technology and Automation – Volume 03 22–27.
10. Monod, J. (1978). On symmetry and function in biological systems. Selected Papers in Molecular Biology by Jacques Monod, 701–713. doi: 10.1016/b978-0-12-460482-7.50061-0.

12 Anaerobic Co-Digestion of Landfill Leachate and Sewage Sludge
Role of Substrate Ratio

N. Anand, Srinjoy Roy, and P. Sankar Ganesh

12.1 INTRODUCTION

Human activities have always resulted in waste generation, and its management has become a significant problem due to industrialisation and the constant increase in the human population (Giusti, 2009). As science and technology evolved, waste management became organised and scientific, especially in affluent countries, to cope with the massive waste generated per capita (OECD, 2003). Unfortunately, the improper management of municipal solid waste (MSW) has led to serious ecological, environmental, and health problems in most third-world countries. Annually, India generates about 68.8 million tonnes of MSW with 500 g/day per capita waste generation (CPCB, 2016). Eighty-seven percent is being dumped openly or landfilled, out of which only 13% undergoes thermal treatment or biological treatment like aerobic or vermicomposting (CPCB, 2016). Unscientific dumping leads to leachate generation, causing surface and groundwater contamination (Mor et al., 2006a, 2006b).

The leachate generated by the decomposition of organic waste and rainfall percolating through the waste materials is a significant issue in landfills. Landfill leachate (LL) is characterised by high chemical oxygen demand (COD), total dissolved solids (TDS), pH, and ammoniacal nitrogen that are majorly determined by the age of landfill, amount and morphology of biodegradable matter, and heavy metals (Malina, 1992; Im et al., 2001). Different methods to treat LL that are currently being practised include biological treatments such as aerobic and anaerobic; physiochemical treatments such as air stripping, oxidation-reduction, pH adjustment, coagulation, and flocculation; and advanced techniques such as carbon adsorption and ion exchange (Inanc et al., 2000; Raghab et al., 2013). Physicochemical processes are highly effective but expensive to use (Del Borghi et al., 2003). Among all the available technologies for treating LL, anaerobic digestion is most promising due to the high organic content in the leachate, which would serve as a valuable substrate for biogas production (Jaroenpoj, 2014). The problem with this process is the presence of a high concentration of refractory organics (such as fulvic acid and humic acid), as well as heavy metals and ammoniacal nitrogen, which hinder the microbial processes and concomitant biochemical processes of anaerobic digestion (Lei et al., 2018; Lin et al., 1999).

Sewage sludge (SS) is the residual, semi-solid material produced as a by-product of aerobic treatment of industrial or municipal wastewater. SS is either disposed of by landfilling or direct application to agricultural lands. It is characterised by a high organic matter content (60–70% TS) and relatively low carbon to nitrogen ratio (C: N), ranging between 6 and 16 (Grobelak et al., 2017). Treatment methods available for SS management include incineration or composting. The negative impacts of the practices are that incineration generates large amounts of bottom ash, and composting leads to foul odour emission. On the other hand, high organic content makes it a suitable substrate for the anaerobic digestion process. SS is subjected to anaerobic treatment in many municipal wastewater treatment plants, leading to biogas production and reduction in sludge volume (Grobelak et al., 2017).

DOI: 10.1201/9781003204435-14

Co-digestion is an established technique used to increase anaerobic digestion efficiency by mixing two or more organic substrates, which may hinder the process if treated alone. The benefits of co-digesting two or more substrates are that they suffice nutrient balance (Siddique and Wahid, 2018); increase syntrophic interaction between microorganisms (Chakraborty et al., 2017); and dilute toxic substances present in any of the substrates (Montusiewicz and Lebiocka, 2011), thereby improving the digestion rate. The addition of suitable organic load can increase biogas production and stabilise the anaerobic digestion process (Cecchi et al., 1996). Various co-digestion studies were performed in the past, such as SS and cow dung/garden waste (Rao and Baral, 2011), dairy cattle manure and pear waste (Dias et al., 2014), food waste and straw (Yong et al., 2015), cattle manure with fruit and vegetable waste, organic household waste and SS (Tufaner and Avşar, 2016), leachate from sanitary landfill and municipal wastewater activated sludge (Del Borghi et al., 2003), animal manure and lignocellulose material (Neshat et al., 2017), and food waste and chemically enhanced primary treated sludge (Chakraborty et al., 2018), among many others.

Montusiewicz and Lebiocka (2011) investigated co-digestion of intermediate landfill leachate and SS using a lab-scale continuous stirred tank reactor in mesophilic conditions. The study was carried out at two different SS: LL ratios, 20:1 and 10:1, with a hydraulic retention time (HRT) of 30 days. The results showed that 10:1 was the best mixing ratio at the mesophilic condition, with a biogas production of 0.0413 m^3/kg VS/day removed. There was an increase in biogas yield by 13% when SS was co-digested with LL. There was also an increase in volatile fatty acid production and decreased N–NH_4^+ and P–PO_4^3 concentrations. Increased methane yields of 6.2% and 16.9% per kg VS removed were observed in the 10:1 and 20:1 ratios, respectively (Christensen et al., 2001). Apart from this research, not much work has been done on these substrates' co-digestion, thus creating a potential field of study concerning industrial-scale and lab-scale research. There is a high scope of improving the process parameters, including substrate ratio, to lower the HRT and increase the biogas yield (Montusiewicz and Lebiocka, 2011).

According to our literature review, co-digestion of MSW LL with SS in thermophilic conditions has not previously been studied. Hence, the current study aims to optimise the substrate ratio, bioconversion, and methane recovery by co-digesting LL and SS in thermophilic conditions. LL and SS were mixed in five different ratios depending on their volatile solid (VS) concentrations. This study was performed in serum bottles incubated at a temperature of 50°C. Physicochemical analyses such as VS, total solids (TS), and COD were performed and daily biogas generation was measured.

12.2 MATERIAL AND METHODS

12.2.1 Substrate and Inoculum

12.2.1.1 Substrate

The two substrates used in this study were LL generated from MSW and SS. LL was collected from an MSW processing and disposal facility located at Jawahar Nagar, Hyderabad, Telangana, India (17.528936, 78.600770).

SS was collected from the sewage treatment plant of Birla Institute of Technology and Science, Pilani, Hyderabad Campus.

The substrates were collected in 10 L ice-packed containers, transported to the laboratory, and stored at 4°C. Before feeding the reactors, the substrates were thoroughly mixed, and the required volume was transferred to a container to equilibrate to room temperature. The substrates were characterised by performing VS, TS, pH, and COD analyses (APHA, 2017).

12.2.1.2 Inoculum

Microorganism-laden slurry from a thermophilic anaerobic digester employed to treat food waste was used as inoculum. The required volume of inoculum was collected in 5 L containers from the digester before starting the experiments.

12.2.2 Optimisation of Substrate Ratio

12.2.2.1 Reactor Configuration

This study was carried out in clear glass serum bottles (33110-U, Sigma Aldrich, Milwaukee, USA) with a total volume of 100 mL and an outer diameter and height of 51.7 mm and 94.5 mm, respectively. Serum bottles have a closure-type crimp-top with a 20 mm diameter opening. For maintaining anaerobic conditions, the bottles were tightly closed with a rubber septum of 20 mm diameter and sealed with an aluminium cap using a hand crimping instrument.

12.2.2.2 Substrate Preparation and Reactor Operation

The substrates were mixed in five different ratios to optimise the substrate ratios based on their VS values (Table 12.1).

The reactors' working volume was 60 mL, out of which mixed substrates comprised 80% (48 mL), and the inoculum with VS of 0.132 g/L constituted the remaining 20% (12 mL). After the addition of substrate and inoculum, serum bottles were tightly closed with a rubber septum and sealed with aluminium caps. Biogas production was measured using a 5 mL syringe with a hypodermic reciprocating pump (Dispo van, Hindustan syringes, and medical devices, Faridabad, Haryana, India) which was injected through the rubber cork. All experiments were done in triplicates. Serum bottles were incubated at 50°C for seven days (HRT) in an incubator (YSI-440, Yorco, New Delhi, India). Daily biogas readings were recorded, and COD reduction was calculated according to *Standard Methods for the Examination of Water and Wastewater*. Based on the COD, TS, and VS reductions and daily biogas production, the substrates' mixing ratio was optimised. The reactors were operated in duplicates (series 1 and series 2).

12.2.3 Analysis

12.2.3.1 Physicochemical Analysis

All the ratios with their triplicates for three cycles were operated at similar conditions. TS, VS, and COD were analysed to track the performance of reactors. A hot air oven (RDHO 50, REMI, Mumbai, Maharashtra, India) was used for TS analysis, and complete moisture removal was carried out by oven-drying the samples at 105°C for 24 hours. A muffle furnace (Unix 96, Fourtech, Mumbai, Maharashtra, India) was used for VS analysis, performed at 550°C for 1 hour of the dried sample obtained from TS analysis. COD digestion was carried out in a COD digester (2015 M, Spectralab, Mumbai, Maharashtra, India) (APHA, 2017).

TABLE 12.1
Mixing Ratios of LL and SS Used in the Experiment

Ratio No.	Ratio (LL%:SS%)	LL (mL)	SS (mL)	VS (g/L) Cycle 1	VS (g/L) Cycle 2	VS (g/L) Cycle 3
1	100:0	48	0	79.33	95.33	52.11
2	75:25	36	12	101.33	100.67	67.40
3	50:50	24	24	126.85	130.60	160.73
4	25:75	12	36	148.40	143.60	208.10
5	0:100	0	48	112.40	94.60	184.140

Abbreviations: LL, landfill leachate; SS, sewage sludge; 12 mL – (20%) inoculum.

12.3 RESULTS AND DISCUSSION

12.3.1 Physicochemical Characteristics of Landfill Leachate and Sewage Sludge

LL and SS were characterised based on their COD, pH, TS, and VS, as seen in Table 12.2. LL was characterised by a high COD value in all the three cycles with pH inclined towards acidic nature. The TS of LL was comparatively higher than TS of SS in the first two cycles. Whereas in the third cycle, the TS was higher in SS than LL. In all the three cycles, the VS of LL was in the range of 62–63% of TS concentration, whereas the VS% of SS was decreased from 98% in cycle 1–34% in cycle 3.

12.3.2 TS Reduction

TS reduction in 25% LL and 75% SS was more than 55% in all three cycles with maximum removal of 72.5% in cycle 2. The least reduction of 10.72% was observed in 100% LL and 0% SS. Comparing our results with the study performed by Montusiewicz and Lebiocka (2011), TS reduction was 30% in the 10:1 ratio of SS and intermediate LL (Figure 12.1).

12.3.3 VS Reduction

VS reduction in 25% LL and 75% SS was above 60% in all three cycles with a maximum reduction of 78.03% in cycle 1. The minimum reduction of 12.34% was observed in 100% LL and 0% SS (Figure 12.2). Compared to this, VS reduction was 40% when Montusiewicz and Lebiocka (2011) co-digested SS and intermediate LL in a 10:1 substrate ratio.

12.3.4 COD Reduction

COD reduction was more than 80% in 25% LL and 75% SS ratio in all three cycles with a maximum reduction of 87.5% in cycle 1. At the same time, 0% LL and 100% SS substrate ratio had the least COD reduction of 50% (Figure 12.3). Comparing this, Montusiewicz and Lebiocka (2011) achieved a COD removal efficiency of 39% when SS and intermediate LL digested in a 10:1 ratio.

12.3.5 Biogas Yield

The biogas yield for 25% LL and 75% SS ratio was higher than the rest of the ratios in all three cycles (Figure 12.4). The maximum biogas yield of 0.145 m^3/kg VS was produced by 25% LL and

TABLE 12.2
Physicochemical Characteristics of Landfill Leachate and Sewage Sludge

	Experimental Cycle					
	1		2		3	
Characteristics	**LL**	**SS**	**LL**	**SS**	**LL**	**SS**
COD (mg/L)	112000	32000	128000	176000	180000	112000
pH	5.86	8.20	5.90	7.90	5.80	8.00
TS (g/L)	126	114.80	151.60	116.60	82.92	534.96
VS (g/L)	79.30	112.40	95.30	94.60	52.11	184.14

Abbreviations: LL, Landfill leachate; SS, Sewage Sludge.

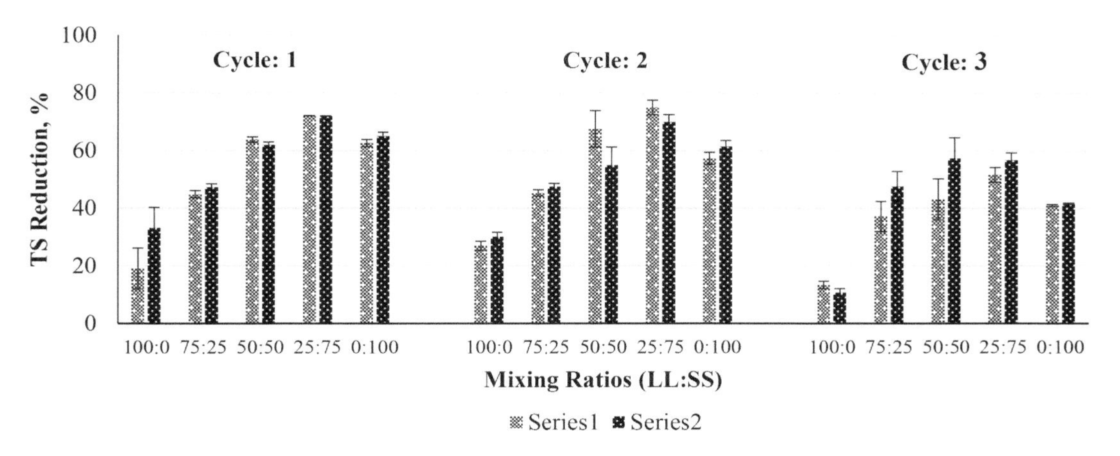

FIGURE 12.1 TS reduction %.

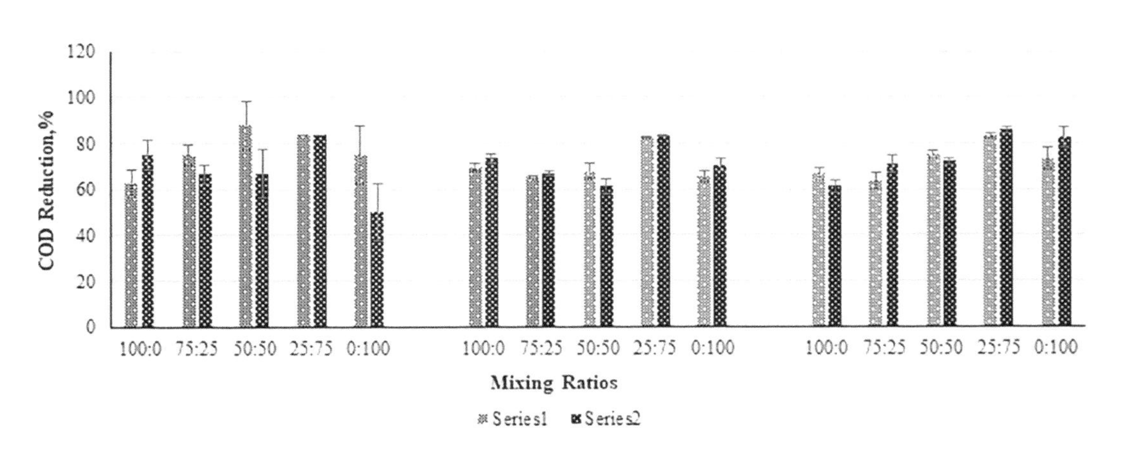

FIGURE 12.2 VS reduction %.

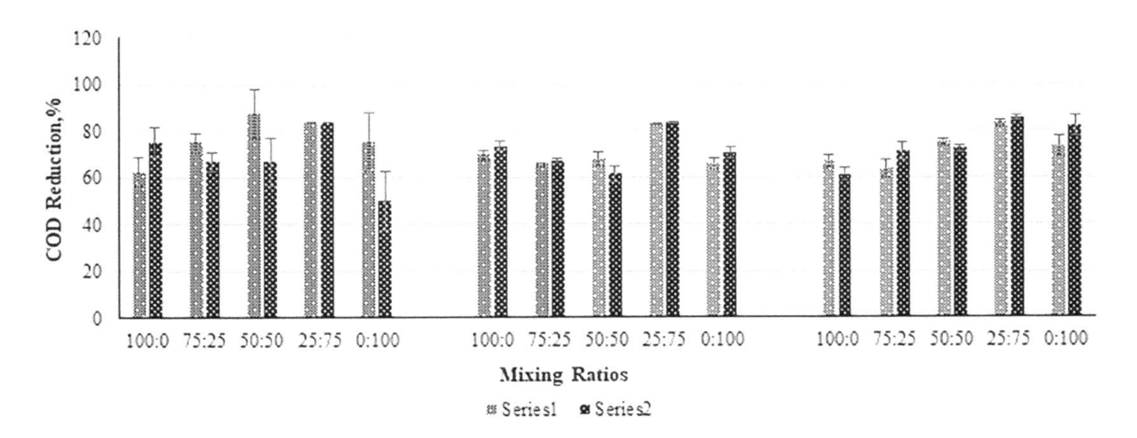

FIGURE 12.3 COD reduction %.

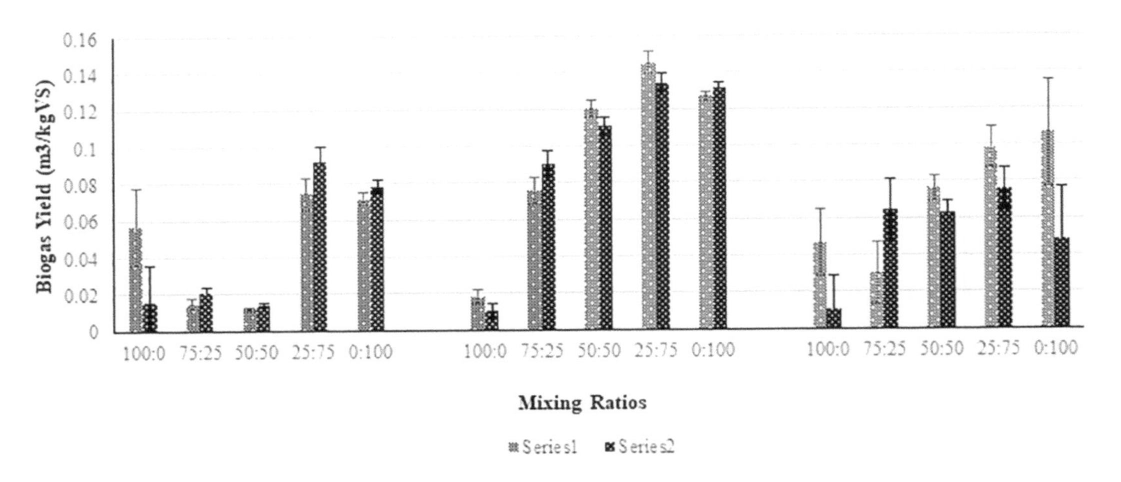

FIGURE 12.4 Biogas yield.

75% SS. Compared to this, the biogas yield for the ratio 10:1 (SS: intermediate LL) in the study performed by Montusiewicz and Lebiocka (2011) was 1.3 m^3/kg VS removed. High biogas yield in 25% LL and 75% SS ratio in this study is because the sewage sludge has higher organic content, which contributed to biogas production.

This study shows that when LL and SS are treated in 25% LL and 75% SS ratio, the COD, VS, and TS reduction were higher. The biogas production was also higher in this ratio. In the study performed by Montusiewicz and Lebiocka (2011), the leachate percentage was significantly less. Compared with the daily generation of leachate, the ratio of 25% LL and 75% SS can be considered an optimal ratio for the co-digestion of LL and SS.

12.4 CONCLUSION

Anaerobic digestion of LL as a sole substrate is complicated because of refractory organics, heavy metals, and high ammoniacal nitrogen content. Refractory organics are not readily biodegradable and thus hinder the hydrolysis process, whereas high ammoniacal nitrogen content inhibits the methanogenesis process of AD. To nullify this inhibition, LL was co-digested with SS in this study. Optimisation of substrate ratio (25% LL and 75% SS) was crucial in achieving improved anaerobic treatment of LL with higher COD reduction, VS reduction, and biogas production. The possible explanation can be the higher organic content of SS and its higher buffer capacity, which stabilises the toxic elements (refractory elements and ammoniacal nitrogen) and high pH of LL, enabling higher nutrition of microorganisms. Therefore, maximum bioresource recovery can be obtained at thermophilic conditions (50°C) with 25% LL and 75% SS as the substrates' mixing ratio, with better biogas yield than anaerobic digestion of sole substrates.

Moreover, the produced digestate after the co-digestion process will be subjected to the composting process. The composting process in aerobic conditions will lead to fractional mineralisation, biomass humification, and oxidation of the organic fraction of the produced digestate. The semi-solid digestate (mainly from 25% LL:75% SS and 0% LL:100% SS substrate ratios) will undergo an aerobic composting. In contrast, the liquid digestate (from 100% LL:0% SS, 75% LL:25% SS, 50% LL:50% SS substrate ratios) will be mixed with the organic fraction of MSW and further subjected to the aerobic composting process.

However, further investigations on the role of HRT, microbial diversity, biochemical processes, pH and biogas composition in the reactors will allow us to understand better the anaerobic co-digestion process, leading to stabilise it with more energy generation.

ACKNOWLEDGEMENTS

PSG, as principal investigator, thanks Biotechnology Industry Research Assistance Council for funding this research work (Project No. BT/SPARSH/0144/03/15). NA and SR thank BITS Pilani, Hyderabad Campus, for providing a PhD fellowship and teaching assistantship, respectively.

ABBREVIATIONS

COD = chemical oxygen demand
HRT = hydraulic retention time
LL = landfill leachate
SS = sewage sludge
VS = volatile solids
TS = total solids

REFERENCES

APHA, 2017. Standard Methods for the Examination of Water and Wastewater, 23rd ed. American Public Health Association, Washington, DC.

Cecchi, F., Pavan, P. and Mata-Alvarez, J., 1996. Anaerobic co-digestion of sewage sludge: application to the macroalgae from the Venice lagoon. Resources, Conservation and Recycling, 17(1), pp. 57–66.

Chakraborty, D., Karthikeyan, O.P., Selvam, A. and Wong, J.W., 2018. Co-digestion of food waste and chemically enhanced primary treated sludge in a continuous stirred tank reactor. Biomass and Bioenergy, 111, pp. 232–240.

Chakraborty, D., Selvam, A., Kaur, B., Wong, J.W.C. and Karthikeyan, O.P., 2017. Application of recombinant *Pediococcus acidilactici* BD16 (fcs+/ech+) for bioconversion of agrowaste to vanillin. Applied Microbiology and Biotechnology, 101(14), pp. 5615–5626.

Christensen, T.H., Kjeldsen, P., Bjerg, P.L., Jensen, D.L., Christensen, J.B., Baun, A., Albrechtsen, H.J. and Heron, G., 2001. Biogeochemistry of landfill leachate plumes. Applied Geochemistry, 16(7-8), pp. 659–718.

CPCB, 2016. NEERI, Waste Generation and Composition.

Del Borghi, A., Binaghi, L., Converti, A. and Del Borghi, M., 2003. Combined treatment of leachate from sanitary landfill and municipal wastewater by activated sludge. Chemical and Biochemical Engineering Quarterly, 17(4), pp. 277–284.

Dias, T., Fragoso, R. and Duarte, E., 2014. Anaerobic co-digestion of dairy cattle manure and pear waste. Bioresource Technology, 164, pp. 420–423.

Giusti, L., 2009. A review of waste management practices and their impact on human health. Waste Management, 29(8), pp. 2227–2239.

Grobelak, A., Placek, A., Grosser, A., Singh, B.R., Almås, Å.R., Napora, A. and Kacprzak, M., 2017. Effects of single sewage sludge application on soil phytoremediation. Journal of Cleaner Production, 155, pp. 189–197.

Im, J.H., Woo, H.J., Choi, M.W., Han, K.B. and Kim, C.W., 2001. Simultaneous organic and nitrogen removal from municipal landfill leachate using an anaerobic-aerobic system. Water Research, 35(10), pp. 2403–2410.

Inanc, B., Calli, B. and Saatci, A., 2000. Characterization and anaerobic treatment of the sanitary landfill leachate in Istanbul. Water Science and Technology, 41(3), pp. 223–230.

Jaroenpoj, S., 2014. Biogas production from co-digestion of landfill leachate and pineapple peel. PhD desertation, Department of Engineering, Griffith Science, Griffith University, Queensland Australia.

Lei, Y., Wei, L., Liu, T., Xiao, Y., Dang, Y., Sun, D. and Holmes, D.E., 2018. Magnetite enhances anaerobic digestion and methanogenesis of fresh leachate from a municipal solid waste incineration plant. Chemical Engineering Journal, 348, pp. 992–999.

Lin, C.Y., Bian, F.Y. and Chou, J., 1999. Anaerobic co-digestion of septage and landfill leachate. Bioresource Technology, 68(3), pp. 275–282.

Malina, J., 1992. Design of anaerobic processes for treatment of industrial and municipal waste (Vol. 7). CRC Press.

Montusiewicz, A. and Lebiocka, M., 2011. Co-digestion of intermediate landfill leachate and sewage sludge as a method of leachate utilization. Bioresource Technology, 102(3), pp. 2563–2571.

Mor, S., Ravindra, K., Dahiya, R.P. and Chandra, A., 2006a. Leachate characterization and assessment of groundwater pollution near municipal solid waste landfill site. Environmental monitoring and assessment, 118(1-3), pp. 435–456.

Mor, S., Ravindra, K., De Visscher, A., Dahiya, R.P. and Chandra, A., 2006b. Municipal solid waste characterization and its assessment for potential methane generation: a case study. Science of the Total Environment, 371(1-3), pp. 1–10.

Neshat, S.A., Mohammadi, M., Najafpour, G.D. and Lahijani, P., 2017. Anaerobic co-digestion of animal manures and lignocellulosic residues as a potent approach for sustainable biogas production. Renewable and Sustainable Energy Reviews, 79, pp. 308–322.

OECD, 2003. OECD Environmental Indicators. Development, Measurement and Use. OECD Environment Directorate, Organisation for Economic Co-operation and Development (OECD), Paris, France. http://www.oecd.org/env/.

Raghab, S.M., El Meguid, A.M.A. and Hegazi, H.A., 2013. Treatment of leachate from municipal solid waste landfill. HBRC Journal, 9(2), pp. 187–192.

Rao, P.V. and Baral, S.S., 2011. Experimental design of mixture for the anaerobic co-digestion of sewage sludge. Chemical Engineering Journal, 172(2-3), pp. 977–986.

Siddique, M.N.I. and Wahid, Z.A., 2018. Achievements and perspectives of anaerobic co-digestion: a review. Journal of Cleaner Production, 194, pp. 359–371.

Tufaner, F. and Avşar, Y., 2016. Effects of co-substrate on biogas production from cattle manure: a review. International journal of environmental science and technology, 13(9), pp. 2303–2312.

Yong, Z., Dong, Y., Zhang, X. and Tan, T., 2015. Anaerobic co-digestion of food waste and straw for biogas production. Renewable energy, 78, pp. 527–530.

Section III

Biomethane: Feedstock, Production, and Application

13 Economic Aspects of Waste Valorisation with the Aid of Anaerobic Digestion and Other Technologies

Sutripta Sarkar and Debaprasad Sarkar

13.1 INTRODUCTION

Any substance that cannot be put to use by man is classified as waste. Depending on its nature, waste can be organic, chemical or industrial, plastic, or e-waste. Rapid and unplanned urbanisation has led to several problems. Managing increasing amount of solid waste in cities and towns has become a major challenge. Most of the waste generated is thrown away or dumped, which causes foul odours and an unhygienic environment. Dumping waste in unplanned landfills and burning those causes emission of gases, which are potentially harmful to mankind. The presence of greenhouse gases (GHGs) like methane and carbon dioxide leads to global warming and climate change. Openly thrown waste becomes a breeding ground for mosquitoes and other pests, which act as vectors or carriers of several diseases. These conditions ultimately lead to economic losses. A lot of food chain supply waste is also being generated in the different stages of supply chain from production to retail (Matharu et al, 2016). In developed countries like the United States and various countries in Europe, food wastage primarily occurs at the consumer end, whereas in case of developing countries, the wastage usually occurs at initial or middle stages of supply chain (Ong et al, 2017). This is primarily due to lack of storage and food processing facilities in these countries.

According to studies, municipal solid waste (MSW) in India comprises 50–55% biodegradables and around 15–20% recyclable materials (Figure 13.1). Similar data on composition of biodegradables has also been reported in Planning Commission Report 2014. However, metropolitan cities like Bengaluru generate around 59–60% organic waste (Naveen and Sivapullaiah, 2020). It is interesting to note that smaller cities like Agartala, which is the capital of the north eastern Indian state Tripura, generates around 55% of biodegradable food waste (Chakraborty et al, 2019). Market waste is composed of leftover raw green leafy vegetables, rotting fruits, straws and other packaging material, plastics, foils, etc.

In India, MSW become an acute problem with rapid urbanisation and growth rate of urban population. Per capita waste generation in India is estimated to be 450 g/day (Sharma and Jain, 2019). It has been anticipated that the population of India will be about 1.823 billion by 2051 and about 300 million tonnes per annum of MSW will be generated, which will require around 1,450 km^2 of land to dispose of it in a systematic manner (Joshi and Ahmed, 2016) if urban local bodies in India continue to rely on landfill route for MSW management (CPCB, 2013, 2000). However, these projections are on conservative side, keeping 1.33% annual growth in per capita generation of MSW (CPCB, 2000; Pappu et al, 2007). Therefore, with 5% annual growth in per capita generation, landfill area required for disposal of waste could be much larger (CPCB, 2013). Details of waste generation, collection, and treatment in India are reviewed in Table 13.1.

Data shows that only 28% of the waste generated is treated. Hence, there is a lot of scope for implementation of valorisation technologies. Studies suggest that waste valorisation attempts to

DOI: 10.1201/9781003204435-16

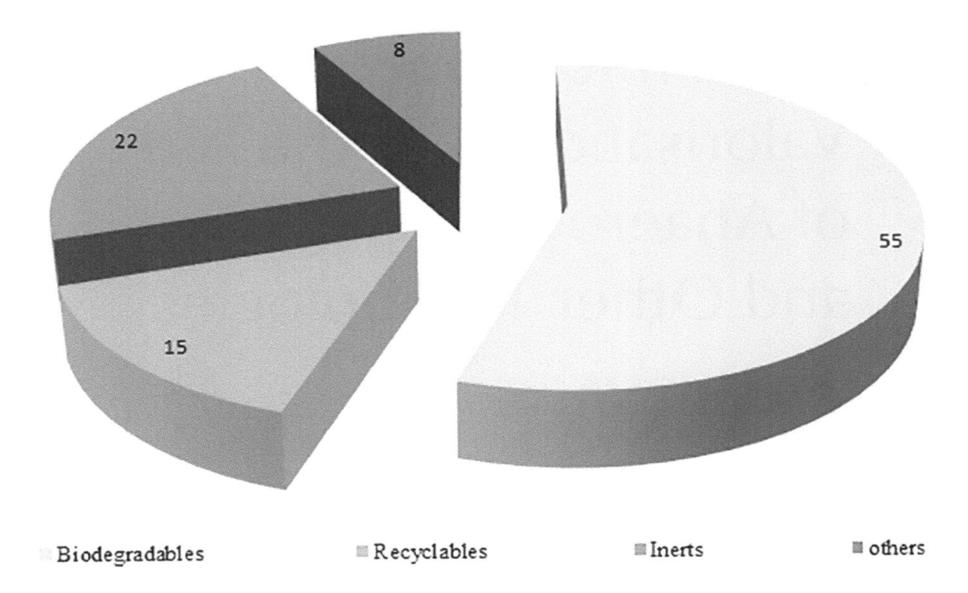

FIGURE 13.1 Composition of municipal solid waste in India.

convert discards into useful products that are not only an economically viable (Aracon et al, 2013) but also socially, economically, and environmentally desirable (Misra and Pandey, 2005; Schoot Uiterkamp et al, 2011). Biochemical (composting, anaerobic digestion, ethanol fermentation, etc.), thermal (incineration and direct combustion), and thermochemical (pyrolysis and gasification) are some of the commonly used techniques to valorise waste (Gumisiriza et al, 2017).

13.2 PYROLYSIS AND GASIFICATION

Pyrolysis is the thermal degradation of organic material in the absence of oxygen. The process requires very high temperature of around 400–900°C (Bosmans et al, 2013). The end product formed is bio-oil, char, and synthesis gas (syngas), which can be used for electricity generation. Gasification is the process of partial oxidation of organic waste at elevated temperatures (500–1800°C) to produce syngas (Gumisiriza et al, 2017). The initial cost of setting up a pyrolysis or gasification plant is usually around Rs. 20 lakhs or USD 28,550 (excluding the cost of energy). The biggest advantage of the pyrolysis and gasification processes is that unsorted or mixed waste can be used. All types of waste, especially plastics, rubber etc. are effectively valorised to generate useful products. However, there are also a lot of disadvantages. Pyrolysis/gasification units release lots of toxic gases like carbon monoxide, methane, hydrogen chloride, and ammonia; heavy metals like mercury and cadmium; and dioxins into the atmosphere (Moukamnerd et al, 2013). This process requires high temperatures, meaning a lot of energy is used, which increases the cost of production drastically. It should also be noted that market waste is comprised of a large faction of wet waste, which is not a very appropriate substrate for pyrolysis and gasification. Thermal drying of wet waste increases energy expenditure.

13.3 COMPOSTING

According to Askarany and Franklin-Smith (2014), composting of organic waste is a commercially viable approach for cleaner production and a valuable alternative to landfilling. Since a large faction of the waste generated is biodegradable, composting seems to be an environmentally friendly and economically viable method to valorise waste.

TABLE 13.1
Current Status of Waste Collection, Disposal, and Treatment in India

Parameter	Status
House-to-house collection of waste	18 states (out of 29)
Segregation of waste at source	5 states (out of 29)
Number of unsanitary unfilled sites identified	1,285
Number of sanitary landfill sites constructed	95
Number of ULBs operating compost/ vermicompost facilities	553
Number of ULBs with under-construction compost/ vermicompost facilities	173
Number of operating pipe composting facilities	7,000
Number of operating RDF facilities	12
Number of operating biogas plants	645
Number of energy generation plants	11 (6 operational)
Waste generation	143,449 Mt/day
Waste collection	117,644 Mt/day (82%)
Waste treated	32,871 Mt/day (28%)

Source: Data from CPCB (2016).
Abbreviations: RDF, refuse derived fuel; ULBs, urban local bodies.

Composting is a method of organic waste stabilisation. It is a self-heating biological conversion that results in useful end products (Sarkar et al, 2016). Composting is an aerobic process of waste treatment where waste material is either piled in heaps or layered in windrows. The time taken for composting depends on the materials to be composted, usually taking 1–3 months to be completed. The addition of microbial inoculum helps speed up the process (Sarkar et al, 2010). The end product can be used for agriculture or horticulture to enhance soil fertility. Factors affecting composting include pH, temperature, particle size, texture of material used, porosity, moisture, and aeration (Rynk, 1992). The carbon and nitrogen ratio is a crucial factor in composting. The initial ratio is set at 30 in order to get compost balanced in carbon and nitrogen. According to a report by the Centre for Integrated Agricultural System, University of Wisconsin (https://www.cias.wisc.edu/windrow-composting-systems-can-be-feasable-cost-effective/), windrows composting systems can be very cost effective for farmers if they are willing to share their equipment. The study states that the annual average cost of composting can range between USD 150/cow and USD 697/cow depending on the method used. But this is applicable only for agricultural fields where the quality and quantity of waste is uniform and regular. Waste generated in cities is usually mixed, and composting plants for waste management and valorisation need to set up where land can be an issue. In India, it is estimated that only 6%–7% of MSW is valorised to compost. Most of the market waste generated is still disposed into landfills unsegregated, as all biodegradables, recyclables, and inert materials are collected together. According to a report by CPCB (2013), India has 59 constructed sites, most of which are overflowing with waste beyond capacity. Around 376 landfill sites are under planning and implementation and more than 1,000 new sites have been identified for future use. Among the states, Maharashtra (125) has the highest number of composting/vermicomposting plants, followed by Gujrat (86) and Andhra Pradesh (32) (CPCB, 2013).

Technically, composting is an ideal process to valorise market waste generated in Indian cities. However, ready-to-use compost has few takers in the cities where, due to lack of space and time, people are unenthusiastic about growing plants. Farmers generally prefer to prepare their own compost for use in fields. Compost quality depends a lot on the input materials, and in this case composition of biodegradable market waste is subject to seasonal changes. Compost quality degrades after long storage, hence unsold compost becomes virtually useless. Aside from these issues, though less than landfills, compost plants also generate a lot of GHGs, which are hazardous to the environment (Andersen et al, 2010).

13.4 ETHANOLIC FERMENTATION

Ethanol production from waste biomass has the attention of researchers worldwide, as it is a renewable fuel that has a negative carbon footprint. The process does not emit any GHGs, having a negative impact on global warming. Fermentation is essentially a microbial process and microbes like *Saccharomyces cerevisiae* have been used extensively (Hossain et al, 2017). The ideal substrate for ethanol production is sugars, and countries like the United States and Brazil use extracts from sugar beets and corn syrup to produce ethanol (McDonald et al, 2001). In India, sugarcane molasses is largely used for ethanol production. Five to ten percent ethanol-blended petrol is sold in India. The government of India intends to increase the blending up to 20% in petrol by year 2030 in order to control petroleum prices. Initially, ethanol was formed from sugar- or starch-rich crops like corn or potatoes. With the debate over food verses fuel gaining momentum, scientist looked for alternatives in waste lignocellulosic biomass.

Ethanol production is perhaps the cheapest among the technologies discussed in this review. It is also a very environmentally friendly process. There are, however, certain limitations in this method. Lignocellulosic biomass needs pretreatment with acid or alkali for production of ethanol from cellulose. Lignin can be removed by microbial degradation, but it is time consuming. During fermentation of ethanol, there are chances of coproduction of other organic solvents like methanol and acetic acid, which need to be removed by distillation. This increases the cost of production. Sometimes the product obtained is very dilute and cannot be used optimally. Ethanol is hygroscopic and absorbs water from air, leading to corrosion in energy-generating engines and power machines (Masjuki and Kalam, 2013).

13.5 BIOMETHANATION OR ANAEROBIC DIGESTION

Biomethanation is a process by which a complex mixture of symbiotic organisms transforms organic material under anaerobic conditions into biogas, which is composed of 60% methane, 40% carbon dioxide, and traces of hydrogen sulphide. Methane gas is captured and used as an alternative to liquified petroleum gas (Chisty et al, 2014). Increases in petroleum and cooking gas prices have led to search for alternate sources of energy. Biogas is a viable alternative. According to the Ministry of New and Renewable Energy website, the cost of setting up a domestic biogas plant can vary between USD 250 and USD 600, depending upon the model (https://mnre.gov.in). The government of India subsidises domestic biogas plants under its National Biogas and Manure Management Programme. This programme claims to have setup 4.31 million family-type biogas plants in the country. Several biogas bottling plants have been set up across the country. Aside from generating gas, the biogas bottling plants also provide organic manure and a solution for managing organic waste. The first bottling plant was set up in Talwade village in Nasik (Maharashtra) in 2010. The biogas generation capacity of the plant is 500 m^3 per day, which can be compressed to 150 bar for filling up cylinders. The biogas produced is 98% pure. Twelve point five MT of cow dung and agricultural wastes are processed per day (MNRE, BGFP Project). Anaerobic digestion (AD) of 20–25 kg of cow dung can generate 1 m^3 of biogas, which can be converted to two units of electricity or 0.4 kg of bio-CNG (https://mnre.gov.in). Bio-CNG is compressed biogas with higher methane content and is used as transportation fuel (Shah et al, 2017). According to a media report, Indian oil companies like HPCL, BPCL, and Indian Oil will invest Rs. 10,000 crores in bio-CNG plants (www.livemint.com, dated April 10, 2018). According to the TERI website (http://terienvis.nic.in), around 21 bottling plants have been sanctioned under Bio-energy Technology Development Group-BGFP Projects. Currently 11 such plants are functional.

Electricity can be generated from biogas through an internal combustion engine. The mechanical energy rotates an electric generator, which produces the electricity. A pilot-scale biogas plant is functional at the Indian Institute of Engineering, Science and Technology, Shibpur, Kolkata. Approximately 500 kg of food waste is added daily, which generates 8–9 m^3 of biogas that can be converted to around 12 kW of electricity per day (Sarkar et al, 2019). According to the CPCB 2013 report, there are 56 biogas-based power plants in India; most of these are located in Maharashtra, Karnataka, and Kerala. The city of Pune processes its organic waste in biogas plants and the electricity generated is used for lighting up street lamps (Kumar and Agrawal, 2020). Similar successfully running biogas plants are in several cities including Bengaluru (Naveen and Sivapullaiah, 2020) and Agartala (Chakraborty et al, 2019).

The four key steps that occur during AD are hydrolysis, acidogenesis, acetogenesis, and methanogenesis (Chisty et al, 2014). A consortium of bacteria and archea contribute in the biomethanation process. MSW, fruits and vegetable solid waste, aquatic weeds, cow dung, pig manure, etc. can be used as feed material in AD systems (Biosantech et al, 2013). However, lignocellulosic wastes need to be pretreated before digestion (Hendricks and Zeeman, 2009), which increases the cost of production. Biogas yield is reduced when substrate is in its diluted form. Cow dung or pig manure is required for good yield of biogas, which are not readily available in cities. Change in substrate composition may hamper the methane to carbon dioxide ratio, thereby decreasing the yield. There are certain operational challenges as well. The biodegradable fraction of market waste or food waste has a very low carbon to nitrogen ratio, so the rate of biodegradation increases, resulting in accumulation of volatile fatty acid and decrease in pH, which ultimately results in the failure of the process (Capson-Tojo et al, 2016; Jabeen et al, 2015; Shen et al, 2013). Methane, which is the primary constituent of biogas, can form explosive mixtures with air, which can be extremely hazardous (Chrebet and Martinka, 2012). Hence safety measures must be maintained wherever a biogas plant is installed. AD systems are closed, meaning GHG emission is largely cut out, which is perhaps the biggest environmental advantage of the process.

13.6 DISCUSSIONS

13.6.1 Cost Analysis

Table 13.2 highlights the cost of collection and disposal of waste in different countries based on income levels. It also depicts the costs incurred in different waste valorisation processes.

Cost comparison of the valorisation techniques (Table 13.2) shows that composting is the cheapest method, but compared to AD and ethanol fermentation, the valorised products of AD systems are more commercially viable. Biogas can be used as fuel and for the generation of electricity. Methane is a very potent GHG and is released from landfills and composting plants. AD is beneficial, as it has zero emissions. Biogas can be used as a cheaper alternative of cooking gas; several biogas bottling plants are being set up in India to exploit this potential. On the flip side, skilled expertise is required for setting up of biogas production, which ultimately leads to increased cost of production. One of the advantages of aerobic digestion is that a lot of heat is produced due to microbial activity, which kills the pathogens present in the waste. If done in large scale, it takes up a lot of land, and there are very few takers for compost made from waste generated in cities. The quality of compost is determined by the raw materials used, and quality changes when there is change in composition of the waste material. Several studies have shown that biogas production has the least adverse effect on the environment (Zulkepli et al, 2017). In towns and cities, the cost of transportation of waste to the composting or biogas plant has to be considered. However, in rural areas, due to availability of raw materials, both the processes can be taken up indigenously.

TABLE 13.2
Cost of Collection, Disposal, and Treatment of Waste (USD/Tonne)

	Collection	Sanitary Landfill	Open Dumping	Composting	Waste-Energy	Anaerobic Digestion	Total Cost 2010 (in billion $)	Total Expected Cost 2025 (in billion $)
Low-income countries	20–50	10–30	2–8	5–30	NA	NA	1.5	7.7
Lower-middle-income countries	30–75	15–40	3–10	10–40	40–100	20–80	20.1	84.1
Upper-middle-income countries	40–90	25–65	NA	20–75	60–150	50–100	24.5	63.5
High-income countries	85–250	40–100	NA	35–90	70–200	65–150	159.3	220.2

Source: World Bank Solid Waste Thematic Group.

Converting waste to energy, burning garbage to produce electricity, is also an alternative. There has been considerable excitement about waste-to-energy (WTE) plants, especially from businesses that want to avail government subsidies. The Ministry of New and Renewable Energy offers interest subsidies of up to Rs 2 crore per megawatt for commercial WTE projects. For demonstration projects, a direct subsidy up to 50% of the capital cost of the plant (up to Rs 3 crore per megawatt) is offered. Municipalities can get incentives of Rs 15 lakh per megawatt for providing garbage free of cost to WTE plants for up to 30 years. State nodal agencies get incentives of Rs 5 lakh per megawatt for promoting and monitoring WTE projects. The biggest cause of concern with the WTE plants (pyrolysis and gasification) is the emissions. The GHGs, dioxanes, heavy metals, and ash that are generated during these processes become an environmental liability. There are, however, no such concerns when waste is converted to energy via AD. Initial start-up cost of a biogas plant is much less compared to that of other WTE projects. Bioethanol production has also gained a lot of importance in recent times. The process is however marred by low product output, which needs distillation and purification. This increases the cost of production to a large extent. Biogas production on the other hand requires very little input cost once the plant has been set up.

13.7 CONCLUSIONS

In this review, a comparison was made of the cost and benefits of different waste valorisation technologies available for market waste in India. The advantages and disadvantages of the available technologies have been discussed in detail. The best technology depends to a large extent on the availability and quality of waste. From this review, we concluded that AD is perhaps the best valorisation method, especially in terms of environmental benefits, fuel, and electricity generation. With the ever-increasing amount of food waste being generated at the consumer end in developed and developing nations, AD seems to be the most viable technology available. Detailed survey-based study needs to be undertaken to develop an understanding regarding the supply chain, expenditure per unit of waste, and circular economy related to waste management in different forms, especially in the Indian context.

ACKNOWLEDGEMENTS

Authors thank Principal, Barrackpore Rastraguru Surendranath College and the International Society of Waste Management, Air and Water for encouragement and support.

REFERENCES

Andersen J K, Boldrin A, Christensen T H, Schuetz C. (2010). Greenhouse gas emissions from home composting of organic household waste. *Waste Management* 30, 2475–2482.

Aracon R A D, Lin C S K, Chan K M, Kwan T H, Luque R. (2013). Advances on waste valorization: New horizons for a more sustainable society. *Energy Science and Engineering* 1(2), 53–71.

Askarany D, Franklin-Smith A W. (2014). Cost benefit analysis of organic waste composting systems through the lens of time driven-based costing. *JAMAR* 12(2), 59–74.

Bosmans A, Vanderreydt I, Geysen D, Helsen L. (2013). The crucial role of waste-to-energy technologies in enhanced landfill mining: A technology review. *Journal of Cleaner Production* 55, 10–23.

Biosantech T A S, Rutz D, Janssen R, Drosg B. (2013). Biomass resources for biogas production. *The Biogas Handbook Science, Production and Applications*, Woodhead Publishing Series in Energy, 19–51.

Capson-Tojo G, Rouze M, Crest M, Steyer J P, Delgenes J P, Escudie R. (2016). Food waste valorization via anaerobic processes: A review. *Reviews in Environmental Science and Bio/Technology* DOI 10.1007/s11157-016-9405-y

Chakraborty S, Majumdar K, Pal M, Roy P K. (2019). Assessment of bio-gas from municipal solid waste for generation of electricity– A case study of Agartala city. *International Journal of Applied Engineering Research* 14(6), 1265–1268.

Chisty P M, Gopinath L R, Divya D. (2014). A review on anaerobic decomposition and enhancement of biogas production through enzymes and microorganisms. *Renewable and Sustainable Energy Reviews* 34, 167–173.

Chrebet T, Martinka J (2012). Assessment of biogas potential hazards. *Annals of Faculty Engineering Hunedoara - International Journal of Engineering* 39–42.

CPCB. (2000). Status of municipal solid waste generation, collection, treatment and disposable in class 1 cities. Central Pollution Control Broad. Ministry of Environment and Forests, Government of India, New Delhi.

CPCB. (2013). Status report on municipal solid waste management. Retrieved from http://www.cpcb.nic.in/divisionsofheadoffice/pcp/MSW_Report.pdfhttp://pratham.org/images/paper_on_ragpickers.pdf

CPCB. (2016). Central pollution control board (CPCB) bulletin, Goverment of India. http://cpcb.nic.in/openpdffile.php?id=TGF0ZXN0RmlsZS9MYXRlc3RfMTIzX1NVTU1BUllfQk9PS19GUy5wZGY.

Gumisiriza R, Hawumba J R, Okure M, Hensel O. (2017). Biomass waste-to-energy valorisation technologies: A review case for banana processing in Uganda. *Biotechnology for Biofuels* 10, 11.

Hendricks A T W M, Zeeman G. (2009). Pretreatments to enhance the digestibility of lignocellulosic biomass. *Bioresource Technology* 100, 10–18.

Hossain N, Zaini J H, Mahlia T M. (2017). A review of bioethanol production from plant-based waste biomass by yeast fermentation. *International Journal of Technology* 8(1), 5–18.

Jabeen M, Yousaf S, Haider M R, Malik R N (2015). High-solids anaerobic co-digestion of food waste and rice husk at different organic loading rates. *International Biodeterioration & Biodegradation* 102, 149–153.

Joshi R, Ahmed S. (2016). Status and challenges of municipal solid waste management in India: A review. *Cogent Environmental Science* 2, 1139434.

Kumar A, Agrawal A. (2020). Recent trends in solid waste management status, challenges, and potential for the future Indian cities – A review. *Current Research in Environmental Sustainability* 2, 100011

McDonald T, Yowell G, McCormack M. (2001). Staff report. US ethanol industry production capacity outlook. California Energy Commission. Available at http://www.energy.ca.gov/reports/2001-08-29_600-01-017.PDF

Masjuki H, Kalam A. (2013) An overview of biofuel as a renewable energy source: Development and challenges. *Procedia Engineering* 56, 39–53.

Matharu, A S, de Melo, E M, Houghton, J A. (2016). Opportunity for high value-added chemicals from food supply chain wastes. *Bioresource Technology* 215, 123–130.

Misra, V, Pandey, S D. (2005). Hazardous waste, impact on health and environment for development of better waste management strategies in future in India. *Environment International* 31, 417–431.

Moukamnerd C, Kawahara H, Katakura Y. (2013). Feasibility study of ethanol production from food wastes by consolidated continuous solid-state fermentation. *Journal of Sustainable Bioenergy Systems* 3, 143–148.

Naveen B P and Sivapullaiah P V. (2020). Solid waste management: Current scenario and challenges in Bengaluru. *Sustainable Sewage Sludge Management and Sustainable Efficiency*. DOI: http://dx.doi.org/10.5772/intechopen.90837

Ong K L, Kaur G, Pensupa N, Uisan K, Lin C S K. (2017). Trends in Food waste valorization for the production of chemicals, materials and fuels: Case study South and Southeast Asia. *Bioresource Technology* doi: http://dx.doi.org/10.1016/j.biortech.2017.06.076

Pappu, A, Saxena, M, Asolekar, S R. (2007). Solid wastes generation in India and their recycling potential in building materials. *Building and Environment* 42, 2311–2320. doi: 10.1016/j.buildenv.2006.04.015

Rynk, R. (1992). *On-Farm Composting Handbook. No.54.* Northeast Regional Agricultural Engineering Service, Ithaca, NY.

Sharma K D, Jain S. (2019). Overview of municipal solid waste generation, composition, and management in India. *Journal of Environmental Engineering* 145(3), 04018143.

Shah M S, Halder P K, Shamsuzzaman A S M, Hossain M S, Pal S K, Sarker E. (2017). Perspective of biogas conversion into bio-CNG for automobile fuel in Bangladesh. *Journal of Renewable Energy* https://doi.org/10.1155/2017/4385295

Sarkar A, Bhattacharjee A, Samanta H, Bhattacharya K, Saha H. (2019). Optimal design and implementation of solar PV-wind-biogas-VRFB storage integrated smart hybrid microgrid for ensuring zero loss of power supply probability. *Energy Conservation and Management* 191, 102–118.

Sarkar S, Banerji R, Chanda S, Das P, Ganguly S, Pal S. (2010). Effectiveness of inoculation with isolated *Geobacillus* strains in the thermophilic stage of vegetable waste composting. *Bioresource Technology* 101, 2892–2895.

Sarkar S, Pal S, Chanda S. (2016). Optimization of vegetable waste composting with an optimum thermophilic phase. *Procedia Environmental Sciences* 35, 435–440.

Shen F, Yuan H, Pang Y et al. (2013) Performances of anaerobic co-digestion of fruit & vegetable waste (FVW) and food waste (FW): Single-phase vs. two-phase. *Bioresource Technology* 144, 80–85.

Schoot Uiterkamp B J, Azadi H, Ho P. (2011). Sustainable recycling model: A comparative analysis between India and Tanzania. *Resources, Conservation and Recycling* 55, 344–335.

Zulkepli N E, Muis Z A, Mahmood N A N, Hashim H, Ho W S. (2017). Cost benefit analysis of composting and anaerobic digestion in a community: A review. *Chemical Engineering Transactions* 56, 1777–1782.

14 Biomass from Agricultural Wastes for Renewable Energy in the Philippines

Aries O. Ativo and Lynlei L. Pintor

14.1 INTRODUCTION

With the current situation of the country in terms of energy usage, it is quite beneficial to use sources of renewable energy like biomass. This type of renewable energy can be considered sustainable, especially as it mostly comes from agricultural wastes. In fact, 14% of the world's energy is provided for by biomass, not only in developing countries, but in industrial countries as well (Hall, 1991).

According to the United States Energy Information Administration (EIA, 2018), biomass is considered an organic and a renewable source of energy since it comes from plants and animals. Through photosynthesis, plants absorb the sun's energy, which makes up the plants' stored energy. Among the organic materials that can be used for biomass energy is agricultural waste. In the Philippines, there are several types of agricultural wastes that can be identified, including rice husk, rice straw, coconut husk, coconut shell, and bagasse, meaning that the country has good potential in terms of producing biomass power (Zafar, 2020).

There are several advantages and benefits, both economically and environmentally, for the Philippines to generate biomass energy. Aside from the fact that it has several sources of biomass energy, the most common of which is agricultural waste, the country has already benefited from using biomass energy for its energy supply. Out of the 100 million people in the Philippines, almost 30% use biomass energy. This is usually utilised for household cooking by those in rural areas. In addition, it was recorded that coconut husks and shell and bagasse comprised 12% of the national energy supply (ASEAN Briefing, 2017).

In the Philippines, biomass as a renewable energy is not yet well studied; generating renewable energy from agricultural waste is still to be explored. But the potential of biomass is quite promising. Go et al. (2019) stated that sugarcane, paddy rice, coconut, and maize are among the widely available residue in the country, and that agricultural and agro-industrial residues are used for liquid-fuel production. However, it was likewise found that consumption of fossil fuel is faster than alternative fuel generation. On the other hand, Grafilo, Gunay, and Malayao (2018) found that biomass can likely be a resource in the future at both small- and large-scale levels. Although biomass is mostly considered a low-status fuel, it is still renewable, unlike fossil fuels. It is a flexible and diverse fuel that can be utilised as feedstock for direct combustion and converted into electricity through gas turbines. Looking into the potential of agricultural wastes as renewable biomass energy is both a challenge and an opportunity because of its numerous uses and benefits. The need to explore this field is important, as the Philippines can be considered highly agricultural and the application of biomass energy may further mean less waste and more sources of energy.

14.2 STATE OF THE ART

The Philippines has abundant biomass-waste resources from numerous agricultural and forestry production operations. Despite the large volume of biomass waste, the utilisation for energy and materials production has been very minimal. There have been studies done in

DOI: 10.1201/9781003204435-17

the past to assess the extent and distribution of biomass wastes around the country, with a common conclusion that these resources have great potential for energy and power production (Capareda, 2012). There are six major biomass wastes in the Philippines: rice hull, animal manure, coconut wastes, sugarcane wastes, urban solid wastes, and forestry residues. Volatile oil price and growing emphasis on environmental conservation have stimulated the development and utilisation of biomass as a vital source of renewable energy (Lim, 2010). This covers three categories of biomass energy: (1) wastes (garbage or animal), (2) residues (plant remains left in the field by forestry or agricultural operations), and (3) fuel crops (biomass cultivated on energy farms specifically for its fuel content).

Several studies conducted (Alawi & Mana, 2012; Abdul & Hashim, 2013; Gallego, Aberilla, & Adizas, 2019; Rao, 2016; Elauria & Castro, 2010; Krista, Aviso, & Eusebio, 2016) explored rice hull, which is the widely abundant agricultural waste from the rice industry. Various researchers (Elan & Rubio, 2014; Malayao, Grafilo, & Gunay, 2018; Zafar, 2020; Diaz & Goluoke, 2017; Fung, 2013) addressed the issues on animal manure as biomass waste, while some researchers (Samson & Zafar, 2020; Pestaño, 2016; Arellano & Kato, 2016) focused on coconut wastes. However, there are studies focusing on sugarcane wastes (Go et al., 2019, Tan & Culaba, 2014; Jamora, Gudia, & Gidoquio, 2015; De Mesa & Soriano, 2019). Forest residues were also given attention by more than a few researchers (Tamayo, 2018; Acda & De Vera, 2014; Pimentel, Moran, & Weber, 2016).

The majority of studies fall under the three categories of biomass energy. Some researchers (Cajes, 2014; Chiu, 2019) reviewed wastes that refer to garbage or animals as biomass energy. Residues of plants were also conducted by multiple researchers (Fernandez, & Costa, 2010). Moreover, recently, fuel crops were the focus of several researchers (Hidon & Enriquez, 2014; Martinez, 2019; Maruyama, 2015; Milla, River, & Huang, 2013).

14.3 RESULTS AND DISCUSSION

Table 14.1 presents the biomass availability from the three major agricultural crops in the Philippines, namely rice, coconut, and sugarcane. The staple food for millions of Filipinos is rice, and most agricultural lands in the country are devoted to rice (4.2×10^6 ha). The area devoted to rice production is 1.17 times larger than coconut production ($3.3–3.6 \times 10^6$ ha) and ten times larger than sugarcane production ($0.38–0.42 \times 10^6$ ha) areas. Though the percentage of land utilisation favours rice production, it yields the lowest biomass among the

TABLE 14.1
Biomass Availability from Three Major Crops in the Philippines

Agricultural Crops	Area Planted (10^6 ha)	Biomass	Annual Yield (tonnes/ha)	Total Yield (tonnes '000)
Rice	4.2	Rice hull	0.95	4,000
		Rice straw	1–1.19	4,200–5,000
Total combined yield			***1.95–2.14***	***8,200–9,000***
Coconut	3.3–3.6	Coconut husk	1.24–1.94	4,100–7,000
		Coconut shell	0.5–0.54	1,800
		Coconut fronds	1.92–2.1	6,900
Total combined yield			***3.66–4.58***	***12,800–15,700***
Sugarcane	0.38–0.42	Cane bagasse	15.2–15.8	6,000–6,400
		Cane trash	14.3–15.8	6,000
Total combined yield			***29.5–31.6***	***12,000–12,400***

Sources: Zubiri (2016), Shead (2017), and Statista (2019).

TABLE 14.2
Electrical Generation and Power Potential of Agricultural Wastes[a]

Agricultural Wastes	Volume (Tonnes)	Electrical Generation[b] (kWh/kg)	Power Potential[c] (MWe)
Plant biomass			
Rice husk	3,767,824	0.627	308
Rice straw	8,813,623	0.774	888
Coco husk	4,143,376	1.398	754
Coco shell	1,970,074	1.758	451
Sugarcane bagasse	6,163,085	0.316	254
Sugarcane trash	1,118,527	0.545	79
Animal biomass			
Chicken manure	46,430,466	0.240	1,451
Pig manure	468,648,600	0.030	1,831

[a] Data cited by Caparino (2018).

[b] Data from energy Efficiency and Power Generation in the Philippine Agro-Industries by Full Advantage Co. Ltd. for the International Finance Corporation; Biomass resource assessment in the Philippines by Philippine Association of Renewable Energy Centers (PAREC) for GEF=UNDP-DOE-CBRED Project.

[c] Based on a 320-day annual operation.

three (rice 2.14 tonne/ha versus coconut 4.58 tonne/ha versus sugarcane 31.6 tonne/ha). Even if the sugarcane production area is ten times smaller than rice, and 8.57 smaller than coconut, sugarcane still yields the highest biomass per hectare (29.5–31.6 tonne/ha versus rice 1.95–2.14 tonne/ha versus coconut 3.66–4.58 tonne/ha). Among the seven identified biomasses, cane bagasse yields the highest volume per hectare at 15.2–15.8 tonnes, followed by cane trash (14.3–15.8 tonne/ha) and coconut fronds (1.92–2.1 tonne/ha). The lowest yield is observed for rice hull at 0.95 tonne/ha. The total combined annual volume biomass ranges from 33–37 million tonnes.

Table 14.2 presents the electrical generation and power potential of agricultural wastes (as cited by Caparino, 2018). The electrical generation was computed based on averages from different sources, which is dependent on the biomass heating value and moisture content, and the electric efficiency of certain technologies. Power potential was based on a 320-day annual operation of the biomass plant. Data shows that on a given volume, agricultural wastes from coconut yield a higher electrical generation rate per kilogram than that of rice or sugarcane. Among all the plant biomass, coco shell yields the highest electrical generation at 1.758 kWh/kg followed by coco husk at 1.398 kWh/kg, with sugarcane bagasse being the lowest at 0.316 kWh/kg. Comparing plant biomass against animal biomass, data shows that plant biomass has a higher electricity generation rate per kilogram than animal biomass. It is interesting to note that the lowest recorded data from plant biomass is still greater than animal biomass's highest electricity generator (sugarcane 0.316 kWh/kg versus chicken manure 0.240 kWh/kg). Recorded power potential data is greater in animal biomass because of the higher initial volume of the waste. This may show that plant biomass is better than animal biomass in terms of electrical generation on a per kilogram basis.

Table 14.3 summarises the cost of operation for rice hull and bagasse biomass plants. Data was taken from the study of the Society for the Advancement of Technology Management in the Philippines in 2001. Table 14.3 shows setups for biomass from rice hull and bagasse and a biodigester on various capacities. Investment-wise, using bagasse costs a lot compared to rice hull, albeit the figures are based on various capacities, thus the best indicator to look at is the levelised cost per kilowatt-hour to determine how agricultural wastes fare compared to one another. Comparing

TABLE 14.3
Economics of Biomass Energy (in Pesos)

	Capacity (kW)	Investment Cost	Investment Cost per kW	Annual Costs	Life-cycle Cost	Levelised Cost per kWh
Rice Hull Biomass						
Rice hull 1	3	311,000,000	103,667	5,107,300	584,229	2.98
Rice hull 2	6	463,500,000	82,750	12,061,812	961,893	2.45
Bagasse Biomass						
Bagasse 1	220	11,475,000,000	52,159	765,966,153	26,255,603,872	1.82
Bagasse 2	94	4,954,000,000	52,702	446,365,292	12,014,324,022	1.95
Biogas Digester						
Commercial	18	865,000	48,056	251,500	2,562,826	5.22
Elevated tank	6	300,000	50,000	86,500	904,122	5.53

Source: SATMP.

the two aforementioned plant biomasses, the levelised cost per kilowatt-hour is lower in bagasse than rice hull (bagasse ₱1.82 kWh, ₱1.95 kWh versus rice hull ₱2.98 kWh, ₱2.45 kWh). Equating the result of plant biomass to a biodigester, which is mainly fuelled by animal biomass, the cost per kilowatt-hour of a biodigester is almost double compared to rice hull and almost triple bagasse. Unfortunately, there is no available data for coconut biomass–fuelled plant.

14.4 CONCLUSION

Although tonnes of agricultural waste are generated in the country, particularly from rice, coconut, and sugarcane, the Philippine government was not able to maximise the production of biomass from these wastes as a source of renewable energy to its fullest potential.

While sugarcane obtained the highest biomass compared with rice and coconut, it was not given utmost research and development attention by the government and researchers. Clearly, there is a paucity of studies done on sugarcane.

Coconut waste, on the other hand, obtained the highest electrical generation rate compared to rice and sugarcane. However, despite its electrical generation rate, coconut waste was made into handicrafts, construction material, tiles, mats, packaging material, and others. Such action is attributed to the lack of awareness on the potential of coconut waste in generating electricity.

Overall, there is a need for further research and development on the potential of sugarcane waste and coconut waste as sources of biomass production in terms of electrical generation from waste.

REFERENCES

Abdul, Z., & Hashim, H. (2013). State of the Art of Biomass Combustion. Retrieved from https://www.tandfonline.com/doi/abs/10.1080/00908319808970051?journalCode=sol9

Acda, A., & De Vera, J. (2014). A Review on Electricity Generation Based on Biomass Residue in the Philippines. Retrieved from https://www.sciencedirect.com/science/article/abs/pii/S1364032112004248

Alawi, S., & Mana, S. (2012). Renewable and Sustainable Energy. Retrieved from https://www.sciencedirect.com/science/article/abs/pii/S1364032114005656

Arellano, G., & Kato, Y. (2016). Evaluation of Fuel Properties of Charcoal Briquettes Derived From Combinations of Coconut Shell, Corn Cob and Sugarcane Bagasse. Retrieved from https://pdfs.semanticscholar.org/6e04/d8d1b18ae89ba53b03d2ef51f10dc77c1d7b.pdf

ASEAN Briefing. (2017). Biomass industry in the Philippines. Retrieved from https://www.aseanbriefing.com/news/biomass-industry-philippines/

Cajes, A. (2014). Philippine Waste Agricultural Biomass: Prospects and Opportunities. Retrieved from https://issuu.com/alancajes/docs/philippine_waste_agricultural_bioma

Capareda, S. C. (2012). State of the Art Assessment of Biomass Waste Resources in the Philippines. Philippines Univ. Los Baños, College, Laguna, Philippines. Institution of Agricultural Engineering. Retrieved from https://agris.fao.org/agris-search/search.do?recordID=PH2002000843

Caparino, O. A. (2018). Status of Agricultural Waste and Utilization in the Philippines. Paper presentation 2018 International Forum on Sustainable Application of Waste-to-Energy in Asia Region, February 22-23, 2018 Novotel Ambassador Hotel Busan, Korea.

Chiu, C. (2019). Philippine Biomass Utilization. Retrieved from http://gec.jp/gec/en/Activities/FY2009/ietc/wab/wab_day2-2.pdf

De Mesa, R., & Soriano, A. (2019). Characterization of Torrefiel Biomass from Sugarcane (Saccharum Officinarum) Bagasse Blended with Semirara Coal. Retrieved from https://www.e3s-conferences.org/articles/e3sconf/abs/2019/46/e3sconf_cgeee2019_02002/e3sconf_cgeee2019_02002.html

Diaz, P., & Goluoke, I. (2017). The Use of Global Positioning System (GPS) to Estimate Point Sources of Selected Biomass Residues in the Philippines. Retrieved from http://citeseerx.ist.psu.edu/viewdoc/summary?doi=10.1.1.592.5293

Elauria, J. C., & Castro, M. (2010). Assessment of Sustainable Energy Potential of Non-plantation Biomass Resources in the Philippines. Biomass and Bioenergy, *29*(3), 191–198. Retrieved from https://doi.org/10.1016/j.biombioe.2005.03.007

Energy Information Administration (EIA). (1998). Biomass explained. Retrieved from https://www.eia.gov/energyexplained/biomass/

Fernandez, A., & Costa, C. (2010). Strategies for Enhancing Biomass Energy Utilization in the Philippines. Retrieved from https://www.nrel.gov/docs/fy02osti/30813.pdf

Fung, W. (2013). Biogas Production from Mixtures of Animal Manure and Fresh Biomass with and without Glucose Addition. Retrieved from https://www.researchgate.net/publication/277327496_Biogas_production_from_mixtures_of_animal_manure_and_fresh_biomass_with_and_without_glucose_addition

Gallego, A., Aberilla, J., & Adizas, A. (2019). Environmental Sustainability of Small-Scale Biomass Power Technologies for Agricultural Communities in Developing Countries. Renewable Energy, *141*, October 2019, 493–506. Retrieved from https://doi.org/10.1016/j.renene.2019.04.036

Go, A. W., Conag, A. T., Igdon, R. M. B., Toledo, A. S., & Malila, J. S. (2019). Potentials of Agricultural and Agro-Industrial Crop Residues for the Displacement of Fossil Fuels: A Philippine Context. Energy Strategy Reviews, *23*, 100–113. Retrieved from https://www.sciencedirect.com/science/article/pii/S2211467X18301226

Grafilo, L. A. D. R., Gunay, C. J. C., & Malayao, V. T. (2018). Renewable Energy Resources in the Philippines: Biomass. Retrieved from https://www.researchgate.net/publication/329191446_Renewable_Energy_Resources_in_the_Philippines_Biomass

Hall, D. O. (1991). Biomass Energy. Energy Policy, *19*(8), 711–737. Retrieved from https://www.sciencedirect.com/science/article/abs/pii/030142159190042M

Hidon, H., & Enriquez, M. (2014). Biomass Energy Projects: *Planning and Management*. Retrieved from https://www.osti.gov/biblio/6328586

Krista, C., Aviso, D., & Eusebio, F. (2016). Biomass Power Development in the Philippines. Retrieved from https://elibrary.asabe.org/abstract.asp?aid=15059

Lim, J. S. (2010). A Review on Utilization of Biomass from Rice Industry as a Source of Renewable Energy. Renewable and Sustainable Energy Reviews, *16*(5), 3084–3094. Retrieved from https://www.sciencedirect.com/science/article/abs/pii/S1364032112001451

Malayao, C., Grafilo, T., & Gunay, B. (2018). A Survey of Waste Management Practices of Selected Swine and Poultry Farms in Laguna, Philippines. Journal of Environmental Science and Management, *13*(2), 44–52 ISSN 0119-1144. Retrieved from https://www.researchgate_A_Survey_of_Waste_Management_Practices_of_Selected_Swine_and_Poultry_Farms_in_Laguna_Philippines/links/0fcfd5101db52cc502000000/A-Survey-of-Waste-Management-Practices-of-Selected-Swine-and-Poultry-Farms-in-Laguna-Philippines.pdf

Martinez, C. (2019). Biomass Resource Facilities and Biomass Conversion Processing for Fuels and Chemicals. Retrieved from https://www.sciencedirect.com/science/article/abs/pii/S0196890400001370

Maruyama, T. (2015). Production from Lignocellulose Agricultural Crop Wastes: *A review in Context to Second Generation of Biofuel Production.* Retrieved from https://www.sciencedirect.com/science/article/abs/pii/S1364032111005818

Milla, M., River, D., & Huang, B. (2013). Biomass and Organic Waste Conversion to Food, Feed, Fuel, Fertilizer, Energy and Commodity Products. Retrieved from https://www.eolss.net/sample-chapters/C17/E6-58-09-04.pdf

Pestaño, W. (2016). Utilization of Waste Coconut Coir Dust as a Source of Fuel. Retrieved form https://www.osti.gov/etdeweb/biblio/6814432

Pimentel, K., Moran, G., & Weber, P. (2016). Wood Energy and Livelihood Patterns: A Case Study from the Philippines. Retrieved from http://www.fao.org/tempref/docrep/fao/005/y4450e/y4450e03.pdf

Rao, S. (2016). Advances in Agricultural Microbiology. Studies in Agricultural & Food Sciences. Retrieved from https://books.google.com.ph/books?hl=en&lr=&id=ho2GDAAAQBAJ&oi=fnd&pg=PP1&dq=State+of+the+Art+on+Renewable+Energy+for+Biomass+from+Agricultural+Wastes+in+the+Philippines&ots=9VJXUtbmax&sig=T7Mx7uc6zs0SzDmRZnx7Chtwp9I&redir_esc=y#v=onepage&q&f=false

Samson, H., & Zafar, D. (2020). Animal Manure as Bioenergy Resources. Retrieved from https://www.bioenergyconsult.com/tag/animal-manure/

Shead, B. (2017). Biomass Industry in the Philippines. ASEAN Briefing 2017. Accessed at https://www.aseanbriefing.com/news/biomass-industry-philippines/#:~:text=The%20majority%20of%20coconut%20waste,estimated%20at%206m%20tons%20annually

Statista. (2019). Philippines: production volume of sugarcane 2018. Statisa.com

Tamayo, L. (2018). Air Pollutant Emissions from Rice Straw Open Field Burning in India, Thailand and the Philippines. Retrieved from https://www.sciencedirect.com/science/article/abs/pii/S0269749109000189

Tan, R., & Culaba, K. (2014). Biomass and Bioenergy: An overview of the Development Potential in the Philippines. Retrieved from https://doi.org/10.1016/j.rser.2014.05.111

Zafar, S. (2020). Agricultural Wastes in the Philippines. BioEnergy Consult Powering a Greener Future. Retrieved from https://www.bioenergyconsult.com/agricultural-resources-in-philippines/#:~:text=The%20most%20common%20agricultural%20wastes,straw%20and%20hulls%20are%20generated

Zubiri, J. M. F. (2016). Promoting agricultural development and inclusive growth through biomass energy development. Speech from 3rd Energy Smart Philippines. Pasay City: SMX Convention Center, SM Mall of Asia.

15 Utilisation of Food Processing Waste for Energy Generation

Jyothilakshmi R., Sumangala Patil,
Hemanth Kumar K. J., Sandhya Jayakumar,
Sadhan Kumar Ghosh, and K. S. Badarinarayan

15.1 INTRODUCTION

Increased demand for renewable energy sources due to scarcity of conventional energy sources has paved the way for innovative energy conversion technologies. Coconut is the major crop in some countries, and coconut industries play an important role in the economy of these countries(Kumar, Senanayake, Visvanathan, & Basu, 2003; Rasyid, 1992). In such countries, wastewater from the coconut industry also creates environmental pollution problems. Coconut-industry wastewater has a high concentration of organic substances, which causes environmental pollution due to high biological oxygen demand (BOD) and chemical oxygen demand (COD). Wastewater from coconut industries have high concentrations of lipids, containing medium-chain saturated fatty acids (up to 55%), long-chain saturated fatty acids (up to 35%), and long-chain unsaturated fatty acids (up to 10%). Desiccated coconut wastewater treated normally consists of a pH of 4.0–5.5, BOD of 1,000–5,000 mg/l, COD of 4,000–8,000 mg/l, and oil and grease at 4,000 mg/l (Kumar et al., 2003). Figure 15.1 shows the flowchart related to the processes involved in the desiccated coconut production.

15.2 OBJECTIVES

1. Utilisation of desiccated coconut wastewater for biogas production
2. Installation of biogas power plant for generation of electricity
3. Cost analysis of biogas power plant

15.3 MATERIALS AND METHODS

Undesirable complex fatty acids are present in both suspension and free-floating forms in desiccated coconut wastewater. Paring and washing processes give rise to particulate kernel debris. This debris and antimicrobial antioxidants from coconut water with short- and long-chain fatty acids are also present. The organic load present in the desiccated coconut wastewater is given in Table 15.1. After this process, the effluents can be easily treated by anaerobic digestion, but anaerobic digestion also faces many hurdles due low pH and high fatty content in the produced wastewater (Salomon & Lora, 2009; Walla & Schneeberger, 2008; Yentekakis, Papadam, & Goula, 2008). Cleaning uncertain restraining materials and biologically treating the effluent stream is the solution to this problem. Various combinations of the filtration, sedimentation, coagulation, and flocculation processes remove the suspension particles. The adsorption, coagulation, and flocculation processes remove the lipids. By adding alkali and bicarbonate ions, acidic wastewater neutralises.

To achieve objectives and improve productivity of the overall anaerobic digestion process, the present research has been taken with central financial assistance from the Ministry of New and Renewable Energy under the public private partnership for installation of biogas plants and

DOI: 10.1201/9781003204435-18

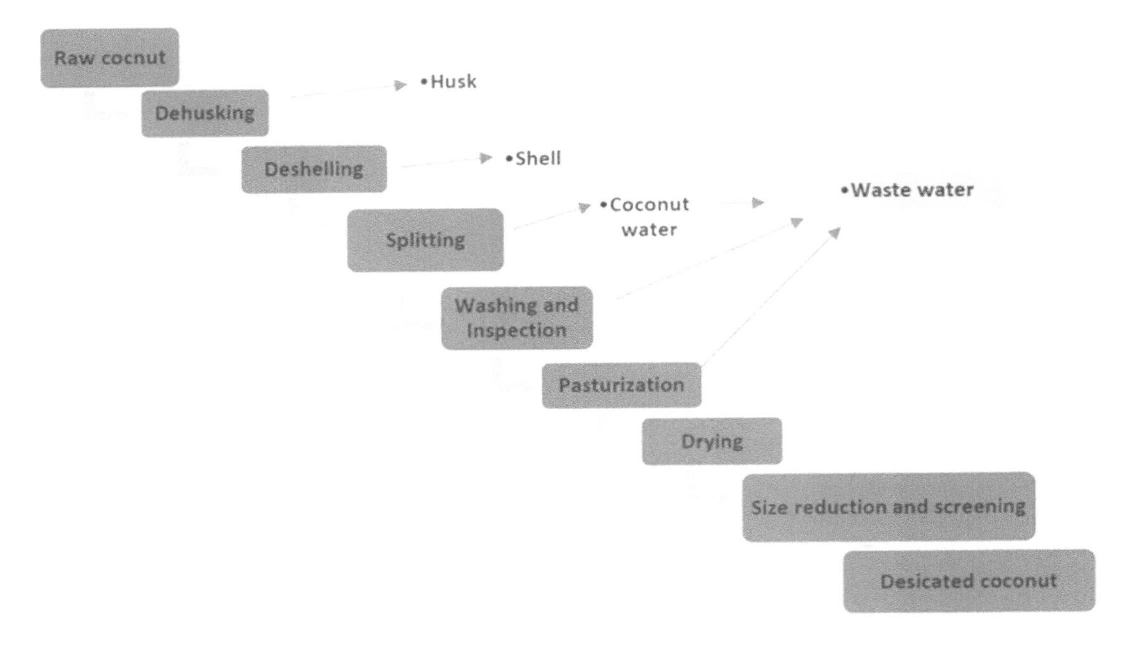

FIGURE 15.1 Wastewater generated in the desiccated coconut production process.

generation of electricity by utilising agro-industrial waste. The 300 m^3 biogas plant has been installed at Indus Bio Products, Puttur, Dakshina Kannada District for performance evaluation.

15.4 RESULTS AND DISCUSSION

The installed biomethanation plant has proved to be an effective source for power generation and biogas. The power thus generated can be utilised for various purposes. Medium- to large-scale biomethanation plants should be installed to control environmental pollution and the greenhouse effect.

TABLE 15.1
Fractionation of the Volumetric Discharge, Organic Load, and Energy Potential of the DC Industry Wastewater

	Small-Scale NMDC	Large-Scale NMDC
Coconut usage/day/industry (in thousands)	10–20	40–50
Coconut kernel water (m^3/day)	1–5	4–12.5
Wash water (m^3/day)	5–20	20–50
Virgin wastewater (m^3/day)	0	0
Total wastewater (m^3/day)	6–25	24–62.5
Total wastewater organic loading rate (kg COD/day)	72.3–340.2	289.2–850.5
Biogas potential (m^3/day)[a]	36.1–170.1	144.4–425.2
Energy generation potential (mJ/day)[a]	794.2–3742.2	3176.8–9354.4
Fuel wood replacement (kg/day)	218.4–1029.1	873.6–2572.4

Source: Bhattacharya & Salam (2002) and Bond & Templeton (2011).

[a] Assumed woodstove efficiency – 10%, biogas stove efficiency – 55%, calorific value of wood – 20 mJ/kg, calorific value of biogas – 22 mJ/m^3.

TABLE 15.2
Basic Information Related to Electricity Generation by the Utilisation of Desiccated Coconut Wastewater

Basic Information	
Process technology	UASB Technology
Design capacity	300 m^3 biogas/day
Segregated organic industrial waste	4–5 tons/day
Electricity generation	350 power units/day 30 kW generator runs for 8–10 h
Energy yield	Equivalent to 160–175 kg LPG/day
Biogas utilisation	Power generation
Organic manure	1.5 tons/day (1.5 tons/day organic manure × Rs. 2000/- = Rs. 3000/-)
Electricity consumption	10–15 electrical units/day
Total project cost	Rs. 38 lakhs
Payback period	2.25 years

Basic information regarding the biogas plant is presented in Table 15.2. Information on the economics of the biogas plant can be found in Table 15.3.

- About 350 electrical units per day are generated by the installed biomethanation plant. A 30 kW biogas generator runs for 8–10 h/day.
- The plant supplies power to the coconut processing unit of Indus Bio Products, Puttur, for running of water lifting from borewell, drying desiccated coconut, feeding unit, running chaff cutter machine, processing machine, and captive use.
- Treatment of 3,240 tonnes of agro-processing waste during the last 36 months generated a total of 3,45,600 units of electrical power.
- The biogas produced is effective and reduces harmful greenhouse gas emissions.
- The biomethanation plant's compact design utilises less floor area and produces biogas.
- Odourless biogas is produced in closed digestor by treatment of agro-industrial waste. Nutrient-rich bio-manure is obtained by recycling the organic matter.

TABLE 15.3
Cost Economics of Biogas Plant and Revenue

	1 Year	3 Years
A) Revenue generation		
Power Cost @ Rs. 7.00 kWh (7 × 350 × 30 days)	882,000.00	2646,000.00
Organic manure cost @ Rs. 2000/tons (1.5 t × 2000 × 30 days)	1,080,000.00	3,240,000.00
Total income	1,962,000.00	5,886,000.00
B) Cost of operation and maintenance		
Organic manure cost @ Rs. 0.50/kg (0.50 × 3500 kg × 30 days × 12 months)	630,000.00	1890,000.00
Cost of labour @ Rs. 6000/month	72,000.00	216,000.00
C) Auxiliary consumption	18,000.00	54,000.00
Total	720,000.00	2,160,000.00
Payback period	2.25 years	

15.5 CONCLUSION

Desiccated coconut wastewater from the industries contains high BOD and COD due to the presence of high organic compounds. After washing the coconut kernel, the particulate solid organic matter is available in higher concentrations in desiccated wastewater. Volatile fatty acids available predominantly in wastewater inhibit the anaerobic bacteria during the combined treatment of desiccated coconut-industry wastewater along with particulate solids. The preliminary treatment helps to remove higher fatty acids content and neutralise wastewater.

REFERENCES

Bhattacharya, S., & Salam, P. A. (2002). Low greenhouse gas biomass options for cooking in the developing countries. *Biomass and Bioenergy*, *22*(4), 305–317.

Bond, T., & Templeton, M. R. (2011). History and future of domestic biogas plants in the developing world. *Energy for Sustainable Development*, *15*(4), 347–354.

Kumar, S., Senanayake, G., Visvanathan, C., & Basu, B. (2003). Desiccated coconut industry of Sri Lanka: opportunities for energy efficiency and environmental protection. *Energy Conversion and Management*, *44*(13), 2205–2215.

Rasyid, F. (1992). *Properties of protein extracted from desiccated coconut*. The Ohio State University.

Salomon, K. R., & Lora, E. E. S. (2009). Estimate of the electric energy generating potential for different sources of biogas in Brazil. *Biomass and Bioenergy*, *33*(9), 1101–1107.

Walla, C., & Schneeberger, W. (2008). The optimal size for biogas plants. *Biomass and Bioenergy*, *32*(6), 551–557.

Yentekakis, I., Papadam, T., & Goula, G. (2008). Electricity production from wastewater treatment via a novel biogas-SOFC aided process. *Solid State Ionics*, *179*(27–32), 1521–1525.

16 A Study of the Processes, Parameters, and Optimisation of Anaerobic Digestion for Food Waste

Jyothilakshmi R., Sumangala Patil, Hemanth Kumar K. J., Sadhan Kumar Ghosh, and Sandhya Jayakumar

16.1 INTRODUCTION

Ancient indication from India designates the practice of biogas for heating bathing water, cooking food as early as the 21st century. The antiquity of biogas invention has humanity having acknowledged of the presence of biogas since the 17th century, and experimentation with the creation of definite biogas production structures and plants were in progress as early as the mid-19th century. Among the oldest biogas systems is the tank, which has been used for the handling of wastewater since the early 19th century and remains used for isolated properties where there are no sewerage systems. In this sort of plant, the biogas is utilised and not stored (Jorgensen, 2009).

The release of flammable gas has been observed within dumping sites due to the decomposition of carbon-based substances (Jorgensen, 2009). Trials have been conducted on flammable gas collected from wetland deposits, creating an immediate association between decaying organic matter and gas production. In 1808, methane (CH_4) was produced, which promoted the formation of flammable gas compost (Karlsson et al., 2014).

Biogas is a flammable blend of gases, consisting largely of CH_4 and carbon dioxide (CO_2), produced from the decaying of organic matters in anaerobic conditions due to the bacterial reactions in the digester. The respiration of decomposer microorganisms forms gases, and the amalgamation of the gases are the product of biomass decay. The CH_4 production is low if the substance contains primarily carbohydrates, like glucose and other simple sugars, and high-molecular compounds (polymers) like cellulose, hemicellulose, and lignin. CH_4 production is high when fat and protein content is high. CH_4 is the third most vital greenhouse gas emission after water vapour and CO_2, with a degree within the troposphere of 1.8 ppm. "Its present concentration is 2.5 times above than the 0.7 ppm detected in ice cores dated to the period of AD 1000–1750" (Van Amstel, 2012). "Methane is an odorless and colorless gas with a boiling point of -162°C and it burns with a blue flame. Methane is also the main constituent (77-90%) of natural gas. Chemically, methane belongs to the alkanes and is the simplest possible form. At normal temperature and pressure, methane has a density of approximately 0.75 kg/m^3. Due to carbon dioxide being somewhat heavier, biogas has a slightly higher density of 1.15 kg/m^3. Pure methane has an upper calorific value of 39.8 MJ/m^3, which corresponds to 11.06 kWh/m^3. If biogas is mixed with 10-20% air and whatever additional hydrogen makes up the combustible part of biogas, which produce flammable air/fuel" (Karlsson et al., 2014).

Existing waste management practices in India cause substantial methane release. Per the 2016 Central Pollution Control Board (CPCB) manual, landfills, wastewater treatment, and animal waste treatment are the main causes of methane emissions, responsible for approximately 45% of methane emissions. The benefit of anaerobic digestion (AD) is not only the decrease in greenhouse gas emissions, it can also produce renewable energy and redirect organic municipal solid wastes from

DOI: 10.1201/9781003204435-19

landfills (Costa et al., 2015). AD of organic fraction municipal solid waste (OFMSW) to produce biogas has systematic method for treatment of waste for the production of methane as a power resource. This is a trend that has arisen not only in India, but globally; in the past 15 years, the economic benefits of biogas have been progressing to the point of currently including aspects for commercial profit. Currently operating biogas plants have initiated efforts to increase methane production, with substantial research and development to enhance AD being carried out within the academic world and the industrial sector, not the smallest amount of which comes from the Biogas Research Center. The main aspects influencing the economy of a biogas plant are related to feedstock supply, price, wanted shipments, digestibility, and the requirements of pre-treatment. The combination of the feedstock affects the procedure plan for digester design considerations, including conceivable organic loading rate (OLR), hydraulic retention time (HRT), total solids (TS) and temperature.

The determination of the hostel mess biogas project is to instigate a small-scale anaerobic digester for an educational institute to digest food waste, which has collected from the hostel kitchen and retrieve biogas to satisfy heat energy requirements in the same hostel kitchen. The food waste degradation is going to be augmented via AD in the current plant, a food waste digester holding the specifications to digest 100 kg of waste per day and produce 15 m^3 of biogas by using following accessories: a food waste crusher, mixing pit, a floating drum for the storage of gas, an aerobic airtight chamber, an outlet pit, a biogas booster pump, a biogas analyser, and a combustion flow meter. The produced biogas is supplied to the gas line via a scrubber and booster pump, and accumulated sludge will be further processed by aerobic composting after production of biogas is decreased. Figure 16.1 shows the important aspects of the plant as far as the production of quality biogas for energy generation. Double ended arrows show cyclic process which has two-way operation, single-ended arrows indicate one-way process.

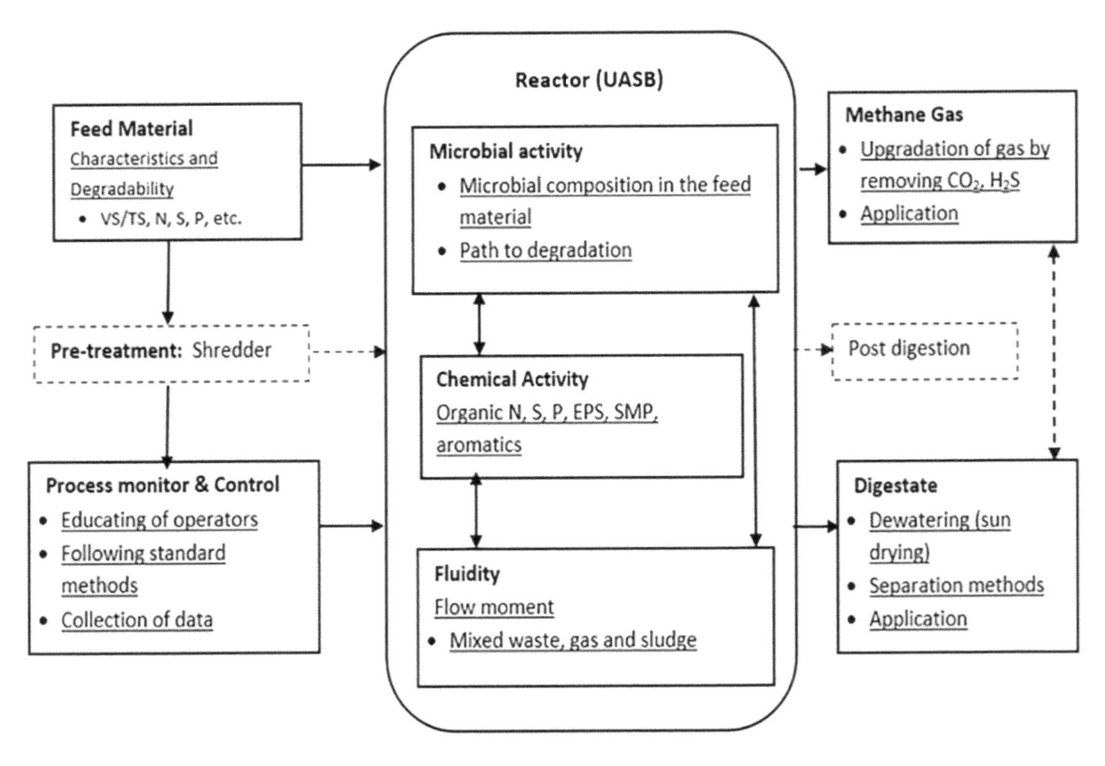

FIGURE 16.1 Identified important areas of this research which affects the plant performance in production of quality methane gas.

As per the publication by the MoHUA "Swachhata Sandesh Newsletter" as of January 2020, 147,613 metric tons of solid waste is generated per day in India (Singh, 2020). At present, the current global waste volume is 64–72 billion tons which is predictable to rise to 125 billion tons by 2031 (Song & Wu, 2021).

Whereas land in Indian cities was initially allotted for emerging landfills for harmless discarding of debris, unprocessed waste is dumped for months and years at landfills and is rising in volume and changing its composition, which is a mix of organic or wet waste, animal waste, water waste, sewage waste, and inorganic waste (Ahluwalia & Patel, 2018).

Biomethanation provides a more progressive process and infrastructure that supports the biochemical alteration of decomposable waste. Though hit is not the simplest process to decrease greenhouse gas discharges, AD can also produce renewable power and divert biodegradable waste from dumpsites to biogas plants (Costa et al., 2015). Nevertheless, these biomethanation plants face several communal and financial problem that stop their complete capability from being leveraged, even though marketplace circumstances are becoming increasingly favourable for the implementation of AD (Voegeli & Zurbrügg, 2008).

This chapter seeks to issue a thorough outline of the biomethanation process and summaries numerous selected subjects regarding biomethanation. Section 16.2 delivers an outline of the four phases of AD and a brief discussion of anaerobic biochemistry. Section 16.3 provides various operating parameters of AD. Section 16.4 gives an outline of numerous pretreatments that may be used to increase system performance and decrease digestion times. Features of this amassed understanding may be carried out to broaden moderate single-level batch digesters. Figure 16.2 provides a schematic illustration of one of these digesters.

16.2 ANAEROBIC DIGESTION STAGES

AD occurs over four successive phases: hydrolysis, acidogenesis, acetogenesis, and methanogenesis. The AD process relies on the interactions of various microorganisms, which are capable of performing these four phases. In single-degree batch reactors, all wastes are loaded immediately, and all four phases occur serially within the reactor. The compost is released after a certain period of storage or the cessation of biogas production (Verma, 2002). Figure 16.3 provides a visual guide to the four digestions phases.

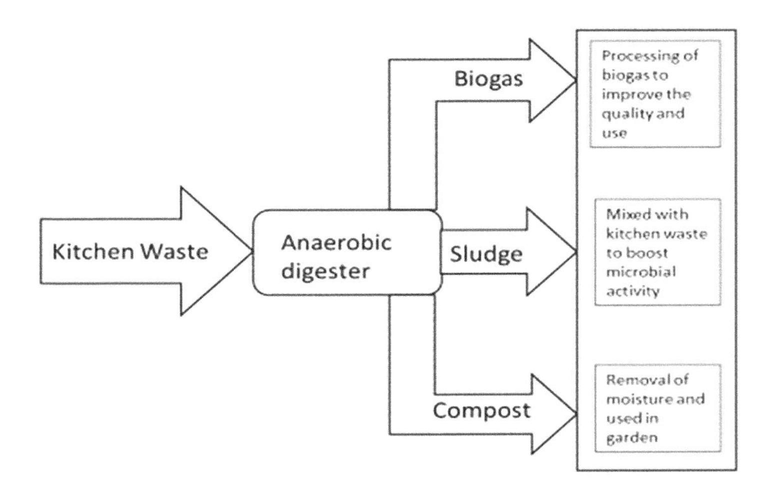

FIGURE 16.2 Anaerobic digestion of canteen waste and processing.

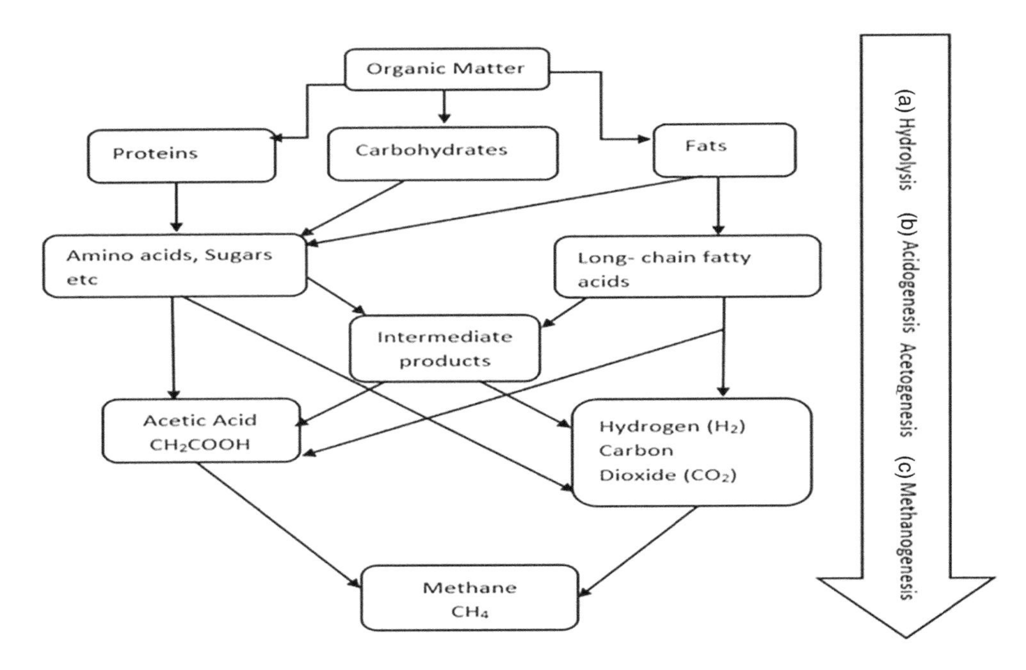

FIGURE 16.3 Organic matter decomposition at anaerobic condition. Contain mainly three phases of anaerobic digestion (a) hydrolysis, (b) acidogenesis, acetogenesis, and (c) methanogenesis.

16.2.1 Hydrolysis

Matter that contains high contents of protein and fat is frequently discarded from the food trade, such as restaurants, slaughterhouses, and food-processing industries. This waste is ideal matter for the generation of biogas, as the high alkalinity makes it capable of producing up to 500–600 dm^3/kg VS (Salminen & Rintala, 2002).

Barlaz, Schaefer, & Ham (1989) stated that "community trash comprises 40–50% of cellulose in the form of plant and paper material". It has been stated that the reaction of cellulose can advance the rate-limiting stage in complete degradation (Lin, Patterson, & Ladisch, 1985).

With regard to the pH and volatile fatty acid (VFA) ranges deliberated, the reaction rate of biowaste relies on pH. Although the hydrolysis rate does not rely on VFA, it might be discovered between pH of 5 and 7 and VFA concentrations up to 30 g COD/L. These findings imply that the pH is that the primary method variable in dominant the hydrolysis rate of the anaerobic solid-state fermentation process, not the VFA concentration (Veeken et al., 2000).

The response rate can be adjusted according to the pH, temperature, and particle size (Li, Park, & Zhu, 2011). At the studied pH and VFA concentrations, the hydrolysis of biological waste was the dependent on pH value. However, the hydrolysis value of VFA between pH 5 and 7 was up to 30 g of COD/L. These findings imply that the pH is the primary variable dominant in the hydrolysis rate of the anaerobic solid-state fermentation process, not the VFA concentration (Veeken et al., 2000). Hydrolysis has an optimum temperature between 30 and 50°C (Van Lier, 1995).

The outcome of hydrolysis aids the process of biogas production by converting organic substances into their reduced elements, which can be used for the bacterial activity in acidogenic phase. Though hydrolysis is an organic or biotic process, it can also be seen as an electrochemical process in AD. "In this hydrolysis process, hydrolytic bacteria are up to excrete exoenzyme that can undergo metamorphosis lipids, carbohydrates and proteins into sugars, long-chain fatty acids(LCFAs), and amino acids, respectively. Subsequently enzymatic breakdown, the outcomes of hydrolysis are up to scatter along the cell membranes of acidogenic microbes" (Van Lier, 1995).

TABLE 16.1
Reactions in Acidogenesis Phase (Arsova, 2010)

Product	Reaction
Ethanol	$C_6H_{12}O_6 \leftrightarrow 2CH_3CH_2OH + 2CO_2$
Propionate	$C_6H_{12}O_6 \leftrightarrow 2CH_3CH_2COOH + 2H_2O$
Acetic acid	$C_6H_{12}O_6 \leftrightarrow 3CH_3COOH$

It is significant to note that specific substances, for instance hemicellulose, cellulose, and lignin, can be difficult to decompose, and they may be unreachable to microorganisms because of the compound forms. To augment the hydrolysis of certain carbohydrates, enzymes are regularly added (Lin et al., 2010). The previous study has confirmed that methanogenesis phase occurs as a rate-determining phase, reliant on the magnitude relation of hydrolytic to methanogenic microorganisms (Luo et al., 2012; Ma et al., 2013).

16.2.2 Acidogenesis

At the stage of acidogenesis, different anaerobic microbes like *lactobacillus* sp. and *Propionibacterium* sp. contribute to break down soluble molecules processed in the hydrolysis phase, long-chain compounds within short intermediate VFAs, and are able to produce additional products like alcohol, H_2S, H_2, and CO_2. These are typical types of acetate-forming fermentative microbials, such as *Acetobacterium*, *Clostridium*, *Eubacterium*, and *Sporomusa* (Amani, Nosrati, & Sreekrishnan, 2010; Li et al., 2011; Kang & Yuan, 2017). The reaction of acidogenesis is shown in Table 16.1. In AD, acidogenesis is a fastest process, as these bacteria produces ten times faster than methanogenic archaebacterium (Saraswat et al., 2019).

As this phase progresses so quickly, the production of VFA generates predecessors for the concluding phase of methanogenesis, and extensive VFA acidification may cause a digester to fail(Akuzawa et al., 2011). A slightly comparable anaerobic method is Bokashi composting, a composting method in which food wastes and a microbic activator are situated to ferment anaerobically. This produces an extremely acidic liquid as the finished product, which is used as a liquid plant food and dry fertiliser for improving soil quality (Yamada & Xu, 2001).

16.2.3 Acetogenesis

In the third AD phase, acetogenesis, higher VFA need to be accessible to methanogenic bacteria. Acetogenesis is the phase when acetogenic microbes convert higher VFAs and alcohols to acetate, in which H_2 and CO_2 are additionally produced (Amani, Nosrati, & Sreekrishnan, 2010; Hansen & Cheong, 2019). Dissimilar families are recognised as acetogens. *Syntrophobacter wolinii* and *Smithella propionica* are recognised microbials that produce acetate by overwhelming butyrate and propionate, respectively. Other species equivalent to *Syntrophobacter fumaroxidans*, *Syntrophomonas wolfei*, *Pelotomaculum thermopropionicum*, and *Pelotomaculum schinkii* have been known as acetogenic bacterium that convert VFAs, H_2, and CO_2. *Clostridium aceticum* may be another familiar bacterium which produces acetate from H_2 and CO_2 (Amani, Nosrati, & Sreekrishnan, 2010).

The hydrogen produced through acetogenesis proposes the concept of a stimulating syntrophic connection that exists within AD, hydrogen interspecific transmission. Whereas acetogenesis is a creator of hydrogen, a maximum partial pressure demonstrates its presence to be harmful to acetogenic bacteria (Kang & Yuan, 2017). At the same time, lipids bear an alternative route of acetogenesis, associated acidogenesis, and β-oxidation, wherever acidogenesis generates glycerol from acetate and β-oxidation produces acetate from long-chain fatty acids(Li, Park, & Zhu, 2011).

Because of the existence of hydrogenotrophic methanogens, hydrogen is ready to be spent apace, upholding partial pressures of hydrogen to acetogenesis by generating an exergonic reaction (Stams & Plugge, 2009).

16.2.4 Methanogenesis

The last stage of AD is methanogenesis, where methane is produced when the products of acetogenesis are consumed by methanogenic bacteria (Ferry, 2010). Methanogens take an essential part as the terminator of acetogenesis products created by acetate-forming microorganisms consuming acetic acid, H_2, and CO_2, transforming them into CH_4 and CO_2 (Luo et al., 2012). Methanogenic microbes signify a collection of anaerobic archaebacteria; as evidence of the methanogenic bacteria's sensitivity to O_2, it was found after ten hours of contact with O_2, 99% of *Methanococcus voltae* and *Methanococcus vannielii* cells had been eliminated (Kiener & Leisinger, 1983). With consideration to the atmosphere required for the methanogenesis process, methanogenic microorganisms need a better pH value compared to earlier phases of AD (Wolfe, 2011). In this methanogenesis phase, there are two important path ways to produce CH_4, processes called hydrogenotrophic methanogenesis and acetoclastic methanogenesis. Hydrogenotrophic methanogenesis can create CH_4 from H_2 and CO_2, with an H_2O by-product (Bapteste, Brochier, & Boucher, 2005), as shown in the following reaction.

$$CO_2 + 4H_2 \rightarrow CH_4 + 2H_2O \text{ (Christy, Gopinath, \& Divya, 2014)}$$

In acetoclastic methanogenesis, CH_4 gains electrons from the oxidation of carbon monoxide released by the breakdown of acetate (Bapteste, Brochier, & Boucher, 2005)

$$CH3COOH \rightarrow CH_4 + CO_2 \text{ (Christy, Gopinath, \& Divya, 2014)}$$

These four phases are conducted in a single reactor, an anaerobic digester is often named as single-phase anaerobic digester, as shown in Figure 16.4 (Christy, Gopinath, & Divya, 2014).

At the same time, methanogens seem to take a notably moderate reproduction period compared other microbes' activity in AD, up to and above 5–16 days (Deublein & Steinhauser, 2011).

FIGURE 16.4 Biomethanation plant (100 kg capacity) installed in MSRIT Hostel, funded by VGST, Govt. of Karnataka.

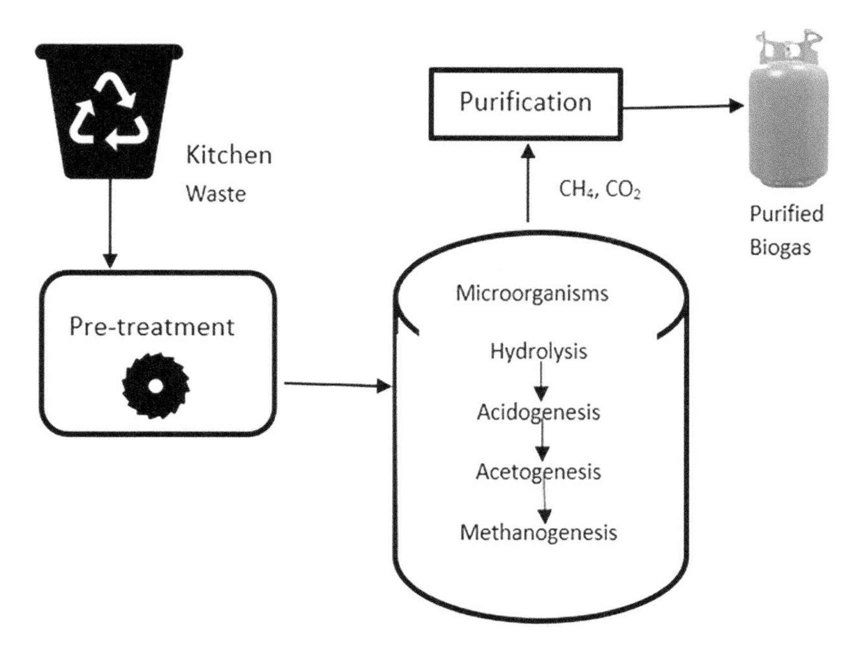

FIGURE 16.5 Schematic diagram of single-phase anaerobic digesterion (SPAD).

The conventional single-phase anaerobic digester (Figure 16.5) transforms pretreated organic waste into biogas, which is rich in methane. After purification, biogas can be utilised as biofuel (Christy, Gopinath, & Divya, 2014).

16.3 OPERATING FACTORS OF AD

The AD of organic waste is an intricate procedure and includes several different phases for degradation. For each degradation step, specific microorganisms participate with requirements of different environmental conditions for bacterial activities (Khalid et al., 2011). To establish the supreme yield of biogas, it is essential to give special considerations to recognised parameters.

The following subcategories outline the frequently utilised factors in quantifiable calculations of the AD method. Factors influencing AD include carbon to nitrogen (C:N) ratio, volatile solids, HRT, temperature, pH, moisture content, OLR, and TS (Noraini et al., 2017).

16.3.1 Temperature

Temperature is a critical operating parameter of AD because it is an important state for the durability and evolution of microorganisms. Problems with retaining the appropriate reaction temperature may lead to reduction in process competence and indirect interruption of the estimated reaction (Noraini et al., 2017). Temperature also governs the standards of the foremost dynamic variables for the progression and, hence, the evaluation of the bacteriological activity. Hypothetically, the amount of reaction will increase with the rise of the ambient temperature. There are two series of temperature that give optimal AD rates in microorganisms' growth rate, gas production rates, and substratum degradation rate. Hence, the generation of biogas will also increase in these temperature ranges (Kang & Yuan, 2017). There are three temperature ranges in AD: psychrophilic (0–15°C), mesophilic (15–45°C), and thermophilic (45–65°C) (Luo et al., 2012). Mesophilic and thermophilic AD are broadly adopted for the production of biogas. These procedures produce individual rewards

TABLE 16.2
Comparisons of Mesophilic and Thermophilic Digestion

Temperature Ranges	Advantages	Disadvantages	References
Mesophilic	System is steadier and easier in controlling of process. Methane production achieved in a longer operating time (40 days). Smaller energy expenses.	Lesser biogas generation. Lower degradation time.	Moset et al. (2015) Fernández-Rodríguez, Pérez, & Romero (2013)
Thermophilic	The values of the higher specific evolution rate of microorganisms. Higher organic substance degradation. Production of methane could be accomplished within a smaller operational period (20 days).	Larger energy expansion. Sensitive to atmospheric variations. Supplementary energy input is essential for reaching higher temperature.	Moset et al. (2015) Veeken et al. (2000)

and downsides, as itemised in Table 16.2. The selection of a temperature range is made based on the substance to be used (Noraini et al., 2017). Mesophilic digestion functions at a lower temperature; at this range of temperature the degradation process is moderate and produces a lesser amount of biogas. Nevertheless, mesophilic digesters remain popular due to their lower energy expenses for heating compared to thermophilic digesters (Moset et al., 2015).

Thermophilic digestion functions at a better temperature range; reaction rates are amplified due to upper matter degradation, adding to augmented biogas generation (Hartmann & Ahring, 2006; Van Lier., 1995). Some digesters are reliant on the atmospheric temperature, needing no heating system. These digesters repeatedly noticed periodic variations in the production of methane (Mao et al., 2015). It should be noted that individual methanogenic microorganisms have their own optimal temperature. Similarly, not all four phases have the same optimal temperature for best procedure in AD. For example, methanogenesis may work favourably at 37°C but the same temperature is not optimal for hydrolysis, acidogenesis, or acetogenesis, which have their own appropriate temperatures (Noraini et al., 2017).

16.3.2 Carbon to Nitrogen Ratio

The magnitude relation of carbon to nitrogen is called the C:N ratio. The C:N ratio influences the capacity of the production of the biogas (Saraswat et al., 2019). OFMSW typically includes protein, starch, and fat. In an AD environment, nitrogen is vital for microorganisms to cultivate and increase; most rich foundations of nitrogen in an anaerobic autoclave facilitate the breakdown of proteins (Chen, Cheng, & Creamer, 2012; Zupančič & Grilc, 2012). It is essential to preserve nitrogen concentration during the process so as not to impede the development of microorganisms. If the nitrogen quantity is restricted, microbic populations may require an extended period to decay the existing carbon (Igoni et al., 2008). Surplus nitrogen, on the other hand, slows down the AD process. It has been alleged that "microorganisms use carbon 25–30 times quicker than nitrogen, a C:N ratio of 20–30:1 was stated as the finest C:N ratio for anaerobic digestion" (Gerlach, Grieb, & Zerger, 2013). A small C:N ratio, or an excess amount of nitrogen, will cause ammonia to gather, which causes the pH to exceed 8.5 (Ostrem & Themelis, 2004).

In order to retain the nutrition and C:N ratios, co-digestion of dissimilar organic combinations can be implemented. A study observed anaerobic degradability of OFMSW and waste activated sludge in a single-stage anaerobic digester functioning at 35°C. The study stated that because the OFMSW percentage of the combination augmented from 10 to 90%, the C:N ratio of the blend enhanced from 6 to 15, and when degradation of the blenders augmented, there was a rise in the generation of methane (Heo, Park, & Kang, 2004).

In AD, nitrogen plays a vital role, as it balances the pH value in the digester, but excess nitrogen can cause difficulties in anaerobic biodegradation due to metabolic products like ammonia (Okonkwo et al., 2018)

16.3.3 pH Value

Another important parameter in AD is operating pH, due to the susceptibility of methanogenic microorganisms to acidic growth during the production of methane. Organic occupation changes the pH level through the various phases of AD (Arsova, 2010). The pH of the biodigester is contingent on CO_2 and VFA. The pH value is additionally lowered with the temperature of reaction phases. A study done by Yadvika et al. (2004) stated that if the pH value is higher than 5, the generation of methane is potentially higher than 75% (Saraswat et al., 2019). Generation of organic acids in the course the acetogenesis may cause the pH value to fall below 5, which is deadly for methanogens and can lead to biodigester failure (Arsova, 2010).

16.3.4 Hydraulic Retention Time

HRT is the time needed for the degradation of biodegradable substances within a reactor (Nijaguna, 2006), which is affected by chemical oxygen demand (COD) or biochemical oxygen demand of the feeder and the waste. HRT affects the complex matter contained in the biodigesters, in interaction with the influent and degraded into metabolic outcomes, such as monosaccharides, polysaccharides, and amino acids (Sreekrishnan, Kohli, & Rana, 2004).

On the other hand, HRT is connected to the loading rate, with a smaller HRT being proportional to a larger loading rate. For this reason, smaller HRTs are related to VFA acidification, which might cause suppression properties (Kim et al., 2013). However, shorter HRT permits augmented procedure competence and diminished principal budgets, even though longer HRTs are essential for the breakdown of lignocellulosic wastes (Shi et al., 2017). Nevertheless, for municipal wastes, the maximum profit is originated for bioreactors functioning on a low loading rate and high HRT (Hartmann & Ahring, 2006). The loading rate of a bioreactor is the quantity of organic waste supplied to an AD daily in continuous digesters. Overcharging of AD may cause let downs, in that feed is rapidly hydrolysed and acidified, therefore generating a higher inflation of VFA and constraining methanogenesis and creating barriers to the AD process(Franke-Whittle et al., 2014)

16.3.5 Total Solids

This the quantity of dry substance in a sludge, regardless of organic or inorganic properties, which contains the amalgamation of liquified matters and suspended matter (Saraswat et al., 2019). The TS content of raw material disturbs the operations of anaerobic digestion, and the modification of TS feedstock will affect the variation of bacteriological activities in methane production in AD. To increase the productivity of AD of OFMSW, it is essential to recognise the effect of the TS contents on the actions of the bacterial environment in the AD of degradation from wet to dry technology (Yi et al., 2014).

TS is a significant characteristic of bioreactor process. High TS in AD is a topic of current consideration, as it is essential for small-scale digesters and lower heating requirements (Yi et al., 2014).

16.4 PRETREATMENT

16.4.1 Mechanical Pretreatments

Reduction of particle size, also referred to as physical pretreatment, is a standard technique for rising the productivity of an AD procedure by decreasing the organic waste particle size. AD progresses in the organic reactions by increasing the surface area accessible to the bacteria and resulting in faster degradation (Sreekrishnan et al., 2004). Mechanical pretreatment crumbles compact assembly (Elliott & Mahmood, 2012). A study proposed that an even bigger particulate size results in less COD in digestion and a lesser amount of methane production; substrate size is reciprocally proportional to the speed of most substrate utilisation. There exist several kinds of mechanical pretreatment methods that may be suitable for individual investigates, like "sonication, lysis-centrifuge, liquid shear, collision, high-pressure homogeniser, maceration, and liquefaction" (Ariunbaatar et al., 2014).

Sonication pretreatment is produced by an automatically moving probe, which interrupts the cell structure. In ultrasonic pretreatment, the processed cavitation is ready to cause cell lysis, the particle size reduces at low frequency sound waves (20 kHz, 80 min) of sonification tends to produce the best generation of biogas.

Rotary drums are common for OFMSW pretreatment and are an alternative method for the productive segregation of OFMSW, which improved the production of biogas by 18–36% (Zhu et al., 2009).

A high-pressure homogeniser raises the compression by several hundred bars and standardises particles below robust depressurisation. The established cavitation encourages internal energy that disturbs the cell membranes (Ariunbaatar et al., 2014).

These pretreatment methods are not common for OFMSW. However, there are a lot of common treatments for different substrates, for example lignocellulosic materials, muck, and wastewater treatment plant (WWTP) sludge. The benefits of mechanical pretreatment are its application without producing smells, an uncomplicated execution, improved dehydration of the resultant anaerobic sediment, and an average power consumption. Drawbacks consist of no substantial outcome on pathogen exclusion and the probability of equipment blockage (Ariunbaatar et al., 2014).

16.4.2 Biological Pretreatments

Biological pretreatments might consist of aerobic and anaerobic preprocessing, along with the addition of explicit enzymes, for example peptidase or carbohydrolase, to the AD system, even though these treatments are generally not functional to OFMSW. This method has been practiced broadly on different forms of organic solid waste, similar to pulp and paper industries and WWTP sludge (Li et al., 2012).

There is a conventional biological treatment called aerobic pretreatment, including things like fertilising or micro-aeration, that oxygenises prior to AD. These vigorous techniques get a high rise in hydrolysis of compound substance because of the upper generation of hydrolytic enzymes, which is incorporated in the raised evolution of particular bacteria (Lim & Wang, 2013).

Few investigators considered the hydrolytic acidogenic phase of a two-stage AD procedure a biological pretreatment methodology (Carrère et al., 2010, Ge, Jensen, & Batstone, 2010; Ge, Jensen, & Batstone, 2011), and additional researchers think about it as a method arrangement of AD, not a pretreatment process (Carballa et al., 2007). Physical breakdown of acetogens from the methanogens can cause an upshot in the generation of a larger amount of methane and eliminate COD regulated at a smaller HRT, as to standard single-phase reactors (Hartmann & Ahring, 2006).

The benefits of such a process consist of amplified constancy with improved pH regulation, elevated loading rate, redoubled precise action of methanogens leading to the next alkane series yield, amplified volatile solids dropping, and high possibility of eliminating pathogens (Bouallagui et al., 2005) The drawbacks consist of hydrogen developing, ensuing in an obstruction of acidic microorganisms, elimination of potential mutually beneficial nutrients needs for the methane developing microorganisms, technical quality, and higher prices (Wang & Zhao, 2009; Ueno et al., 2007).

16.4.3 Chemical Pretreatments

Chemical pretreatment is the elimination of the organic compounds using robust acids, alkalis, or oxidants. AD normally needs a modification of the pH value by raising alkalinity, therefore alkaline pretreatment is the preferred chemical pretreatment process (Li et al., 2012). Chemical pretreatment involves lignocellulosic substrates being broken down into their individual monosaccharides (Ariunbaatar et al., 2014). Acidic pretreatment helps with the decaying of substrates reducing the time needed for breakdown, but the cost prevents it from being as economically worthwhile as using alkaline pretreatment (Couger, 2015).

Through alkaline pretreatment, the primary method where the best response takes place is saponification, which brings the increase of solids particles (Carlsson, Lagerkvist, & Morgan-Sagastume, 2012), and the substrates are readily available to anaerobic microorganisms (Hendriks & Zeeman, 2009; Torres & Lloréns, 2008). This process is appropriate for matter with high lignocellulosic content, as its breaks down the lignin, and because of the ability of hydrolytic microorganisms to adapt to acidic environments (Mussoline et al., 2013).

Strong acidic pretreatment is finished with dilute acids mixed by a thermal process. It ends in the assembly of repressive by-products, like aldehyde and hydroxymethyl furfural (Hendriks & Zeeman, 2009; Mussoline et al., 2013). Chemical pretreatment is not appropriate for simply perishable substrata holding large quantities of carbohydrates because of their enhanced degradation and subsequent gathering of VFA, which ends up in failure of the methanogenesis phase (Ariunbaatar et al., 2014).

16.4.4 Thermal Pretreatments

Thermal pretreatment is one of the foremost pretreatment approaches, and has been practiced with success at an industrial scale (Carrère et al., 2010, Carlsson, Lagerkvist, & Morgan-Sagastume, 2012; Cesaro & Belgiorno, 2014). Thermal pretreatment conjointly leads to infectious agent reduction, progresses dehydrating performance, and decreases viscousness of the sludge and subsequent sweetening of digestate (Carlsson, Lagerkvist & Morgan-Sagastume, 2012; del Río et al., 2011; Liu et al., 2012). Thermal pretreatment includes exposing wastes to higher temperatures and pressure to persuade hydrolysis while avoiding evaporation (Ariunbaatar et al., 2014). Thermal pretreatments consolidated by amplified OLR will be applied to the digester (Barber, 2016). Additionally, cell breakdown and hydrolysis are accomplished to make a sludge that is more decomposable and permits for supplementary steady digestion (Keppet al., 2000).

The different temperature levels of thermal pretreatment have a relevance on the productivity of the pretreatment process. In addition, an extreme temperature might affect in the breakdown of volatile solids, reducing the accessible substratum for AD (Ariunbaatar et al., 2014). Though it has been suggested that the process by which lower temperatures at the thermal pretreatment process occur is through enzymatic hydrolysis (Ferrer et al., 2008). Studies have recommended that thermal pretreatment at high temperatures of more than 70°C forced the formation of chemical bonds and outcome in the accumulation of the elements (Bougrier et al., 2006). Temperatures from 50 to 250°C have been deliberated to improve the AD of diverse organic solid waste, largely WWTP sludge and lignocellulosic substrates.

16.5 CONCLUSION

The rising worldwide concerns about the growing volume of waste, energy demands, and global warming have inspired investigation into the improvement of the anaerobic process. Multiple treatment techniques have been evaluated for batch scale to estimate the biogas efficiency from pretreated organic waste. It has been observed that there is the possibility of increased biogas production from OFMSW by controlling functioning parameters. The important roles of C:N ratio, pH value, temperature, HRT, volatile solids, and TS content in AD were observed. Several researchers have completed projects to discover the best scope of these operating parameters. In this paper, varied diversity of feeder pretreatment methodologies before feeding to the digester was also studied. Among the briefly stated pretreatment technologies, partial mechanical, biological, chemical, and thermal pretreatment technologies remained efficiently functional at full scale to benefit the AD process, such as high rise in biogas production, decrease in retention time, and reduction in digestate amount. The result of this investigation sought to raise biogas production and to handle the issues related with the municipal solid waste management in rural areas and urban cities.

ACKNOWLEDGEMENT

Our sincere thanks to VGST, Government of Karnataka for funding this project under RGS.

REFERENCES

Ahluwalia, I. J., & Patel, U. (2018). "Solid waste management in India: An assessment of resource recovery and environmental impact". Indian Council for Research on International Economic Relations. Working Paper 356.

Akuzawa, M., Hori, T., Haruta, S., Ueno, Y., Ishii, M., & Igarashi, Y. (2011). Distinctive responses of metabolically active microbiota to acidification in a thermophilic anaerobic digester. *Microbial Ecology, 61*(3), 595–605.

Amani, T., Nosrati, M., & Sreekrishnan, T. R. (2010). Anaerobic digestion from the viewpoint of microbiological, chemical, and operational aspects—a review. *Environmental Reviews, 18*, 255–278.

Ariunbaatar, J., Panico, A., Esposito, G., Pirozzi, F., & Lens, P. N. (2014). Pretreatment methods to enhance anaerobic digestion of organic solid waste. *Applied Energy, 123*, 143–156.

Arsova, L. (2010). Anaerobic digestion of food waste: Current status, problems and an alternative product. Department of Earth and Environmental Engineering Foundation of Engineering and Applied Science Columbia University.

Bapteste, É., Brochier, C., & Boucher, Y. (2005). Higher-level classification of the Archaea: evolution of methanogenesis and methanogens. *Archaea, 1*(5), 353–363.

Barber, W. P. F. (2016). Thermal hydrolysis for sewage treatment: a critical review. *Water Research, 104*, 53–71.

Barlaz, M. A., Schaefer, D. M., & Ham, R. K. (1989). Bacterial population development and chemical characteristics of refuse decomposition in a simulated sanitary landfill. *Applied and Environmental Microbiology, 55*(1), 55–65.

Bouallagui, H., Touhami, Y., Cheikh, R. B., & Hamdi, M. (2005). Bioreactor performance in anaerobic digestion of fruit and vegetable wastes. *Process Biochemistry, 40*(3–4), 989–995.

Bougrier, C., Albasi, C., Delgenès, J. P., & Carrère, H. (2006). Effect of ultrasonic, thermal and ozone pre-treatments on waste activated sludge solubilisation and anaerobic biodegradability. *Chemical Engineering and Processing: Process Intensification, 45*(8), 711–718.

Carballa, M., Manterola, G., Larrea, L., Ternes, T., Omil, F., & Lema, J. M. (2007).Influence of ozone pre-treatment on sludge anaerobic digestion: removal of pharmaceutical and personal care products. *Chemosphere, 67*(7), 1444–1452.

Carlsson, M., Lagerkvist, A., & Morgan-Sagastume, F. (2012). The effects of substrate pre-treatment on anaerobic digestion systems: a review. *Waste Management, 32*(9), 1634–1650.

Carrère, H., Dumas, C., Battimelli, A., Batstone, D. J., Delgenes, J. P., Steyer, J. P., & Ferrer, I. (2010). Pretreatment methods to improve sludge anaerobic degradability: a review. *Journal of Hazardous Materials, 183*(1–3), 1–15.

Cesaro, A., &Belgiorno, V. (2014). Pretreatment methods to improve anaerobic biodegradability of organic municipal solid waste fractions. *Chemical Engineering Journal, 240*, 24–37.

Chen, Y., Cheng, J. J., & Creamer, K. S. (2008). Inhibition of anaerobic digestion process: a review. *Bioresource Technology, 99*(10), 4044–4064.

Christy, P. M., Gopinath, L. R., & Divya, D.(2014). A review on anaerobic decomposition and enhancement of biogas production through enzymes and microorganisms. *Renewable and Sustainable Energy Reviews, 34*, 167–173.

Costa, A., Ely, C., Pennington, M., Rock, S., Staniec, C., &Turgeon, J. (2015). *Anaerobic digestion and its applications.* US Environmental Protection Agency: Washington, DC, 24.

Couger, M. B. (2015). Genomic and Transcriptomic Analysis of the Anerobic Fungus Orpinomyces Strain C1a, a Versatile Biodegrader of Plant Biomass (Doctoral dissertation).

del Río, A. V., Morales, N., Isanta, E., Mosquera-Corral, A., Campos, J. L., Steyer, J. P., & Carrère, H. (2011). Thermal pre-treatment of aerobic granular sludge: Impact on anaerobic biodegradability. *Water Research*, *45*(18), 6011–6020.

Deublein, D., &Steinhauser, A. (2011). *Biogas from waste and renewable resources: an introduction*. John Wiley & Sons.

Elliott, A., & Mahmood, T. (2012). Comparison of mechanical pretreatment methods for the enhancement of anaerobic digestion of pulp and paper waste activated sludge. *Water Environment Research*, *84*(6), 497–505.

Fernández-Rodríguez, J., Pérez, M., & Romero, L. I. (2013). Comparison of mesophilic and thermophilic dry anaerobic digestion of OFMSW: Kinetic analysis. *Chemical Engineering Journal*, *232*, 59–64.

Ferrer, I., Ponsá, S., Vázquez, F., & Font, X. (2008). Increasing biogas production by thermal (70 C) sludge pre-treatment prior to thermophilic anaerobic digestion. *Biochemical Engineering Journal*, *42*(2), 186–192.

Ferry, J. G. (2010). The chemical biology of methanogenesis. *Planetary and Space Science*, *58*(14–15), 1775–1783.

Franke-Whittle, I. H., Walter, A., Ebner, C., & Insam, H. (2014). Investigation into the effect of high concentrations of volatile fatty acids in anaerobic digestion on methanogenic communities. *Waste Management*, *34*(11), 2080–2089.

Ge, H., Jensen, P. D., & Batstone, D. J. (2010). Pre-treatment mechanisms during thermophilic–mesophilic temperature phased anaerobic digestion of primary sludge. *Water Research*, *44*(1), 123–130.

Ge, H., Jensen, P. D., & Batstone, D. J. (2011). Temperature phased anaerobic digestion increases apparent hydrolysis rate for waste activated sludge. *Water Research*, *45*(4), 1597–1606.

Gerlach, F., Grieb, B., & Zerger, U. (2013). *Sustainable biogas production: a handbook for organic farmers*. Frankfurt, Germany: FiBL Projekte.

Hansen, C. L., & Cheong, D. Y. (2019). Agricultural waste management in food processing. In *Handbook of farm, dairy and food machinery engineering* (pp. 673–716). New Delhi, India: Academic Press.

Hartmann, H., & Ahring, B. K. (2006). Strategies for the anaerobic digestion of the organic fraction of municipal solid waste: an overview. *Water Science and Technology*, *53*(8), 7–22.

Hendriks, A. T. W. M., & Zeeman, G. (2009). Pretreatments to enhance the digestibility of lignocellulosic biomass. *Bioresource Technology*, *100*(1), 10–18.

Heo, N. H., Park, S. C., & Kang, H. (2004). Effects of mixture ratio and hydraulic retention time on single-stage anaerobic co-digestion of food waste and waste activated sludge. *Journal of Environmental Science and Health, Part A*, *39*(7), 1739–1756.

Igoni, A. H., Ayotamuno, M. J., Eze, C. L., Ogaji, S. O. T., & Probert, S. D. (2008). Designs of anaerobic digesters for producing biogas from municipal solid-waste. *Applied Energy*, *85*(6), 430–438.

Jørgensen, P. J. (2009). *Biogas-green energy: process, design, energy supply, environment*. Researcher for a Day. Agricultural Sciences, Aarhus University, Denmark, ISBN 978-87-992243-2-1.

Kang, A. J., & Yuan, Q. (2017). Enhanced anaerobic digestion of organic waste. *Solid Waste Management in Rural Areas. Croatia: InTechopen*, 123–142.

Karlsson, A., Bjorn, A., Yekta, S. S., & Svensson, B. H. (2014). Improvement of the biogas production process. Biogas Research Center (BRC) Report.

Kepp, U., Machenbach, I., Weisz, N., & Solheim, O. E. (2000). Enhanced stabilisation of sewage sludge through thermal hydrolysis-three years of experience with full scale plant. *Water Science and Technology*, *42*(9), 89–96.

Khalid, A., Arshad, M., Anjum, M., Mahmood, T., & Dawson, L. (2011). The anaerobic digestion of solid organic waste. *Waste Management*, *31*(8), 1737–1744.

Kiener, A., & Leisinger, T. (1983). Oxygen sensitivity of methanogenic bacteria. *Systematic and Applied Microbiology*, *4*(3), 305–312.

Kim, W., Shin, S. G., Lim, J., & Hwang, S. (2013). Effect of temperature and hydraulic retention time on volatile fatty acid production based on bacterial community structure in anaerobic acidogenesis using swine wastewater. *Bioprocess and Biosystems Engineering*, *36*(6), 791–798.

Li, H., Li, C., Liu, W., & Zou, S. (2012). Optimized alkaline pretreatment of sludge before anaerobic digestion. *Bioresource Technology*, *123*, 189–194.

Li, Y., Park, S. Y., & Zhu, J. (2011). Solid-state anaerobic digestion for methane production from organic waste. *Renewable and Sustainable Energy Reviews*, *15*(1), 821–826.

Lim, J. W., & Wang, J. Y. (2013). Enhanced hydrolysis and methane yield by applying microaeration pretreatment to the anaerobic co-digestion of brown water and food waste. *Waste Management*, *33*(4), 813–819.

Lin, K. W., Patterson, J. A., & Ladisch, M. R. (1985). Anaerobic fermentation: microbes from ruminants. *Enzyme and Microbial Technology*, *7*(3), 98–107.

Lin, L., Yan, R., Liu, Y., & Jiang, W. (2010). In-depth investigation of enzymatic hydrolysis of biomass wastes based on three major components: cellulose, hemicellulose and lignin. *Bioresource Technology*, *101*(21), 8217–8223.

Liu, X., Wang, W., Gao, X., Zhou, Y., & Shen, R. (2012). Effect of thermal pretreatment on the physical and chemical properties of municipal biomass waste. *Waste Management*, *32*(2), 249–255.

Luo, K., Yang, Q., Li, X. M., Yang, G. J., Liu, Y., Wang, D. B., ... & Zeng, G. M. (2012). Hydrolysis kinetics in anaerobic digestion of waste activated sludge enhanced by α-amylase. *Biochemical engineering journal*, *62*, 17–21.

Ma, J., Frear, C., Wang, Z. W., Yu, L., Zhao, Q., Li, X., & Chen, S. (2013). A simple methodology for rate-limiting step determination for anaerobic digestion of complex substrates and effect of microbial community ratio. *Bioresource Technology*, *134*, 391–395.

Mao, C., Feng, Y., Wang, X., &Ren, G.(2015). Review on research achievements of biogas from anaerobic digestion. *Renewable and Sustainable Energy Reviews*, *45*, 540–555.

Moset, V., Poulsen, M., Wahid, R., Højberg, O., & Møller, H. B. (2015). Mesophilic versus thermophilic anaerobic digestion of cattle manure: methane productivity and microbial ecology. *Microbial Biotechnology*, *8*(5), 787–800.

Mussoline, W., Esposito, G., Giordano, A., & Lens, P. (2013). The anaerobic digestion of rice straw: a review. *Critical Reviews in Environmental Science and Technology*, *43*(9), 895–915.

Nijaguna, B. T. (2006). *Biogas technology*. New Delhi, India: New Age International.

Noraini, M., Sanusi, S., Elham, J., Sukor, Z., & Halim, K. (2017). Factors affecting production of biogas from organic solid waste via anaerobic digestion process: A review. *Solid State Science and Technology*, *25*(1), 29–39.

Okonkwo, U. C., Onokpite, E., & Onokwai, A. O. (2018). Comparative study of the optimal ratio of biogas production from various organic wastes and weeds for digester/restarted digester. *Journal of King Saud University-Engineering Sciences*, *30*(2), 123–129.

Ostrem, K., & Themelis, N. J. (2004). Greening waste: Anaerobic digestion for treating the organic fraction of municipal solid wastes. *Earth Engineering Center Columbia University*, 6–9.

Salminen, E. A., & Rintala, J. A. (2002). Semi-continuous anaerobic digestion of solid poultry slaughterhouse waste: effect of hydraulic retention time and loading. *Water Research*, *36*(13), 3175–3182.

Saraswat, M., Garg, M., Bhardwaj, M., Mehrotra, M., & Singhal, R. (2019, November). Impact of variables affecting biogas production from biomass. In *IOP Conference Series: Materials Science and Engineering* (Vol. 691, No. 1, p. 012043). New Delhi, India: IOP Publishing.

Shi, X. S., Dong, J. J., Yu, J. H., Yin, H., Hu, S. M., Huang, S. X., & Yuan, X. Z. (2017). Effect of hydraulic retention time on anaerobic digestion of wheat straw in the semicontinuous continuous stirred-tank reactors. *BioMed Research International*, *17*, 1–6.

Singh, S. (2020). Decentralized solid waste management in India: A perspective on technological options. *New Delhi: National Institute of Urban Affairs*, 290–304.

Song, Y., & Wu, P. (2021). Earth observation for sustainable infrastructure: a review. *Remote Sensing*, 13(8), 1528.

Sreekrishnan, T. R., Kohli, S., & Rana, V. (2004). Enhancement of biogas production from solid substrates using different techniques—a review. *Bioresource Technology*, *95*(1), 1–10.

Stams, A. J., & Plugge, C. M. (2009). Electron transfer in syntrophic communities of anaerobic bacteria and archaea. *Nature Reviews Microbiology*, *7*(8), 568–577.

Torres, M. L., & Lloréns, M. D. C. E. (2008). Effect of alkaline pretreatment on anaerobic digestion of solid wastes. *Waste Management*, *28*(11), 2229–2234.

Ueno, Y., Tatara, M., Fukui, H., Makiuchi, T., Goto, M., & Sode, K. (2007). Production of hydrogen and methane from organic solid wastes by phase-separation of anaerobic process. *Bioresource Technology*, *98*(9), 1861–1865.

Van Amstel, A. (2012). Methane: A review. *Journal of Integrative Environmental Sciences*, *9*(Sup 1), 5–30.

Van Lier, J. B. (1995). *Thermophilic anaerobic wastewater treatment: temperature aspects and process stability*. Julius Bernardus van Lier, Wageningen, Netherlands: Van Lier.

Veeken, A., Kalyuzhnyi, S., Scharff, H., & Hamelers, B. (2000). Effect of pH and VFA on hydrolysis of organic solid waste. *Journal of Environmental Engineering*, *126*(12), 1076–1081.

Verma, S. (2002). Anaerobic digestion of biodegradable organics in municipal solid wastes. Columbia University, *7*(3), 98–104.

Voegeli, Y., & Zurbrügg, C. (2008). Decentralised anaerobic digestion of kitchen and market waste in developing countries-'state of the art' in south India. *Proceedings Venice.*

Wang, X., & Zhao, Y. C. (2009). A bench scale study of fermentative hydrogen and methane production from food waste in integrated two-stage process. *International Journal of Hydrogen Energy*, *34*(1), 245–254.

Wolfe, R. S. (2011). Techniques for cultivating methanogens. *Methods in Enzymology*, *494*, 1–22.

Yadvika, Santosh, Sreekrishnan, T., Kohli, S. & Rana, V. (2004). Enhancement of biogas production from solid substrates using different techniques—a review. *Bioresource Technology*, 95, 1–10.

Yamada, K., & Xu, H. L. (2001). Properties and applications of an organic fertilizer inoculated with effective microorganisms. *Journal of Crop Production*, 3(1), 255–268.

Yi, J., Dong, B., Jin, J., & Dai, X. (2014). Effect of increasing total solids contents on anaerobic digestion of food waste under mesophilic conditions: performance and microbial characteristics analysis. *PloS One*, *9*(7), e102548.

Zhu, B., Gikas, P., Zhang, R., Lord, J., Jenkins, B., & Li, X. (2009). Characteristics and biogas production potential of municipal solid wastes pretreated with a rotary drum reactor. *Bioresource Technology*, *100*(3), 1122–1129.

Zupančič, G. D., & Grilc, V. (2012). Anaerobic treatment and biogas production from organic waste. *Management of Organic Waste*, 1–28.

17 Assessment of Biogas Generation Potential of Mixed Fruits Solid Waste

David O. Olukanni, Gbemisola I. Megbope, and Oluwatosin J. Ogundare

17.1 INTRODUCTION

Management of a large number of solid wastes generated by an ever-growing population is a challenge several government agencies are contending with. A significant amount of this waste is fruit waste, which is being generated without any appropriate environmental control for its processing and disposal. This problem is more prominent in developing countries, like Nigeria, where proper waste-management practices and techniques are still picking up. Food waste and agriculture materials make up 64% of recyclable waste (Olukanni et al., 2013; Olukanni and Aremu, 2017) and fruit waste deposits by households and fruit processing factories account for a major portion. This is because fruit provides humans with vitamins and other essential minerals for good health and living, and has thus risen vastly to be a prominent food for the global population (Schieber et al., 2001). Similarly, Jahid et al. (2018) observed that the rejected portion of fruits such as mango, banana, pineapple, and orange ranges from 20% to 50%. According to UNEP (1995), vegetable processing and preservation of vegetables can see up to one-third of the total quantity of raw materials being rejected. The heaps of this rotten organic refuse pose a danger, as they can be a breeding ground for disease vectors, pollute underground and surface water resource, and give offensive stench, thereby reducing quality of life (Owens et al., 1951; Wikandari et al., 2014).

With the continuing increase in population and the problem of overcrowded cities, governments spend huge sums of money managing the waste generated in their cities (Abalo et al., 2018). Cost of effective waste-management systems and implementation of waste-management practices is high (Joshi and Ahmed, 2016; Pap and Maróti, 2016). To reduce these expenses and facilitate better use of funds, the concept of waste to energy continues to be explored by researchers all over the world, investigating organic and inorganic disposed materials for energy production (Adeoti et al., 2000; Wamuyu, 2014; Nkoi et al., 2018). The significance of this energy source cannot be overemphasised, as it is a key factor for growth, development, and the building of sustainable cities and societies. Various energy sources ranging from conventional to renewable energy sources are being assessed, improved, and utilised to meet the numerous energy-based desires of the world's teeming population. The newly found focus on sustainable development, of which environmental stability and sustainability is a key factor, has greatly intensified the search for alternative, environmentally friendly energy sources (Kwon et al., 2012).

Reports have shown that the world energy consumption trends and household energy consumption accounts for 25% and 40% in developed and developing countries, respectively, of which cooking gas is the main contributor. Nigeria's demand for cooking gas is increasing because of city population explosion and accidents caused by cooking with kerosene. This has led to over-exploitation of cooking gas price by marketers who run an almost unregulated market (World LP Gas Association, 2001; Rapu et al., 2015). Consequently, Nigeria is a country with serious waste management and energy sustainability problems (Olukanni and Olatunji, 2018). Essentially, it is necessary to harness the biogas generation potential of organic wastes to achieve efficient waste management and optimal supply of energy needs.

DOI: 10.1201/9781003204435-20

The chemical composition of fruit waste consists of high concentrations of soluble sugars (Marın et al., 2007; Rivas-Cantu et al., 2013) and this makes it a suitable source of cellulosic biomass for biofuel (bioethanol and biogas) generation (Forgacs, 2012). Conversion of biomass to biogas is done through an anaerobic digestion process, under mesophilic or thermophilic conditions, which sustainably utilises the potential of fruit wastes (Zema et al., 2018). Biogas serves as an alternative to fossil fuels and can be utilised for cooking, electricity, lighting, and vehicle fuel with no net carbon to the earth and a by-product that can be used as fertiliser (Forgacs, 2012). Anaerobic digestion of fruit waste to produce biogas therefore serves as a cheap alternative that combines as a sustainable means to reduce environmental pollution impact and produce clean energy (Forgacs, 2012; Wikandari et al., 2014).

This study focused on the assessment of the biogas generation potential of mixed-fruit solid waste comingled with cow dung as a means of providing cheap and clean energy, while also providing an effective waste-management technology. Fruit waste selected for co-digestion with cow dung was limited to banana, watermelon, pineapple, and orange wastes. The study does not consider the design of the co-digester in details, neither does it evaluate the economic viability of the bio-digester and the economics of biogas generated. It deals mainly with assessing the biogas generation potential of the selected fruit wastes. The specific objectives of our study were to assess the generation of biogas using selected mixed-fruit wastes, evaluate the result of retention time on biogas production, and evaluate the effect of substrate composition on biogas generation.

17.2 MATERIALS AND METHODS

In this work, a solution to provide affordable waste-management technology at both household and at the municipal levels was investigated. Solid fruit waste was collected from the waste container of the Covenant University Student Cafeteria II. The collected waste was sorted into pineapple peels, orange peels, watermelon peels, and banana peels and kept in different containers, as shown in Figure 17.1. The weight of each waste sample was measured, and their moisture content determined of using thermo-gravimetric analysis as described by Fujishima et al. (2000).

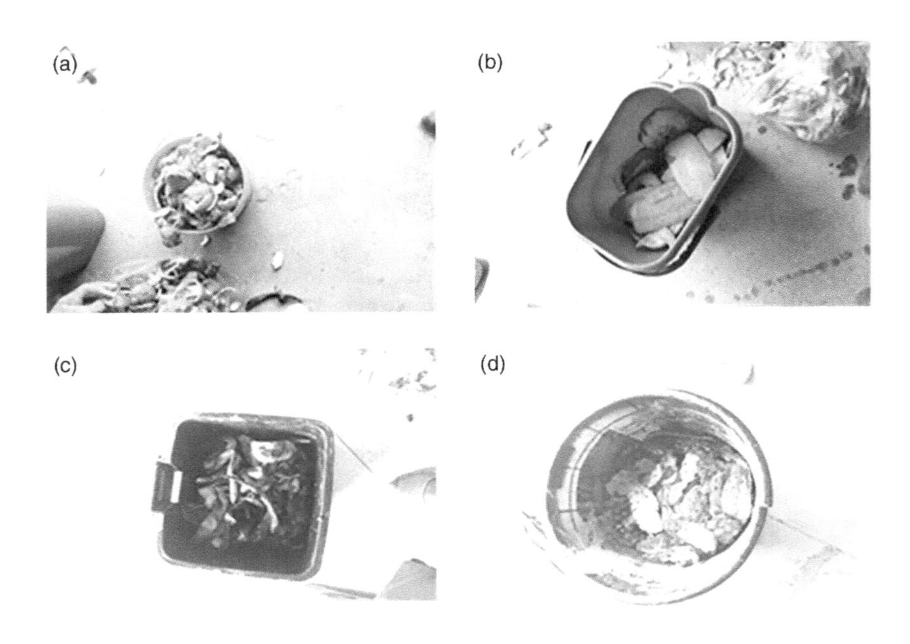

FIGURE 17.1 (a) Orange peels, (b) watermelon peels, (c) banana peels, and (d) pineapple peels.

FIGURE 17.2 Cow dung.

Cow manure was mixed with the fruit waste for co-digestion due to its high biogas generation potential and its capacity to ensure that biodegradation of the fruit waste increased uniformly (Sagagi et al., 2009; Majeed and Malik, 2018). Cow manure was gotten from an abattoir within Ota municipality. The cow dung is shown in Figure 17.2.

17.3 PREPARATION OF SUBSTRATES

The fruit waste samples were chopped and blended to increase surface area to improve the anaerobic digestion process. Two distinct substrate samples, Substrate A and Substrate B, were prepared from the combination of base materials of fruit waste, cow dung, and distilled water in similar proportions. The substrate composition and number of materials used are presented in Tables 1a and 1b, respectively.

17.4 DIGESTER

Locally sourced plastic containers, pipes, and valves were used to construct a digester. A schematic diagram of a typical digester is shown in Figure 17.3.

17.4.1 Digester Construction

Two digesters were constructed for this experiment. The digester tanks were constructed using two sterile 30-litre rubber tanks; two ¾ inch valves, two ½ inch hoses, four ¾ inch elbows, a 3-inch

TABLE 17.1a
Substrate Composition

Sample	Composition of Mixture
Substrate A	Pineapple peels + banana peels + cow dung + water
Substrate B	Watermelon peels + orange peels + cow dung + water

TABLE 17.1b
Substrates A and B Composition by Weight

Substrate A	
Base Material	**Amount**
Cow dung	1000 g
Pineapple peels	1000 g
Banana peels	1000 g
Water	1.5 l
Substrate B	
Base Material	**Amount**
Cow dung	1000 g
Watermelon peels	1000 g
Orange peels	1000 g
Water	1.5 l

valve, ¾ inch PVC pipes and two funnels made up the piping and valve control systems of the biogas production setup for each tank. Tyre tubes were used as gas storage device. Other miscellaneous materials used were sand, gum, paper tape, hand gloves, nose mask, soldering iron, black paint, and brush. The digester was painted black to absorb as much heat as possible and to improve microbial action on substrate. The construction of the setup took place in the environmental Laboratory of Covenant University and is shown in Figure 17.4.

17.4.2 Anaerobic Digestion of Substrates

Anaerobic digestion was done as batch experiments over a period of six weeks under mesophilic conditions. The digesters were maintained in a room at ambient temperature. Biogas generated was collected through the gas outlet of the digester into the airtight tubes. The amount of gas generated was determined as the difference in weight of tubes between measurements. The initial weight of tube was recorded, and the tubes weighed weekly, hence an increase in the weight of the tube indicated the generation of biogas and the quantity of biogas produced in kg/week.

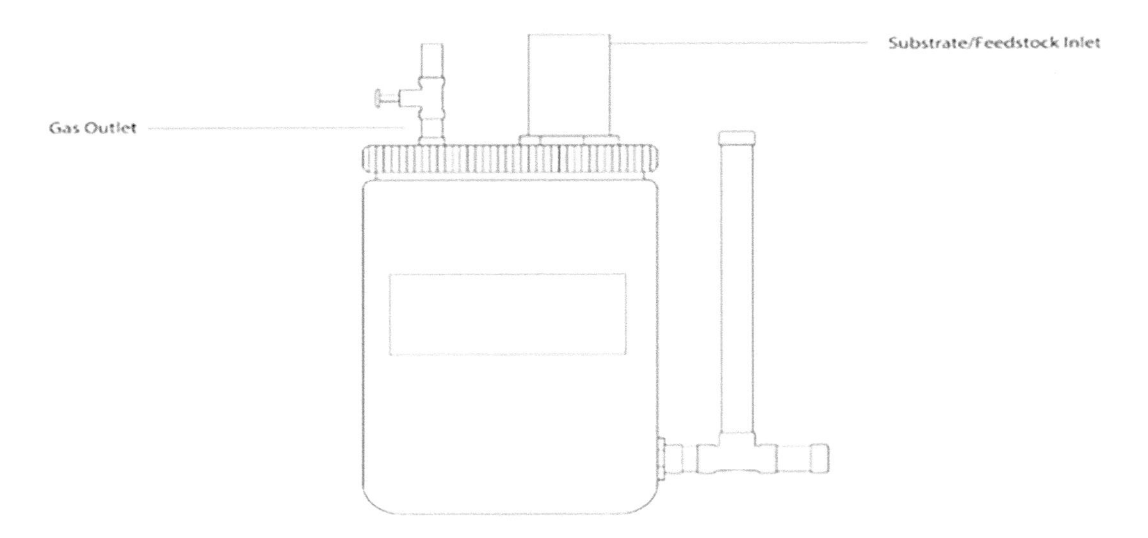

FIGURE 17.3 Digester schematic drawing.

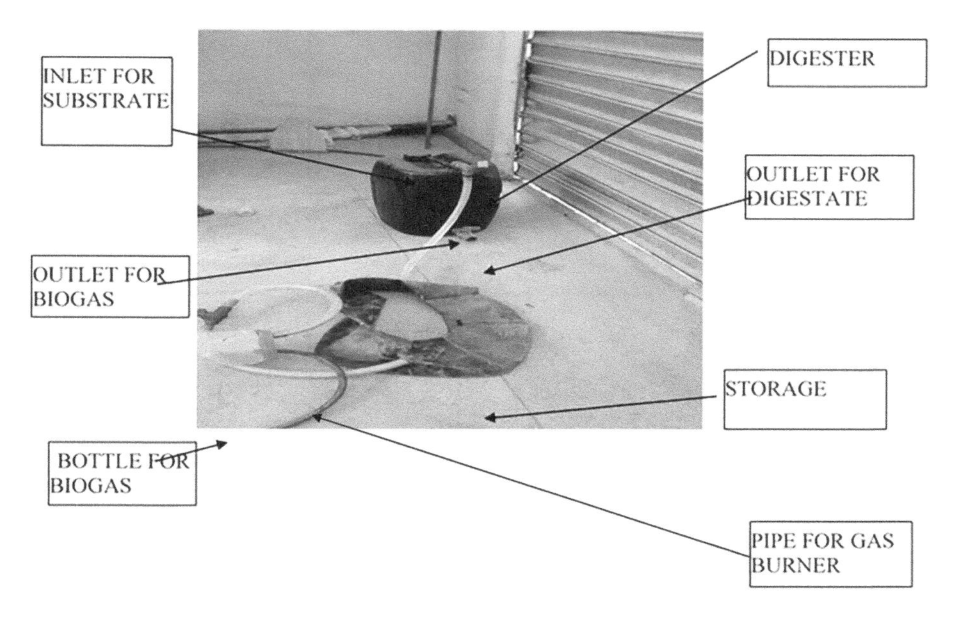

FIGURE 17.4 Constructed digester.

17.4.3 Characterisation of Biogas Production

The following parameters were employed in measuring biogas production in the bio-digester used in this study: daily gas production and specific gas production.

17.4.4 Daily Gas Production (G)

This is the measure of biogas produced per day based on the specific gas yield (Gy) of the biomass (feedstock) and the substrate input (S_d):

The estimation has its bases on the following models.

17.4.5 The Volatile Solids Content (VS)

$$G\left(\frac{m^3}{d}\right) = VS\left(kgm^3\right) \times Gy(m^3/d \times kg)\ (SW) \tag{17.1}$$

17.4.6 The Measure of the Moist Mass (Substrate) "B"

$$G\left(\frac{m^3}{d}\right) = B\ \left(kgm^3\right) \times Gy(m^3/d \times kg) \tag{17.2}$$

17.4.7 Average Gas-Yield Values Per Livestock Unit LSU

$$G\left(\frac{m^3}{d}\right) = number\ \ of\ \ LSU\ \ (number) \times Gy\ \ (species)\ (m^3/d \times number) \tag{17.3}$$

17.4.8 Detailed Gas Production (Gp)

This indicates the day-to-day gas discharge in m^3 of digester volume V_d.

$$specific\ gas\ production = daily\ gas\ production \div digester\ volume$$

$$Gp = G\left(\frac{m^3}{d}\right) \div Vd(m^3) \tag{17.4}$$

17.4.9 Efficiency of the Biogas Generation Process

The effectiveness of the biogas generation process is the fraction of the theoretical daily gas production.

$$\text{Efficiency} = \frac{Actual\ daily\ gas\ production}{theoretical\ daily\ gas\ production}$$

$$\eta_{bgp} = \frac{G_a}{G_e} \tag{17.5}$$

17.5 RESULTS AND DISCUSSION

17.5.1 Total Solid Composition of Fruit Wastes

The total solid content (TS) of the solid fruit waste used in this study was measured using thermogravimetric analysis and the results are presented in Table 17.2 and Figure 17.5.

Figure 17.5 shows that orange has the highest TS content of 40%, followed by pineapple (24%), banana (16%), and watermelon (8%). The moisture contents ranged from 60% to 92% and this indicates that the selected solid fruit wastes are viable substrates for the generation of biogas, mainly because of higher biodegradability and moisture value (Misi and Forster, 2002).

17.5.2 Parameters of Biogas Production Process

Parameters such as the theoretical and actual daily gas production, specific gas production, and digester conversion efficiency were calculated and presented in Tables 17.3a and Table 17.3b. A gas yield of 0.083 m^3/kg for wet food waste was assumed based on estimation retrieved from http://www.biteco-energy.com/biogas-yield and was used in calculating the theoretical biogas production for the solid fruit waste used in this study.

TABLE 17.2
Total Solid Content of Fruit Wastes

Fruit Waste	W_W(g)	W_d(g)	$T_S = W_d/Ww \times 100$
Watermelon	100	8	8%
Pineapple	100	24	24%
Banana	100	16	16%
Orange	100	40	40%

Abbreviations: T_S, total solids (%); W_W, weight when wet (g); W_d, weight when dry (g).

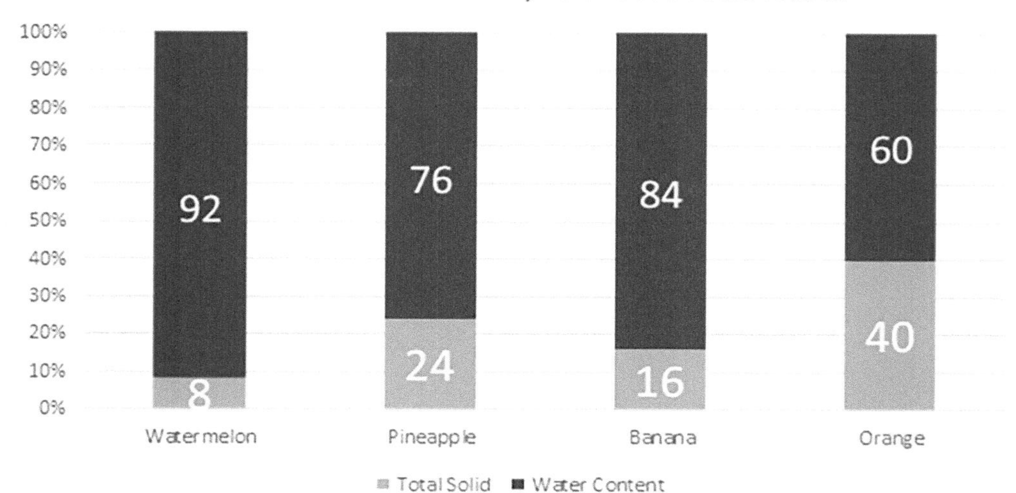

FIGURE 17.5 Total solid content of solid fruit wastes.

TABLE 17.3a
Parameters of Biogas Production

	Substrate A			
Wk	Average Actual Biogas Production (Ga)[kg/d]	Theoretical Biogas Production (Ge)[kg/d]	Specific Gas Production (Gp)[kg/m³]	Digester Conversion Efficiency (bgp) [%]
1	0.114	0.3	3.8	38
2	0.128	0.3	4.27	42.7
3	0.15	0.3	5.0	50.0
4	0.185	0.3	6.17	61.7
5	0.193	0.3	6.43	64.3
6	0.2	0.3	6.67	66.7

TABLE 17.3b
Parameters of Biogas Production

	Substrate B			
Wk	Average Actual Biogas Production (Ga)[kg/d]	Theoretical Biogas Production (Ge)[kg/d]	Specific Gas Production (Gp)[kg/m³]	Digester Conversion Efficiency (Bgp) [%]
1	0.100	0.3	3.33	33.3
2	0.128	0.3	4.27	42.7
3	0.128	0.3	4.27	42.7
4	0.128	0.3	4.27	42.7
5	0.135	0.3	4.50	45.0
6	0.143	0.3	4.77	47.7

TABLE 17.4
Cumulative Quantity of Biogas Produced in (kg) from Solid Fruit Waste

	Substrate	
No of Weeks	Substrate A Pineapple, Banana, and Cow Dung (kg)	Substrate B Watermelon, Orange, and Cow Dung (kg)
Week 1	0.80	0.70
Week 2	0.90	0.90
Week 3	1.05	0.90
Week 4	1.30	0.90
Week 5	1.35	0.95
Week 6	1.40	1.00

The results indicate that the average daily biogas generated for both substrates A and B are favourable and within operating standards when compared to the research done by Islam et al., (2012) and Deressa et al., (2015).

17.5.3 Biogas Production from Substrate

The quantity of biogas produced from this study at a retention time of 6 weeks is indicated in Table 17.4.

The biogas production was measured weekly; the result showed that Substrate A produced biogas measuring 1.4 kg at the 6-week retention time, while the maximum weekly yield of biogas for Substrate A came within the first week measuring 0.8 kg.

Substrate B produced biogas measuring 1.0 kg over the 6-week retention time, while the maximum weekly biogas yield of Substrate B was also recorded at the first week measuring 0.7 kg. It was determined that the temperature of the plant within the 6-week retention period ranges from 26 °C to 27 °C, which is sufficient for the anaerobic digestion process (Wang et al., 2019). Figure 17.6 presents the weekly trend in biogas yield within the 6-week retention time.

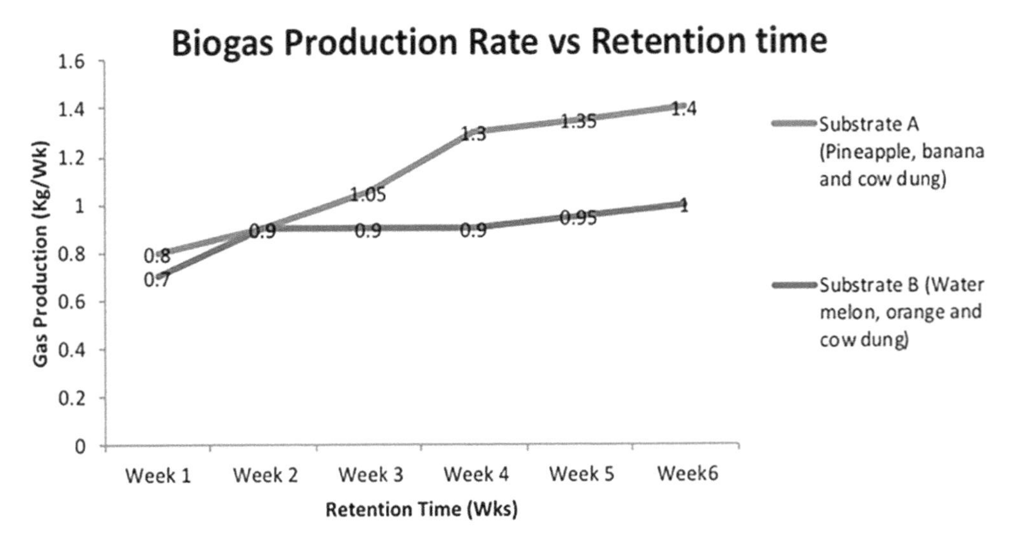

FIGURE 17.6 Biogas production yield within the retention time.

17.5.4 Effect of Substrate on Biogas Production

Findings from the study have indicated that the constituents of the biomass used have an important effect on the rate of biogas production. The gas yield of Substrate A (pineapple, banana, and cow dung) was higher than Substrate B (watermelon, orange, and cow dung). The reduced gas generation in substrate B can be attributed to the presence of D-limonene, a major constituent of orange peel oils, which is known to hamper the anaerobic digestion process (Wikandari et al., 2015; Martin et al., 2018; Zema et al., 2018). D-limonene is an anti-microbial agent that inhibits microbial actions and can cause total failure of anaerobic digestion systems (Lohrasbi et al., 2010; Martín et al., 2010; Forgacs et al., 2012). Pourbafrani et al. (2010) suggested that the threshold level of D-limonene for inhibiting anaerobic digestion under thermophilic conditions is between 450 and 900 μL/L, while the threshold level under mesophilic conditions is 400 μL/L (Mizuki et al., 1990). This emphasises that the substrate type is a determinant of the rate of biogas generation rate in a bio-digester.

17.6 CONCLUSION

The study focused on biogas generation from fruit waste as a waste-management technique and clean energy source. A simple laboratory-scale biogas production unit built using readily available household materials was able to produce substantial amounts of gas. One substrate sample was able to discharge 1.4 kg of gas, while the other discharged 1.0 kg, over a retention period of six weeks. Although there is a possibility of optimising the digestion process to increase biogas production rate, the substrate used, and bio-digester unit shows good prospect for commercialisation and meeting small-scale energy demand and waste-management issues. However, there is need for more studies to be done to optimise generation potential and ascertain the economic feasibility of the digestion process. Households and city governments can generate an alternative energy source that enables environmental sustainability and the judicious use of natural gas reserves.

17.6.1 Recommendation

To improve on this study, the pretreatment of orange waste by leaching should be done to reduce the amount of D-limonene. Due to the high lignocellulose content of watermelon, it should also be pretreated for higher biogas production. Biogas collection should be executed by the water displacement method, which is a more effective method when compared to the less expensive tyre collection method employed in this study. Parameters such as pH of digester, temperature, chemical oxygen demand, and biological oxygen demand should be considered. The effect of the ambient digester temperature and the effect of mechanical stirring on biogas production rate should be considered for optimisation purposes.

REFERENCES

Abalo, E. M., Peprah, P., Myonyo, J., Ampomah-Sarpong, R. and Agyemang-Duah, W. (2018). A review of the triple gains of waste and the way forward for Ghana. Hindawi Journal of Renewable Energy, Doi: http://doi.org/10.1155/2018/9737683.

Adeoti, O., Ilori, M. O., Oyebisi, T. O. and Adekoya, L. O. (2000). Engineering design and economic evaluation of a family-sized biogas project in Nigeria. Technovation, 20(2): 103–108. Doi: 10.1016/S0166-4972(99)00105-4.

Deressa, L., Libsu, S., Chavan, P. B., Manaye, D. and Dabasa, A. (2015). Production of biogas from fruit and vegetable wastes mixed with different wastes. Environment and Ecology Research, 3(3): 65–71.

Forgacs, G. (2012). Biogas Production from Citrus Wastes and Chicken Feather: Pre-treatment and Co-digestion" Unpublished Ph.D. thesis, Chalmers University of Technology, Göteborg.

Forgacs, G., Pourbafrani, M., Niklasson C., Taherzadeh M.J. and Hovath I.S. (2012). Methane production from citrus wastes: Process development and cost estimation. Journal of Chemical Technology and Biotechnology, 87(2):250–255.

Fujishima, S., Miyahara, T. and Noike, T. (2000). Effect of moisture content on anaerobic digestion of dewatered sludge: Ammonia inhibition to carbohydrate removal and methane production. Water Science and Technology: A Journal of the International Association on Water Pollution Research, 41(3): 119–127. Doi: 10.2166/wst.2000.0063.

Islam, M. N., Park, K. J. and Yoon, H. S. (2012). Methane production of food waste and food waste mixture with swine manure in anaerobic digestion. Journal of Biosystems Engineering, 37(2): 100–105.

Jahid, M., Gupta, A. and Sharma, D. K. (2018). Production of bioethanol from fruit wastes (banana, papaya, pineapple and mango peels) under milder conditions. Journal of Bioprocessing & Biotechniques, 8(3):1–11.

Joshi, R. and Ahmed, S. (2016). Status and challenges of municipal solid waste management in India: A review. Cogent Environmental Science, 2(1):1–18.

Kwon, E. E., Kim, S., Jein, Y. J. and Yi, A. (2012). Biodiesel production from sewage sludge: new paradigm for mining energy from municipal hazardous material. Environmental Science and Technology, 46 (18): 10227–10228.

Lohrasbi, M., Pourbafrani, M., Niklasson C. and Taherzadeh M. J. (2010). Process design and economic analysis of a citrus waste biorefinery with biofuels and limonene as products. Bioresource Technology. 101(19):7382–7388.

Majeed, A. and Malik, S. R. (2018). Enhancement of biogas production by co-digestion of fruit and vegetable waste with cow dung and kinetic modeling. International Research Journal of Engineering and Technology (IRJET), 5(12):1238–1246.

Marín, R. F., Cristina, S., Obdulio, B., Julian, C. and Jose, A. P. (2007). By-products from Different Citrus Processes as a Source of Customized Functional Fibres. Journal of Food Chemistry, 100(2):736–741.

Martín, M. A., Siles, J. A., Chica, A. F. and Martín, A. (2010). Biomethanization of orange peel waste. Bioresource Technology, 101(23): 8993–8999.

Martin, M.A., Fernandez, R., Gutierrez, M.C. and Siles, J.A. (2018). Thermophilic anaerobic digestion of pretreated orange peel: Modelling of methane production. Journal of Process Safety and Environmental Protection, 117:245–253.

Misi, S. N. and Forster, C. F. (2002). Semi-continuous anaerobic co-digestion of agrowastes. Environmental Technology, 23(4):445–451.

Mizuki, E., Akao, T. and Saruwatari, T. (1990). Inhibitory effect of citrus unshu peel on anaerobic digestion. Biological Wastes, 33(3):161–168.

Nkoi, B., Lebele-Alawa, B. T. and Odobeatu, B. (2018). Design and fabrication of a modified portable biogas digester for renewable cooking gas production. European Journal of Engineering Research and Science, 3(3):21–29.

Olukanni, D. O., Ede, A. N., Akinwumi, I. I. and Ajanaku, K. O. (2013). Environment, Health and Wealth Issues of Municipal Solid Waste Management in Southwest Nigeria: The Example of Ota, Ogun State. Nigeria. The 22nd, IBIMA Conference. Rome, Italy: International Business Information Management Association (IBIMA).

Olukanni, D. O. and Aremu, O. D. (2017). Provisional evaluation of composting as priority option for sustainable waste management in South-West Nigeria. Pollution, 3(3): 417–428.

Olukanni, D. O. and Olatunji, T. O. (2018). Cassava waste management and biogas generation potential in selected local government areas in Ogun state, Nigeria. Recycling, 3(58):1–12.

Owens, S. H., Veldhuis, M. K. and Maclay, W. D. (1951). Making Use of Tons of Citrus Waste. 1950-1951 Yearbook of Agriculture. Available online: https://naldc.nal.usda.gov/download/IND43894068/PDF (accessed on 14 February 2020).

Pap, B. and Maróti, G. (2016). Diversity of microbial communities in biogas reactors. Current Biochemical Engineering, 3(3):177–187.

Pourbafrani, M., Forgács, G., Sárvári H. I., Niklasson, C. and Taherzadeh, M. J. (2010). Production of biofuels, limonene and pectin from citrus wastes. Journal of Bio-Resource Technology, 101:4246–4250.

Rapu, C. S., Adenuga, A. O. Kanya, W. J., Golit, P. D., Hilili, M. J., Uba, I. A. and Ochu, E. R. (2015). Analysis of energy market conditions in Nigeria, Occasional Paper No 55, Central Bank of Nigeria.

Rivas-Cantu, R. C., Jones, K.D. and Mills, P.L. (2013). A citrus waste-based bio-refinery as a source of renewable energy: Technical advances and analysis of engineering challenges. Journal of Waste Management & Research, 31 (4):413–420.

Sagagi, B. S., Garba, B. and Usman, N. S. (2009). Studies on biogas production from fruits and vegetable waste. Bayero Journal of Pure and Applied Sciences, 2(1):115–118.

Schieber, A., Stintzing, F. C. and Carle, R. (2001). By-products of plant food processing as a source of functional compounds: Recent developments. Trends in Food Science & Technology, 12:401–413.

UNEP. (1995). Industry and environment: Food Processing and the Environment (Paris) 18:4

Wamuyu, M. S. (2014). Analysis of biogas technology for household energy, sustainable livelihoods and climate change mitigation in Kiambu County, Kenya. An unpublished Thesis, Kenyatta University, Kenya.

Wang, S., Ma, F., Ma, W., Wang, P., Zhao, G. and Lu, X. (2019). Influence of temperature on biogas production efficiency and microbial community in a two-phase anaerobic digestion system. Water, 11(133):1–13.

Wikandari, R., Youngsukkasem, S., Millati, R. and Taherzadeh M.J. (2014). Performance of semi-continuous membrane bioreactor in biogas production from toxic feedstock containing d-limonene. Bioresource Technology, 170:350–355.

Wikandari, R., Claes, N., Nguyen, H., Ria, M. and Taherzadeh, M. J. (2015). Improvement of biogas production from orange peel waste by leaching of limonene. BioMed Research International, Doi: 10.1155/2015/494182.

Zema, A. D., Adele, F., Giovanni, Z., Paolo, S. C., Vincenzo, T. and Santo, M. Z. (2018). Anaerobic digestion of orange peel in a semi-continuous pilot plant: An environmentally sound way of citrus waste management in agro-ecosystems. Science of the Total Environment, 630: 401–408.

18 Agricultural Waste to Biogas Energy

Design and Simulation Using the Anaerobic Digestion Model No. 1

Preseela Satpathy, Frank Uhlenhut, and Chinmay Pradhan

18.1 INTRODUCTION

India is essentially an agricultural land and remains one of the largest producers of rice, wheat, millets, cotton, jute, pulses, groundnut, sugarcane, etc. This presents the challenge of managing the several million tonnes of agricultural residues produced after the harvest. Statistics indicate that nearly 550 million tonnes of agricultural crop residues are generated each year in the country, the majority of which are openly burnt on farms (Devi et al., 2017). Recent studies report that from rice alone, 90–140 tonnes of crop residue is openly burnt every year as a cheaper, faster alternative to prepare for the next crop and get rid of pests (Bisen and Rahangdale, 2017). Although such practices are favourable for the farmers to save time and money, from the environmental point of view, crop residue burning (CRB) is a grave concern. Burning crop residues releases pollutants like carbon dioxide, carbon monoxide, methane, nitrous oxide, nitric oxide, sulphur dioxide, black carbon, volatile organic compounds, and particulate matter into the atmosphere (Singh et al., 2015). The International Food Policy Research Institute (IFPRI) reports that due to stubble burning, the particulate matter of Delhi's air reached 20 times more than the limits prescribed by the World Health Organization (Sample, 2019). Exposure to the air pollutants released from CRB is linked with acute respiratory infections, which could prove to be fatal. Even worse, poor air quality is also harmful, especially for children less than five years of age. The detrimental effects of the open-burning practices also extend to soil as well. Ash formed from CRB stays in the surface of the crops, which has been found to immobilise pesticides in the soil (Yang and Sheng, 2003). Such harmful chemicals further find their way into the human food chain, leading to severe health impacts. In addition to these environmental damages, one cannot overlook the economic impacts. A recent study by IFPRI estimates a loss of nearly USD 30 billion, or 2 lakh crores, each year due to the CRB practices in India (Sample, 2019).

The issue with CRB could be addressed when the agricultural residues could be treated as resources and not just waste. Numerous studies recommend utilisation of such crop residues to generate energy. Biogas is one such solution, with the added benefit of attaining a clean and renewable energy source. It could be utilised as a cooking fuel, converted to electricity, and compressed and utilised as vehicle fuel. Furthermore, the residues or the digestate from the digesters could also find applications as rich organic fertilisers, thus qualifying biogas systems as a zero-waste concept (Seadi et al., 2008). Decentralised biogas systems have also been determined to be cost-effective and sustainable for farmers, compared to other methods like gasification (Gumisiriza et al., 2017).

This study suggests utilising residues from rice crops as substrates for biogas generation. Biogas formation occurs by anaerobic digestion, which involves conversion of organic material to methane and carbon dioxide in the absence of oxygen. A virtual biogas plant was designed, and the set of operational parameters were determined by employing a mathematical model. The Anaerobic

DOI: 10.1201/9781003204435-21

Digestion Model No. 1 (ADM1) was utilised to optimally design and derive the ideal operating conditions for a cost-effective biogas plant, and simulations were performed for 600 days to observe the digester behaviour over a longer duration. The goal remained to suit the needs and conditions of the farmers in India and other lesser-developed nations.

18.2 MATERIALS AND METHODOLOGY

The ADM1 implemented in the simulation software package of MATLAB/Simulink was utilised for the mathematical modelling and simulations. While dynamic modelling of the various physicochemical processes involved during anaerobic digestion were performed by MATLAB, graphic representation of the whole process and designing of a biogas plant were performed with SIMBA software, developed by ifak system GmbH. With the building blocks available in the SIMBA library, a static biogas plant can be designed that can perform real-time dynamic simulations in the modelling environment. The software also facilitates virtually preparing biogas systems operating in differing conditions and situations, along with selecting and sizing the various reactors and equipment involved in a biogas plant. The SIMBA 6.6 version incorporated into MATLAB was employed and the processes and reactions described in the ADM1 model were studied. A simple, easy to construct and maintain, cost-effective biogas plant was constructed on this simulation platform.

18.3 THE ANAEROBIC DIGESTION MODEL (ADM1)

The ADM1 model was developed by the Task Group of the International Water Association in 2002 and remains the most advanced, widely applicable and competent mathematical model to simulate complex biogas systems (Batstone et al., 2002). The model is powerful and considers 31 total processes with 19 biochemical and 2 physicochemical processes to interpret the several dynamic processes, including the gas-liquid mass transfers involved during the conversion of organic materials to biogas (Satpathy et al., 2016). The ADM1 model considers the five major processes, namely disintegration, hydrolysis, acidogenesis, acetogenesis, and methanogenesis, involved during conversion of complex organic substrate (Schoen et al., 2009). Metabolic intermediates like hydrogen, acetic acid, propionic acid, butyric acid, and valeric acid are also considered in the model. Complex organic substrates (represented as XC fraction in the model) are considered to degrade majorly into four major fractions: carbohydrates (XCh), proteins (XPr), lipids (XLi), and an inert fraction (XI) to ultimately form methane after several consecutive processes. The inclusion of the inert fraction, which remains unaffected by any of the biological or physicochemical reactions, is identified to enhance the prediction capability to the ADM1 model. Enzymatic decomposition and reaction kinetics like substrate uptake, biomass decay, and disintegration constants are vital components of this model. The ADM1 model has remained an effective tool to investigate the static and dynamic behaviour of a biogas system involving minimal experimental work and has been found reliable in other parts of the world for a broad range of organic substrates like energy crops and farm wastes, including crop residues, sewage sludge, and industrial wastes (Biernacki et al., 2013).

18.4 DESIGN OF THE FARM-SCALE BIOGAS PLANT

A decentralised biogas plant was designed utilising MATLAB/Simulink based on the SIMBA 6.6 version to graphically design a biogas plant and combine numerical equations to describe the processes involved in the whole setup. SIMBA incorporates the ADM1 model for calculating the several biokinetic processes, including biomass decay, at each stage involved during anaerobic digestion (Schoen et al. 2009). A 60 m^3 anaerobic digester was designed in a virtual setup where 40 m^3 was liquid volume and the remaining 20 m^3 was collecting biogas (Figure 18.1). The input substrates for the biogas plant were rice straw co-digested with animal manure. Co-digestion has been determined to enable stability inside biogas digesters, along with providing enhanced

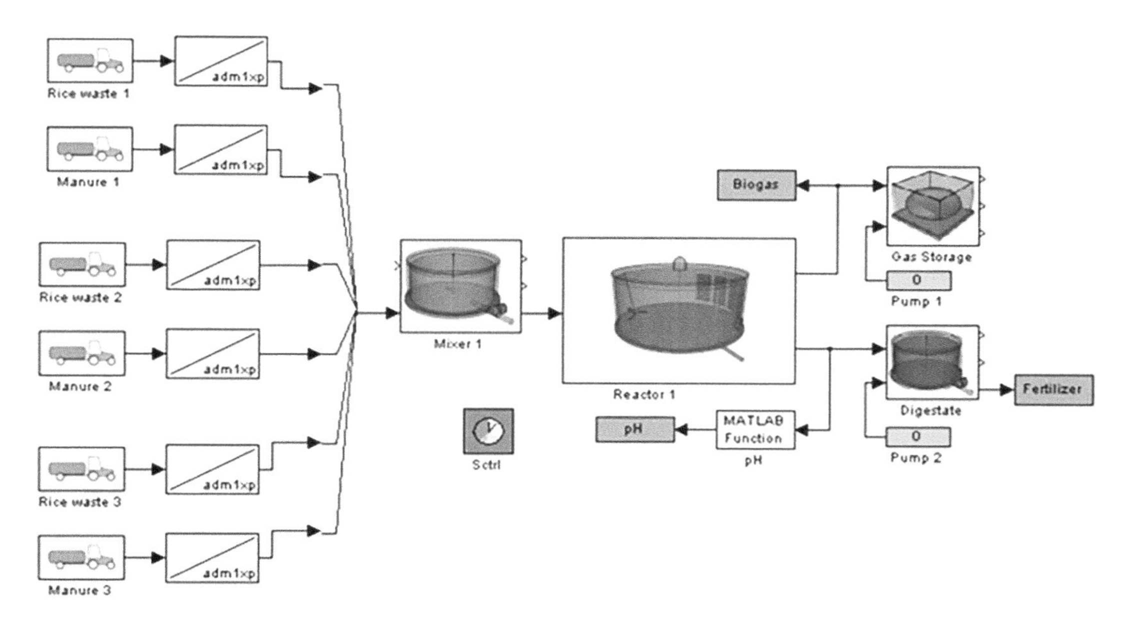

FIGURE 18.1 Model description of a rice crop residue-based biogas plant designed using SIMBA 6.6.

methane production. Other benefits of co-digesting include enhanced buffer capacity, diverse microbial consortia involved during anaerobic digestion, and dilution of toxic elements (Li et al., 2010). A mixing tank was prepared in the model to mix these co-substrates and was then fed into the anaerobic digester. A separate gas storage tank was arranged to collect the biogas, and the digestate was collected in a model block to store as fertiliser.

Simulations were performed with the help of SIMBA 6.6 based on MATLAB 2013a to perform the mathematical calculations (The MathWorks, 2011). The biogas system was simulated for 600 days to observe the digester's performance and understand the behaviour of such crop residue–based biogas system in a longer run.

18.5 MODEL INPUT PARAMETERS

Mathematical models describe a process and provide accurate predictions when provided with characterised data specific for the substrates. The biogas system designed in this study considered two substrates, rice straw and animal manure, which were fed thrice daily. Previously described experimental data from Daiem et al. (2018) and Biernacki et al. (2013) was considered for the rice straw and animal manure, respectively (Table 18.1). Substrate parameters such as the carbohydrate (cellulose, hemicellulose, lignin, etc.), protein, and fat content for the two co-substrates were introduced to the model. This step remarkably helps in improving the model's sensitivity to the specific substrates.

18.6 RESULTS AND DISCUSSIONS

Mathematical simulations performed for this cost-effective, farmer-oriented biogas setup demonstrated that crop leftovers like rice straw could be effectively used in combination with animal manure to generate biogas. Model predictions indicated the need for considering at least 2 m^3 of animal manure inside the biogas digester, failing which the model displayed digester instabilities. This is understood to be due to the high buffering capacity in manures, which enables stable operation inside a biogas digester (Li et al., 2010). Animal manure was kept constant at the aforesaid volume of 2 m^3, and different volumes of rice straw were mixed to determine the optimal feed composition

TABLE 18.1
Substrate Characterisation of Rice Straw and Animal Manure Considered in This Study

Parameter	Rice Straw[a] (%)	Animal Manure[b] (%)
Dry matter (DM)	93.63	9
Organic dry matter (ODM)	69.38	80
Nitrogen	0.52	3.5
Ammonium nitrogen	0.01	2.2
Raw protein	0.04	0.74
Raw lipid	0.085	0.17
Raw fibre	0.6	1.15
Inert fraction	0.3	0.509

[a] Daiem et al. (2018).
[b] Biernacki et al. (2013).

to enable maximum biogas and methane production and functions under stable conditions for a longer duration. As the rice straw volumes fed into the biogas reactor increased, simulations showed an increase in the total biogas yield (Figure 18.2). While the total biogas production gradually enhanced with the increased feed, simulations revealed an abrupt increase in the biogas production when the virtual reactor was fed at and above 0.43 m^3 of rice straw.

Interestingly, further analysis revealed that such improved total biogas yields with the increased feed volumes were not necessarily healthy for the biogas reactor. Although the high biogas production is encouraging, additional investigations into the dynamic processes occurring inside this biogas digester in the model implied the sudden enhanced biogas yields (after feeding 0.43 m^3/d) were due to the increased carbon dioxide and hydrogen production that eventually decreased the methane content in biogas to 0% with time. The overall digester performance and stability were also at stake as the increasing organic load beyond 0.42 m^3/d resulted in a sudden pH drop to a value of 4 (Figure 18.3). Further examinations of the model parameters and kinetic equations within the model matrix

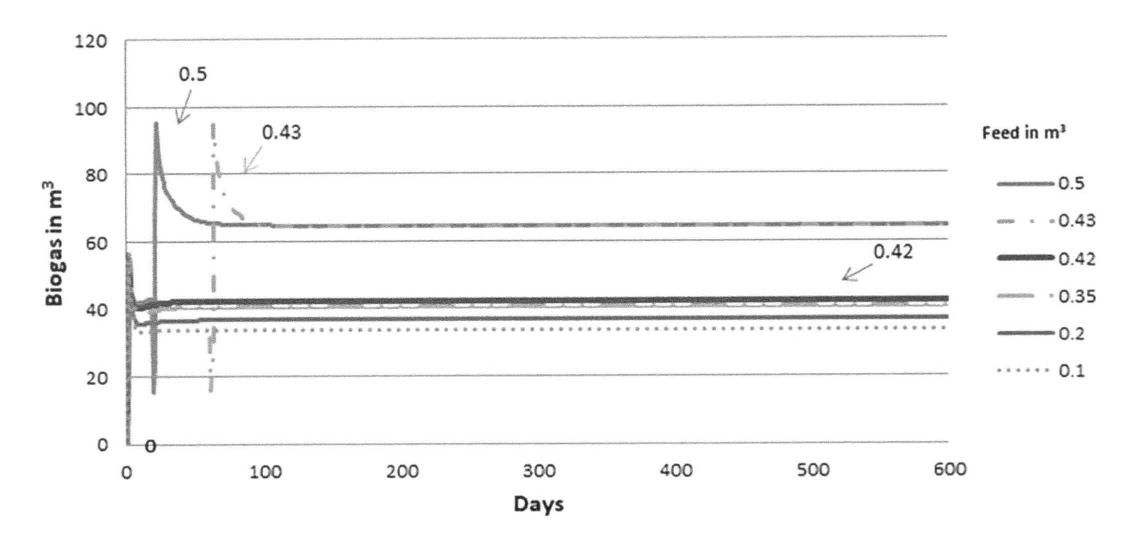

FIGURE 18.2 Total biogas production in the biogas digester when different volumes of rice crop residues are mixed with 2 m^3 of animal manure.

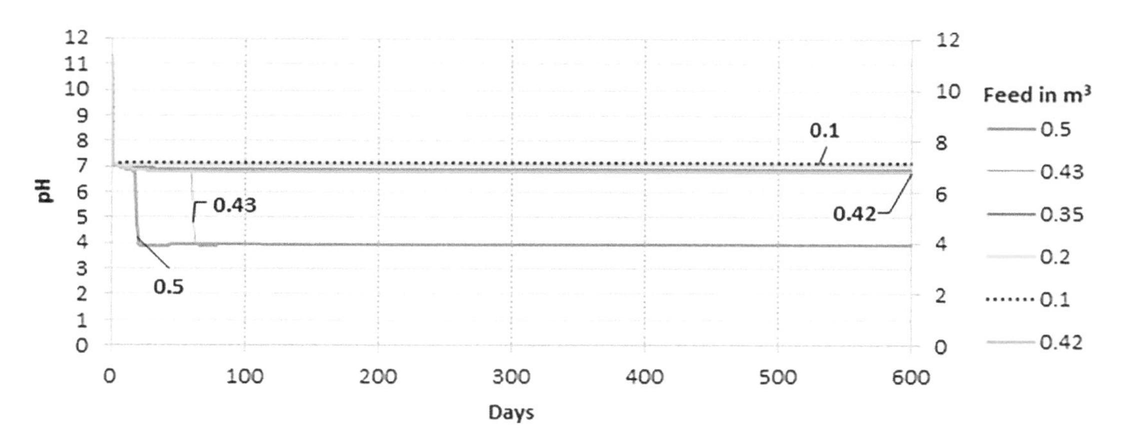

FIGURE 18.3 pH values inside the biogas digester predicted by the ADM1 model when different volumes of rice crop residues mixed with 2 m³ of animal manure.

demonstrated the sudden acidification inside the digester to be the reason for the pH drop. The model indicated the accumulation of intermediate organic acids, majorly acetic, propionic acid and lactic acid, inside the digester to be the major factor causing the abrupt pH drop. Volatile fatty acid accumulation has been recognised to play a significant role in influencing the overall performance of the biogas digester as they inhibit microbial activities and can even lead to complete digester breakdown (Zhang et al., 2018). These intermediate organic acids result in process imbalances and disturb the microbial populations, especially the methane-producing bacteria, which further disrupt the digester's performance and stable operation. Hence, a pH in the range of 6.5–8.0 has been widely recommended for stable operation and optimal performance of a biogas digester. Following the model predictions for the pH, a feed volume of 0.42 m³/d was thus found to be optimal for maintaining stability in the biogas plant for a longer duration.

When checked for the energy-rich methane yields in the biogas system, simulations clearly indicated the ideal feed to be 2 m³ of animal manure mixed with 0.42 m³ of rice straw. This co-digestion mixture ratio demonstrated stable conditions (with a pH maintained at 7), generating the maximum methane yield throughout the simulation run of nearly 2 years. A maximum methane yield of 21.6 m³ (constituting 50% of the total biogas yield) was predicted by the model in the designed farm-scale biogas digester (Figure 18.4). The results also clearly demonstrated how feeding rice

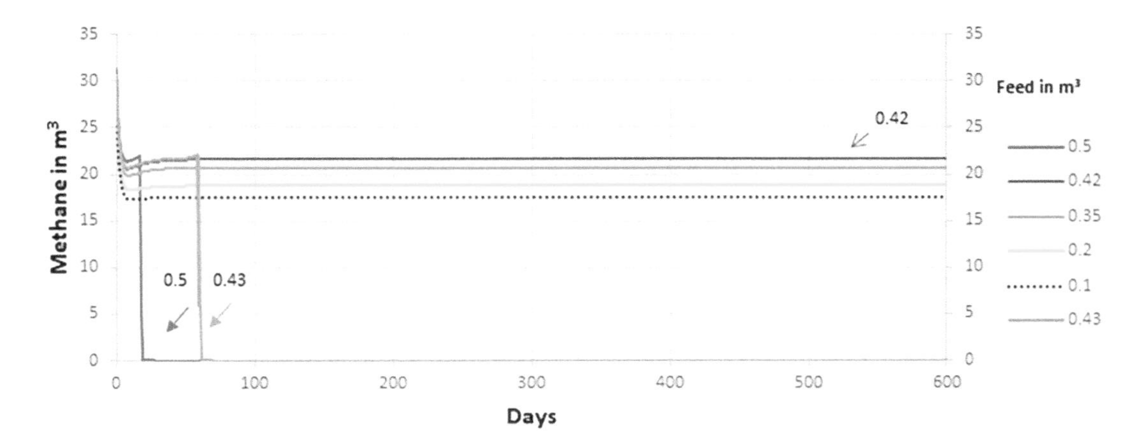

FIGURE 18.4 Methane production in the biogas digester when different volumes of rice crop residues were mixed with 2 m³ of animal manure.

straw even slightly above this volume to 0.43 m^3 showed an abrupt digester breakdown with methane yield reaching zero after few days of operation. As previously discussed, the accumulation of organic acids causing acidification in the reactor could be directly observed to play an influential role in decreasing the methane yields. Carbohydrate fermentation results in production of metabolic intermediates like acetic acid, lactic acid, and butyric acid, and the accumulation of these acids could explain the decrease in methane yields while the pH reduced. Clearly, the maximum permissible feed for such a system was determined to be 0.42 m^3, beyond which there is acidification resulting in the failure of the whole biogas process.

The ADM1 model could thus indicate the ideal design and feed volumes to attain optimal biogas production and limit inhibitions. The model also recommended feeding the digester equally at least three times in order to efficiently operate the biogas plant without the risk of reactor failures, especially in the longer run. Feeding the reactor just once showed substrate overload that eventually resulted in a digester breakdown due to acidification. The design proposed in this study is a farm-scale, decentralised biogas plant. One could also opt for the widely popular fixed dome or floating drum digesters that are historically trusted for their low investment cost requirements and easy design (Martins et al., 2014). The design proposed in this study could help in utilisation of crop residues and generate biogas to either find applications as cooking fuel or be further converted into electricity. Considering the calculations that 1 m^3 biogas can provide 1.7 kWh electricity (Shokri, 2011), the biogas generated in a farm-scale decentralised agriculture waste biogas plant could generate nearly 73 kWh electricity, energy sufficient to power 73 bulbs of 100 W for 10 hours each day. The digestate or bio-slurry could also be utilised as fertilisers that provide readily available nutrients to plants, offering benefits like improving the soil quality, unlike chemical fertilisers.

18.7 CONCLUSION

Stubble burning is a serious concern in India, damaging the ecosystem beyond repair while the government fails to curb these practices. Despite all the ill effects, CRB saves time, money, and efforts for farmers, and this nuisance largely continues. Generating biogas instead could offer a solution to treat different kinds of crop wastes in an affordable manner, along with providing an extra source of revenue for farmers and stakeholders. Biogas could be utilised as a source of renewable energy, and the by-products like the high-quality digestate from the biogas digesters can recycle nutrients back into the soil when utilised as organic fertilisers (Luostarinen et al., 2011). Producing biogas can thus be validated as a sustainable, effective waste management method, offering a possibility to generate energy from waste.

The mathematical ADM1 model in combination with the SIMBA simulation platform is a powerful and competent tool that describes the ideal design suiting the requirements and optimal operating conditions, along with demonstrating the behaviour of a biogas system under different desired scenarios. The ADM1 mathematical model provides a benefit as a time- and cost-effective tool that predicts the performance of a biogas plant with limited experiments and investments, thus favouring farmers in countries like India. This study demonstrated the ideal design and operational conditions to run a farm-scale biogas digester of 60 m^3, which could be fed with rice straw co-digested with animal manure. Co-digestion has been confirmed to generate more biogas due to the carbon to nitrogen ratio and trace metals required by the microorganisms involved during biogas formation (Schunurer and Jarvis, 2009). From an environmental perspective, increased generation of biogas is a sustainable way out for limiting the greenhouse gas emissions that would pollute the atmosphere both from agricultural wastes and animal manures.

ACKNOWLEDGEMENTS

The financial contribution to the first author was awarded by the Dr. D.S. Kothari Postdoctoral Fellowship, UGC India and is highly acknowledged. The authors extend their thanks to the Department of Science and Technology, UGC, Government of India for the support.

REFERENCES

Batstone, D., Keller, J., Angelidaki, I., Kalyuzhnyi, S.V., Pavlostathis, S.G., Rozzi, A., Sanders, W.T., Siegrist, H. and Vavilin, V.A. (2002) Anaerobic Digestion Model No. 1. International Water Association Water Science Technology, 45 (10): 65–73.

Biernacki, P., Steinigeweg, S., Borchert, A. and Uhlenhut, F. (2013) Application of Anaerobic Digestion Model No. 1 for Describing Anaerobic Digestion of Grass, Maize, Green Weed Silage, and Industrial Glycerine. Bioresource Technology [online], 127 (1): 188–194. http://dx.doi.org/10.1016/j.biortech.2012.09.128.

Bisen, N. and Rahangdale, C. (2017) Crop Residues Management Option for Sustainable Soil Health in Rice-Wheat System: A Review. International Journal of Chemical Studies [online], 5 (4): 1038–1042. Available at: https://www.researchgate.net/publication/318959582_Crop_residues_management_option_for_sustainable_soil_health_in_rice-wheat_system_A_review.

Daiem, M., Said, N. and Negm, A.M. (2018) Potential Energy from Residual Biomass of Rice Straw and Sewage Sludge in Egypt. Procedia Manufacturing [online], 22: 818–825. Available from https://www.sciencedirect.com/science/article/pii/S2351978918304128.

Devi, S., Gupta, C., Jat, S.L. and Parmar, M.S. (2017) Crop Residue Recycling for Economic and Environmental Sustainability: The Case of India. Open Agriculture [online], 2 (1): 486–494. Available from http://www.degruyter.com/view/j/opag.2017.2.issue-1/opag-2017-0053/opag-2017-0053.xml

Gumisiriza, R., Funa, J., Okure, M. and Hensel, H. (2017) Biomass Waste-to-Energy Valorisation Technologies: A Review Case for Banana Processing in Uganda. Biotechnology for Biofuels, 10 (11): 1–29. doi: 10.1186/s13068-016-0689-5.

Li, R., Chen, S. and Li, X. (2010) Biogas Production from Anaerobic Co-Digestion of Food Waste with Dairy Manure in a Two-Phase Digestion System. Applied Biochemistry and Biotechnoogy, 643–654.

Luostarinen, S., Normak, A. and Edström, M. (2011) Overview of Biogas Technology. Baltic MANURE WP6 Energy potentials [online], pp. 41–42. Available from http://www.build-a-biogas-plant.com/PDF/baltic_manure_biogas_final_total.pdf

Martins, M.R., Schleder, A.M. and Droguett, E.L. (2014) A Methodology for Risk Analysis Based on Hybrid Bayesian Networks: Application to the Regasification System of Liquefied Natural Gas Onboard a Floating Storage and Regasification Unit. Risk Analysis [online], 34 (12): 2098–2120. http://doi.wiley.com/10.1111/risa.12245.

Sample, D. (2019) Air Pollution from India's Stubble Burning Leads to USD 35 Billion Economic Losses, Poses Significant Health Risk. International Food Polciy Research Institute [online], Available from https://www.ifpri.org/news-release/new-study-air-pollution-indias-stubble-burning-leads-usd-35-billion-economic-losses

Satpathy, P., Biernacki, P., Cypionka, H. and Steinigeweg, S. (2016) Modelling Anaerobic Digestion in an Industrial Biogas Digester: Application of Lactate-including ADM1 Model (Part II). Journal of Environmental Science and Health, 51 (14): 1226–1232.

Schoen, M.A., Sperl, D., Gadermaier, M., Goberna, M., Franke-Whittle, I., Insam, H., Ablinger, J. and Wett, B. (2009) Population Dynamics at Digester Overload Conditions. Bioresource Technology, 100 (23), 5648–5655.

Schunurer, A. and Jarvis, A. (2009) Microbiological Handbook for Biogas Plants. Swedish Waste Management, Swedish Gas Centre, Malmö, pp. 1–74.

Seadi, T.A., Rutz, D., Prassl, H., Köttner, M., Finsterwalder, T., Volk, S. and Janssen, R. (2008) 'Biogas Handbook'. in Big East Project [online], Esbjerg, Denmark: University of Southern Denmark, Esbjerg, pp. 1–126. Available from www.lemvigbiogas.com.

Shokri, S. (2011) Biogas Technology, Applications, Perspectives and Implications. International Journal of Agricultural Science and Research, 2 (3): 53–60.

Singh, R., Chanduka, L. and Dhir, A. (2015) Impacts of Stubble Burning on Ambient Air Quality of a Critically Polluted Area– Mandi-Gobindgarh. Journal of Pollution Effects & Control, 03 (02): 1–7. doi: 10.4172/2375-4397.1000135.

The MathWorks (2011) MATLAB – Optimization ToolboxTM 6, User's Guide. Natick, MA.

Yang, Y. and Sheng, G. (2003) Enhanced Pesticide Sorption by Soils Containing Particulate Matter from Crop Residue Burns. Environmental Science & Technology, 37 (16): 3635–3639.

Zhang, W.L., Xing, W.L. and Li, R.D. (2018) Real-time Recovery Strategies for Volatile Fatty Acid-Inhibited Anaerobic Digestion of Food Waste for Methane Production. Bioresource Technology, 265: 82–92.

19 A Review of Operating Parameters, Pretreatment Process, and Digester Design Effect on Biogas Production in Anaerobic Digestion

Amit Kumar Sharma, Ravi K. Dwivedi, and Bharat Modhera

19.1 INTRODUCTION

Anaerobic digestion (AD) is a process in which complex biochemical reactions are involved to decompose biodegradable waste like manure, food waste, and sewage sludge into biogas, which can be used as fuel to fulfil various energy demands like heating, electric power, and engine fuel. (Deepanraj, Sivasubramanian and Jayaraj, 2017). AD occurs in the absence of oxygen and includes four stages by which waste is decomposed and converted in biogas. Stages of the AD process are hydrolysis, fermentation, acetogenesis, and methanogenesis. In hydrolysis, carbohydrates, fats, and proteins are converted to sugar, amino acids, and fatty acids. In fermentation, the products of hydrolysis convert into volatile fatty acid and subproducts like acetic acid, hydrogen, and carbon dioxide, due to the biochemical reactions of available bacteria. In acetogenesis, acetic acid, carbon dioxide, and hydrogen produce acetate, which is further used in the methanogenesis stage by bacteria to generate biogas (Kondusamy and Kalamdhad, 2014). Biogas is composed of methane, carbon dioxide, and trace amounts of hydrogen sulfide and ammonia. Calorific value of biogas is around 21.48 MJ/m^3 (Monnet, 2003). A comparison of various gas compositions is shown in Table 19.1.

19.2 AD PROCESS STAGES

Figure 19.1 shows the stages and activities involved in the AD process (Jain et al., 2015).

19.3 OPERATING PARAMETERS

There are some design parameters that are very important for the AD process. By changing these parameters, the methane production rate may be changed. For designing an AD plant, optimised value of these parameters are required for efficient operation. The details of some important parameters are further discussed.

19.3.1 Total Solid Content

Total solids (TS) content affects the volume of the digester. As the percentage of TS in waste increases, the required amount of water decreases. The net volume of the digester will also decrease (Monnet, 2003). Percentage of TS also affects methane output. Its optimum percentages are different for different materials. For food waste, if the percentage of TS increases, the methane output will decrease (Liotta et al., 2014). According to Riya et al., 18% to 27% TS will increase the methane

DOI: 10.1201/9781003204435-22

TABLE 19.1
Comparison of Biogas, Natural Gas, and Landfill Gas (Monnet, 2003)

Component	Units	Biogas	Landfill Gas	Natural Gas
Methane	Vol%	55–70	45–58	91
Carbon dioxide	Vol%	30–45	32–45	0.61
Hydrogen sulphide	ppm	~500	~10–200	~1
Nitrogen	Vol%	0–2	0–3	0.32

yield in thermophilic AD of rice straw and pig manure. Further increase TS will start volatile fatty acid accumulation in the digester, which lowers the methane yield (Riya et al., 2018). Aboudi et al. reported that co-digestion of cow manure with sugar beet by-products gives maximum methane yield at 8% TS (Aboudi, Álvarez-Gallego and Romero-García, 2017). It is noted that optimum TS to enhance methane production in AD is different for different feedstock, so for proper operation of digester with optimum output, TS needs to be analysed regularly and maintained accordingly.

19.3.2 Organic Loading Rate

Organic loading rate (OLR) measures the amount of input to the anaerobic digester. Too much input can lead to the accumulation of the acids, which leads to low biogas yield. If OLR is too high, this will lead to system failure (Monnet, 2003). Per the experiment conducted by Elsayed et al. (Elsayed et al., 2016), maximum methane yield observed for primary sludge and wheat straw at optimum OLR (7.50 gVS/L) was 345.5 mL/g VSadd. It was also observed that minimum methane yield occurred at an OLR of 3 and 12 gVS/L. It is clear that AD should be done at optimum OLR. At low OLR, the microbial activity in the digester is reduced and leads to low methane yield. At high OLR, volatile fatty acids increase in the digester, which take time to be consumed and toxic condition formation occurs in the digester, leading to low methane yield.

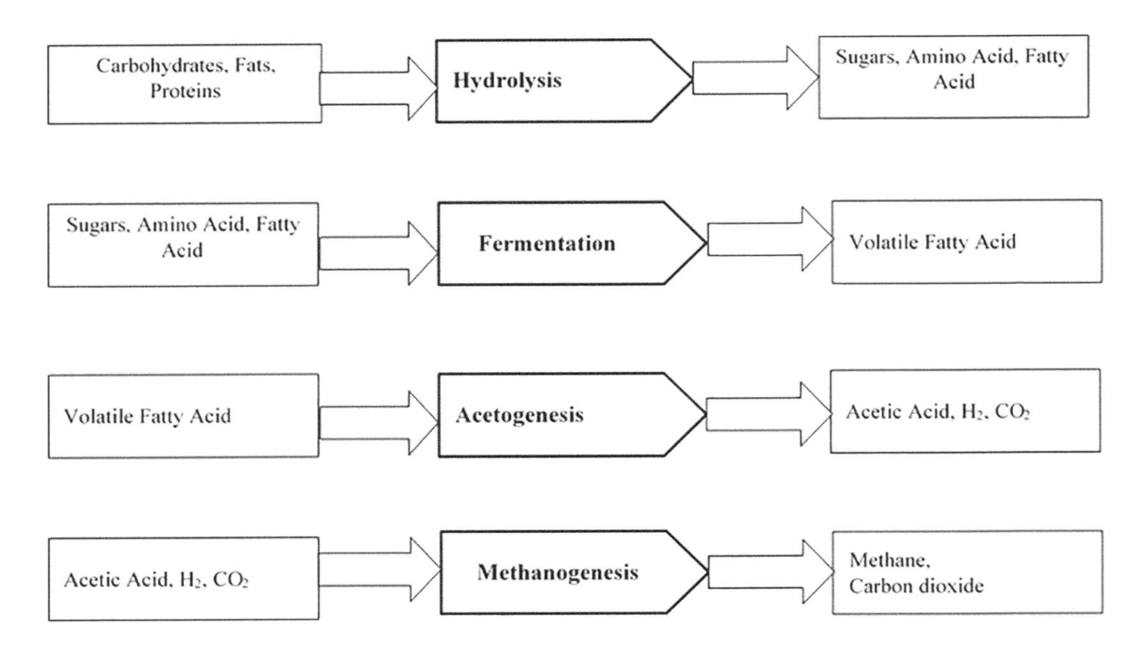

FIGURE 19.1 AD process stages.

19.3.3 Temperature

Temperature is a very important design parameter because it affects the enzyme and bacteria growth and reaction, which leads to the methane production rate. There are two optimum conditions for temperature: mesophilic range (35–40°C) and thermophilic range (50–65°C) (Kondusamy and Kalamdhad, 2014). According to Chae et al. (Chae et al., 2008), AD of swine manure shows high methane yield at 30°C and 35°C, as compared to 25°C. However, methane yield is not linear with temperature increase. According to Kim et al. (Kim et al., 2006), AD of food waste at a temperature range from 40 to 55°C gave maximum methane yield at 50°C for 10 days retention time. Because higher temperatures in AD lead to a higher reaction rate, fast feedstock decomposition and the fast growth of bacteria are responsible for methane in the thermophilic range, as compared to the mesophilic temperature range.

19.3.4 pH

As there are various biochemical reactions associated with AD, optimal pH varies in different stages of the AD process. An AD digester will work efficiently between a pH of 6.5 and 7.5. In this range, required bacteria for the AD stages are very active and maximum methane output has been observed (Jain et al., 2015). For food waste, pH is very sensitive to acid accumulation because a higher concentration of volatile fatty acids in the acetogenesis stage reduces the pH level to 5 or below. This leads to the destruction of the methanogenic bacteria responsible for methane production and hampers the bacterial activity, which can be a reason for digester failure. To maintain the pH level in a digester, alkaline chemical like sodium hydroxide can be used to maintain smooth operation (Mir, Hussain and Verma, 2016). According to Zhang et al., pH adjustment to 7 for AD of kitchen waste increases the volatile fatty acid rate and the solubilisation rate, which further enhances the hydrolysis and acidogenesis process, leading to an increase in methane output (Zhang et al., 2005).

19.3.5 Retention Time

Retention time is the total time required to complete decomposition of feedstock. The optimal retention time depends on the temperature of the digester. For the mesophilic range, retention time varies from 15 to 30 days, and for the thermophilic range it varies from 12 to 14 days (Monnet, 2003). According to Elsayed et al., during experimentation of anaerobic co-digestion of primary sludge and wheat straw, it was observed that maximum methane yield occurred in the first 20 days under mesophilic conditions (Elsayed et al., 2016). In batch-type anaerobic digesters, retention time is per the ambient temperature and digester temperature, but for semi-continuous and continuous digesters, it varies with loading rate. If loading rate is very frequent, the retention time is lower. Very low retention time leads to incomplete decomposition of biodegradable material and a decrease in methane output, so there should be a balance between retention time and loading rate for optimum output and proper digester operation (Meegoda et al., 2018).

19.3.6 C:N Ratio

The carbon to nitrogen (C:N) ratio indicates the proportion of carbon and nitrogen in the organic feedstock for anaerobic digesters. C:N ratio affects methane production. High C:N ratio causes the immediate consumption of nitrogen by the bacteria, which produces lower methane yield. High C:N ratio increases the pH of the system, which leads to accumulation of volatile fatty acids and low methane yield. The optimal C:N ratio suggested by various research is 20 to 30 (Monnet, 2003; Li, Park and Zhu, 2011; Puyuelo et al., 2011). Zhang et al. observed that different feedstocks give optimum result of methane production at different C:N ratio, and that optimal C:N ratio also depends on inoculum (Zhang et al., 2014). For enhanced production of methane in AD, the correct ratio of carbon and nitrogen is required in the digester.

19.4 PRETREATMENT

Pretreatment of input feedstock for a digester is required to remove the non-biodegradable material, which does not contribute to methane production. Pretreatment should also reduce the particle size to increase surface area for faster disintegration of fats, carbohydrates, and proteins. There are various pretreatment processes available to enhance biogas production for various types of feedstock. Some of the important pretreatment processes are listed and further discussed in the following sections (Kondusamy and Kalamdhad, 2014; Zhang et al., 2014). In the AD process, hydrolysis is the first step in the disintegration of feedstocks, breaking complex organic matter like carbohydrates and protein into soluble molecules like amino acids, fatty acids, and sugar, which are further processed during fermentation. Hydrolysis is the rate-limiting process of AD, because if completed at a faster rate, all subsequent processes will also start early and methane production will enhance with less retention time (Shefali and Themelis, 2002; Kondusamy and Kalamdhad, 2014). By using various pretreatment methods, the rate of hydrolysis increases and methane production is enhanced (Zhang et al., 2014).

Methods of pretreatment are:

1. Mechanical
2. Ultrasonic
3. Microwave
4. Thermal
5. Chemical
6. Biological

19.4.1 Mechanical Pretreatment

In mechanical pretreatment, lager solid particles are disintegrated into smaller particles by means of mechanical processes like grinding or shearing. In this way, the surface area of the substrate is increased, which helps in faster reaction with bacteria for enhanced biogas production in AD (Ariunbaatar et al., 2014). According to Rodriguez et al., when a hollander beater is used for pretreatment of paper pulp, in which the beater uses shear action, high pressure, and fast blades for feedstock processing, methane production increased 21% compared to untreated feedstock (Rodriguez et al., 2017). According to Kang et al., pretreatment of hybrid pennisetum incresed the methane production when size of particle ground up to 0.25 mm (Kang et al., 2019). Dahunsi et al. conculded that mechanical pretreatment of lignocelluloses decresed the lag time during the hydrolysis process by the degradation of structural material and increased biogas production up to 20% (Dahunsi, 2019).

19.4.2 Ultrasonic Pretreatment

Ultrasonic pretreatment is used to shatter the bacterial cell and floc structure of material. Hydrolytic bacteria react faster with the intercellular organic substrate of feedstock, which increases the hydrolysis rate and enhances methane generation (Wang et al., 1999). Xu et al. concluded that ultrasonic pretreatment of municple sludge decreases the retention time and increases the methane production due to an increase in the hydrolysis rate for the AD process (Xu et al., 2019). Lizama et al. concluded that ultrasonic pretreatment of waste-activated sludge with increased OLR enhances methane production without acidification in the anaerobic digester due to the higher OLR, as compared to untreated feedstock. Due to the sonication of waste-activated sludge, organic matter reduction increased; at higher OLR there was no pH variation from optimum range, which enhanced biogas production (Lizama et al., 2018).

19.4.3 Microwave Pretreatment

Microwave interaction affects enzyme activity in food, breaking the hydrogen bonds in protein due to the thermal energy of irradiation (Ponne and Bartels, 1995). According to Marin et al., microwave pretreatment of kitchen waste increases the release of bound water from feedstock, which helps in dewatering. This pretreatment also increases the solubilisation rate, which helps in the hydrolysis stage and enhances biogas production (Marin, Kennedy and Eskicioglu, 2010). Zhang et al. observed that microwave pretreatment breaks down microbial cells and allows the intercellular material to react with bacteria in the hydrolysis stage and increases the hydrolysis rate, which further improves the AD process (Zhang et al., 2016).

19.4.4 Thermal Pretreatment

Thermal pretreatment is mostly done in a temperature range of 50°C to 250°C to break the cell structure and increase the solubilisation rate, decreasing the retention time and facilitating a faster AD process with enhanced biogas generation (Ariunbaatar et al., 2014). McVoitte and Clark showed that thermal pretreatmnet of cow manure does not provide significant improvement in biogas production (McVoitte and Clark, 2019). According to Barua, Goud, and Kalamdhad, enhanced biogas generation occcured by using thermally pretreated water hyacinth co-digested with banana peel. This incresed the hydrolysis rate of AD (Barua, Goud and Kalamdhad, 2018). Li et al. concluded that high methane yield and decrease in oil, fat, and grease occurred in AD of thermally pretreated kitchen waste. Thermal pretreatment increases in the speed of the disintegration of organic substrate of feedstock and the solubilisation rate of organics (Li et al., 2018).

19.4.5 Chemical Pretreatment

Chemical pretreatment is applied to feedstock to degrade the material quickly by breaking the covalent bond of the material structure. This allows faster disintegration of organic material, which increases the hydrolysis rate and the reaction between substrate and bacteria increases, reducing retention time and enhancing biogas generation in AD (Hendriks and Zeeman, 2009; Yu, Liu, Li and Ma, 2019). The prevalent chemicals used in pretreatment are alkali, acidic, or oxidised. In alkali pretreatment, the chemical reaction causes an enlargement of solid substrate, which increases the surface area of substrate available to react with microbes, enhancing the AD process (Ariunbaatar et al., 2014). According to Calabro et al., alkali pretreatment of tomato-processing waste does not enhance methane production, but in some cases, due to increase in pH level, reduces the process rate and lowers methane yield (Calabrò et al., 2015). The effect of alkali pretreatment with extrusion on rice straw shows an increase in methane generation compared to untreated feedstock (Zhang et al., 2015). As the result of chemical pretreatment varies with feedstock, for enhanced biogas generation, material behaviour should be studied with the pretreatment method chosen.

19.4.6 Biological Pretreatment

In biological pretreatment, an incolumn of bacteria is made, and the bacteria are helpful in the disintegration of organic substrate by breaking the cell structure and exposing the cellular organics to cause a faster reaction with the bacteria responsible for AD. In this, the proper bacteria should be selected to disintegrate the feedstock faster, help in the hydrolysis and solubilisation processes, and increase their population in the system (Yu, Liu, Li and Ma, 2019). Barua, Goud, and Kalamdhad observed that biological pretreatment of water hyacinth with selected bacterial cultures accelerated the hydrolysis process and degradability of organic substrate by breaking the lignocellulosic complex structure of water hyacinth and enhanced the methane generation rate and solubilisation as compared to untreated feedstock (Barua, Goud and Kalamdhad, 2018). Biological pretreatment of corn straw also showed increase in TS and decrease in the volatile solid and hydraulic retention time, which enhanced the biogas production up to 33.07% as compared to untreated substrate. It also increased the speed of the degradability of complex organics, which plays an important role in the AD process (Zhong et al., 2011).

19.5 DIGESTER DESIGN

There are various designs available for AD for different types of feedstock inputs. In digester design, various design parameters are kept in mind for efficient operation of the digester, including location, size, operating methods, and feedstock properties. Some digester types are listed and discussed in this section (Wilkie, 2005).

1. Covered lagoon
2. Complete mix
3. Plug-flow
4. Fixed-film

Covered lagoon digesters are used where large quantities of manure or sludge with less than 2% solid content are used. In this, a cover is used on the lagoon, which is used to store the biogas that is generated. It operates on the environment temperature, so it requires high hydraulic retention time varying from 35 to 60 days, depending upon ambient temperature. It is difficult to maintain the temperature range of mesophilic AD. In winter, biogas yield is very low and retention time increases to 60 days. In two-cell covered lagoon digesters, the first cell is covered and the second is open to the environment; solid sludge in the first cell has a long solid retention time. As the cost of this type of system is low, it is widely used in the agriculture and dairy industries (Wilkie, 2005; Hamilton, 2014; Mutungwazi, Mukumba and Makaka, 2018). Sharpe and Harper reported that methane yield from swine manure lagoon decreased in winter and pH of liquid and solid sludge also decreased. Ambient temperature has a critical role in lagoon-type anaerobic digesters (Sharpe and Harper, 1999).

A complete-mix digester is a closed-tank digester with intermittent mixing provisions. It operates in mesophilic conditions or thermophilic conditions by maintaining the temperature through a heating system, using either a hot water jacket or heat exchanger coil. Insulation is also provided on the digester for constant temperature operation. For mesophilic conditions, retention time varies from 21 to 35 days, which can be lowered in thermophilic conditions to 12 to 20 days. Mixing can be done either by a mechanical mixer or by recirculation of biogas in the system. There is one inflow port for feedstock input and one outflow port from which an equal quantity of input material flows out from the digester. In this digester, feedstock with a TS content from 3% to 10% can be used (Wilkie, 2005; Hamilton, 2014; Mutungwazi, Mukumba and Makaka, 2018).

A plug-flow digester is a horizontal tubular digester made of concrete or steel with insulation for temperature control and a cover for biogas storage. Digester content is not mixed in this system. This type of digester supports the solid content from 11% to 14%, with a retention time of 20 to 30 days under mesophilic conditions. In this system, there is an input port at one end and an output port at opposite end from which slurry will flow. Cost associated with this type of digester is low due to its simple design. This type of digester is used in industrial- or farm-scale AD (Hamilton, 2014; Mutungwazi, Mukumba and Makaka, 2018).

Fixed-film digesters are used to handle solid content from 1% to 3%, with a retention time of only 3 to 5 days. In this digester, methane-forming bacteria immobilise on supporting media like wood chips or plastic and grows on this, making a thin coating called biofilm. The biofilm formation prevents washout of bacteria, and they are retained in the system for a longer period of time. Biogas is generated in this digester when feedstock passes through the media. This type of digester is best suited for flushed manure. Biogas generation rate depends on the ambient temperature. For efficient operation, the biofilm has to be changed periodically (Wilkie, 2005; Ghosh and Bhattacherjee, 2013; Mutungwazi, Mukumba and Makaka, 2018).

19.6 CONCLUSION

This review specifies various key operating parameters, pretreatments that are useful in preprocessing of feedstock, and designs of anaerobic digesters. In this chapter, various recent and past literature was reviewed, and it can be concluded that the study of operating parameters and various pretreatment methods is essential, since the application of these on AD is very important for efficient operation and enhanced biogas yield. The survey of various types of physical designs of anaerobic digesters, including their type, operation, required feedstock properties, and working principles are also in the scope of discussion in this chapter. It is noteworthy to mention that anaerobic digester design can be selected for maximum biogas generation and efficient operation without failure.

REFERENCES

Aboudi, K., Álvarez-Gallego, C. J. and Romero-García, L. I. (2017) 'Influence of total solids concentration on the anaerobic co-digestion of sugar beet by-products and livestock manures', *Science of the Total Environment*, 586, pp. 438–445. doi: 10.1016/j.scitotenv.2017.01.178.

Ariunbaatar, J., Panico, A., Esposito, G., Pirozzi, F. and Lens, P. N. L. (2014). 'Pretreatment methods to enhance anaerobic digestion of organic solid waste', *Applied Energy*, 123, pp. 143–156. https://doi.org/10.1016/j.apenergy.2014.02.035

Barua, V. B., Goud, V. V. and Kalamdhad, A. S. (2018) 'Microbial pretreatment of water hyacinth for enhanced hydrolysis followed by biogas production', *Renewable Energy*, 126, pp. 21–29. doi: 10.1016/j.renene.2018.03.028.

Calabrò, P. S., Greco, R., Evangelou, A. and Komilis, D. (2015). 'Anaerobic digestion of tomato processing waste: Effect of alkaline pretreatment', Journal of Environmental Management, 163, pp. 49–52

Chae, K. J. *et al.* (2008) 'The effects of digestion temperature and temperature shock on the biogas yields from the mesophilic anaerobic digestion of swine manure', *Bioresource Technology*, 99(1), pp. 1–6. doi: 10.1016/j.biortech.2006.11.063.

Dahunsi, S. O. (2019) 'Mechanical pretreatment of lignocelluloses for enhanced biogas production: Methane yield prediction from biomass structural components', *Bioresource Technology*, 280, pp. 18–26. doi: 10.1016/j.biortech.2019.02.006.

Deepanraj, B., Sivasubramanian, V. and Jayaraj, S. (2017) 'Effect of substrate pretreatment on biogas production through anaerobic digestion of food waste', *International Journal of Hydrogen Energy*, 42(42), pp. 26522–26528.

Elsayed, M. *et al.* (2016) 'Effect of VS organic loads and buckwheat husk on methane production by anaerobic co-digestion of primary sludge and wheat straw', *Energy Conversion and Management*, 117, pp. 538–547. doi: 10.1016/j.enconman.2016.03.064.

Ghosh, R. and Bhattacherjee, S. (2013) 'A review study on anaerobic digesters with an Insight to biogas production', *International Journal of Engineering Science Invention ISSN* (Online): 2319–6734, 2(3), pp. 8–17.

Hamilton, D. W. (2014) 'Anaerobic digestion of animal manures : Types of digesters', *Oklahoma Cooperative Extension Service*, BAE-1750, pp. 1–4.

Hendriks, A. T. W. M. and Zeeman, G. (2009) 'Pretreatments to enhance the digestibility of lignocellulosic biomass', *Bioresource Technology*, 100(1), pp. 10–18. doi: 10.1016/j.biortech.2008.05.027.

Jain, S. *et al.* (2015) 'A comprehensive review on operating parameters and different pretreatment methodologies for anaerobic digestion of municipal solid waste', *Renewable and Sustainable Energy Reviews*, 52, pp. 142–154. doi: 10.1016/j.rser.2015.07.091.

Kang, X., Zhang, Y., Song, B., Sun, Y., Li, L., He, Y., Kong, X., Luo, X. and Yuan, Z. (2019). 'The effect of mechanical pretreatment on the anaerobic digestion of Hybrid Pennisetum', *Fuel*, 252, pp. 469–474. https://doi.org/10.1016/j.fuel.2019.04.134

Kim, J. K. *et al.* (2006) 'Effects of temperature and hydraulic retention time on anaerobic digestion of food waste', *Journal of Bioscience and Bioengineering*, 102(4), pp. 328–332. doi: 10.1263/jbb.102.328.

Kondusamy, D. and Kalamdhad, A. S. (2014) 'Pre-treatment and anaerobic digestion of food waste for high rate methane production–A review', *Journal of Environmental Chemical Engineering*, 2(3), pp. 1821–1830.

Li, Y. *et al.* (2018) 'Corrigendum to "Effects of thermal pretreatment on degradation kinetics of organics during kitchen waste anaerobic digestion" [Energy 118 (2017) 377–386] (S0360544216318461) (10.1016/j.energy.2016.12.041))', *Energy*, 153, p. 1089. doi: 10.1016/j.energy.2018.04.162.

Li, Y., Park, S. Y. and Zhu, J. (2011) 'Solid-state anaerobic digestion for methane production from organic waste', *Renewable and Sustainable Energy Reviews*, 15(1), pp. 821–826. doi: 10.1016/j.rser.2010.07.042.

Liotta, F. *et al.* (2014) 'Effect of total solids content on methane and volatile fatty acid production in anaerobic digestion of food waste', 32(10), pp. 947–953.

Lizama, A. C. *et al.* (2018) 'Effect of ultrasonic pretreatment on the semicontinuous anaerobic digestion of waste activated sludge with increasing loading rates', *International Biodeterioration and Biodegradation*, 130, pp. 32–39. doi: 10.1016/j.ibiod.2018.03.013.

Marin, J., Kennedy, K. J. and Eskicioglu, C. (2010) 'Effect of microwave irradiation on anaerobic degradability of model kitchen waste', *Waste Management*, 30(10), pp. 1772–1779. doi: 10.1016/j.wasman.2010.01.033.

McVoitte, W. P. A. and Clark, O. G. (2019) 'The effects of temperature and duration of thermal pretreatment on the solid-state anaerobic digestion of dairy cow manure', *Heliyon*, 5(7), p. e02140. doi: 10.1016/j.heliyon.2019.e02140.

Meegoda, J. N. *et al.* (2018) 'A review of the processes, parameters, and optimization of anaerobic digestion', *International Journal of Environmental Research and Public Health*. Multidisciplinary Digital Publishing Institute, 15(10), p. 2224.

Mir, M. A., Hussain, A. and Verma, C. (2016) 'Design considerations and operational performance of anaerobic digester: A review', *Cogent Engineering*. Taylor & Francis, 3(1), p. 1181696.

Monnet, F. (2003) 'An introduction to anaerobic digestion of organic wastes EXECUTIVE SUMMARY', 379, p. 45. Available at: https://www.cti2000.it/Bionett/BioG-2003-002%20IntroAnaerobicDigestion.pdf

Mutungwazi, A., Mukumba, P. and Makaka, G. (2018) 'Biogas digester types installed in South Africa: A review', *Renewable and Sustainable Energy Reviews*, 81, pp. 172–180. doi: 10.1016/j.rser.2017.07.051.

Ponne, C. T. and Bartels, P. V. (1995) 'Interaction of electromagnetic energy with biological material - relation to food processing', *Radiation Physics and Chemistry*, 45(4), pp. 591–607. doi: 10.1016/0969-806X(94)00073-S.

Puyuelo, B. *et al.* (2011) 'Determining C/N ratios for typical organic wastes using biodegradable fractions', *Chemosphere*, 85(4), pp. 653–659. doi: 10.1016/j.chemosphere.2011.07.014.

Riya, S. *et al.* (2018) 'The influence of the total solid content on the stability of dry-thermophilic anaerobic digestion of rice straw and pig manure', *Waste Management*, 76, pp. 350–356. doi: 10.1016/j.wasman.2018.02.033.

Rodriguez, C. *et al.* (2017) 'Mechanical pretreatment of waste paper for biogas production', *Waste Management*, 68, pp. 157–164. doi: 10.1016/j.wasman.2017.06.040.

Sharpe, R. R. and Harper, L. A. (1999) 'Methane emissions from an anaerobic swine lagoon', *Atmospheric Environment*, 33(22), pp. 3627–3633. doi: 10.1016/S1352-2310(99)00104-1.

Shefali, V. and Themelis, N. J. (2002) 'Anaerobic digestion of biodegradable organics in municipal solid waste', Doctor of. Available at: http://www.seas.columbia.edu/earth/vermathesis.pdf.

Wang, Q. *et al.* (1999) 'Upgrading of anaerobic digestion of waste activated sludge by ultrasonic pretreatment', *Bioresource Technology*. Elsevier, 68(3), pp. 309–313.

Wilkie, A. C. (2005) 'Anaerobic digestion of dairy manure: Design and process considerations', Dairy Manure Management: Treatment, Handling, and Community Relations, p. 301–312. Available at: https://biogas.ifas.ufl.edu/publs/nraes176-p301-312-mar2005.pdf.

Xu, X. *et al.* (2019) 'Study on ultrasonic treatment for municipal sludge', *Ultrasonics Sonochemistry*, 57, pp. 29–37. doi: 10.1016/j.ultsonch.2019.05.008.

Yu, Q., Liu, R., Li, K. and Ma, R. (2019) 'A review of crop straw pretreatment methods for biogas production by anaerobic digestion in China', *Renewable and Sustainable Energy Reviews*, 107, pp. 51–58. doi: 10.1016/j.rser.2019.02.020.

Zhang, B. *et al.* (2005) 'The Influence of pH on Hydrolysis and Acidogenesis of Kitchen Wastes in Two-phase Anaerobic Digestion', *Environmental Technology. Taylor & Francis*, 26(3), pp. 329–340. doi: 10.1080/09593332608618563.

Zhang, C. *et al.* (2014) 'Reviewing the anaerobic digestion of food waste for biogas production', *Renewable and Sustainable Energy Reviews*, 38, pp. 383–392.

Zhang, J. *et al.* (2016) 'Optimization and microbial community analysis of anaerobic co-digestion of food waste and sewage sludge based on microwave pretreatment', *Bioresource Technology*, 200, pp. 253–261. doi: 10.1016/j.biortech.2015.10.037.

Zhang, Y. *et al.* (2015) 'A physicochemical method for increasing methane production from rice straw: Extrusion combined with alkali pretreatment', *Applied Energy*, 160, pp. 39–48. doi: 10.1016/j.apenergy.2015.09.011.

Zhong, W. *et al.* (2011) 'Effect of biological pretreatments in enhancing corn straw biogas production', *Bioresource Technology*, 102(24), pp. 11177–11182. doi: 10.1016/j.biortech.2011.09.077.

20 Biogas Potential of Kitchen Waste at Visva-Bharati, Santiniketan

Aman Basu, Ankita Laha, Indranil Bhui, Anita Biswas, Krishanu Sarkar, Shibani Chaudhury, and Srinivasan Balachandran

20.1 INTRODUCTION

The amount of food produced by humans for human consumption amounts to approximately 3.9 billion tonnes per year, one-third of which, approximately 1.3 billion tonnes, gets lost or wasted (Food and Agriculture Organization of the United Nations, 2017). In India, 40% of the food produced is wasted, which costs 1 lakh crore rupees per year (CSR, 2018). Not all of this waste amounts to actual wastage, but instead produces unaccounted economic and ecosystem benefits for other forms of life associated with humans. As economic benefits, a good amount of waste is utilised to feed livestock. Fertilisers are made for food crops through vermicomposting. An amount sustains scavengers like dogs, cats, crows, and pigs. The cyclicity of an ecosystem and balanced nutrient flow through the cycle is an indication of a steady ecosystem. Human presence in the food web is immense, and our actions of managing the food source and waste can exert a major influence on energy and nutrient-flow dynamics, which can disrupt the cyclicity of ecosystem functions (Milner-Gulland, 2012). It is often important to understand the existing system of waste management, especially in case of food waste, which gets assimilated into the food chain, for uninterrupted nutrient and energy flow within the ecosystem. From a nutrient flow perspective, the close loop of the flow of nutrients from biomass through the soil is shown in Figure 20.1. Among the described processes, the process which shows the human excreta fraction going back to the soil as nutrient (3) has good biogas potential but cannot be used as biogas feedstock because of social stigma. Energy produced from human faeces is still undesired among the Indian population.

In this study, it was also observed that not all kitchen produce is fated to be actual waste; the majority of the waste is incorporated into the ecosystem by active human actions. The total amount of kitchen waste is not discarded, but utilised as compost and cow feed, which is incorporated directly inside the profit margin of the kitchen and canteen management. By supporting a large amount of macrofauna and flora with food supply and compost, the majority of the waste produced from kitchens, hostels, and canteens maintains an ecosystem. This means the total amount of waste generated from the canteen is not readily available, only a portion of it.

Around 55% of the total waste generated from the university kitchen are readily utilised by the kitchen management as vermicompost and cow feed, with some residuals as crow feed and bones as dog feed. The 45% of waste that is the actual discarded amount is dumped at different locations and not properly managed. This waste can be procured easily without any need to interfere with kitchen and canteen management, because of its apparent lack of usability or potential for profit. The completely discarded kitchen wastes are estimated for the generation of biogas, which is a clean source of energy. In a biogas plant, a mixture of both methane and carbon dioxide is created during anaerobic digestion and serves as a high-energy renewable fuel that can be used as a substitute for fossil fuels. After the waste is treated, a lumpsum amount of slurry is also produced, which can be used as fertiliser, further increasing economic and ecological productivity (Figure 20.2).

DOI: 10.1201/9781003204435-23

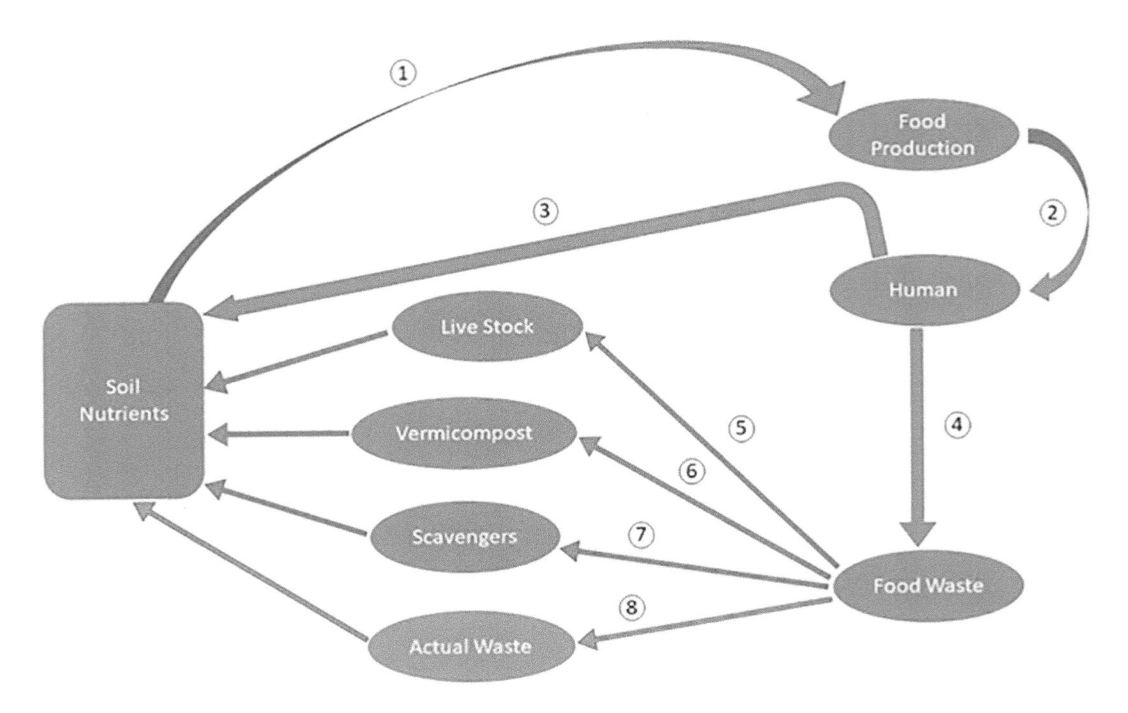

FIGURE 20.1 (1) Nutrient flow from soil to food crop. (2) Food consumption by humans. (3) Human excreta. (4) Food waste produced by humans. (5) Food waste used for feeding livestock. (6) Food waste used for vermicomposting. (7) Food waste consumed by scavengers. (8) Actual discarded food waste.

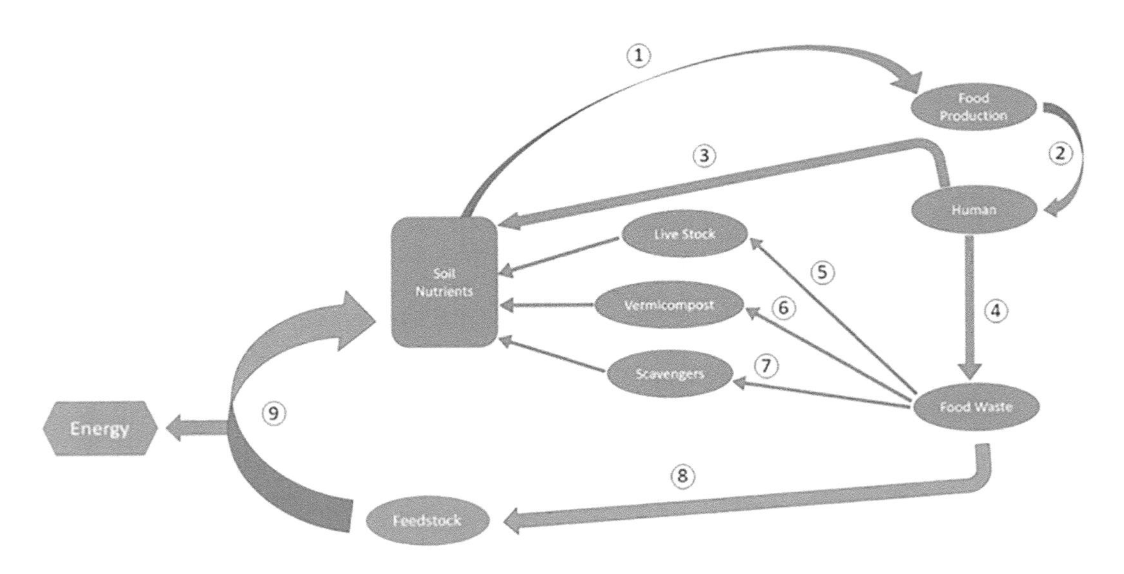

FIGURE 20.2 (1) Nutrient flow from soil to food crop. (2) Food consumption by humans. (3) Human excreta. (4) Food waste produced by humans. (5) Food waste used for feeding livestock. (6) Food waste used for vermicomposting. (7) Food waste consumed by scavengers. (8) Discarded food waste used for feedstock at anaerobic digester. (9) Energy recovery and fertiliser replenishing soil nutrient.

20.2 DATASET

To estimate the amount of usable waste generated in the university, two senior boys' hostels with a total capacity of 548 and two senior girls' hostels with a total capacity of 197 were surveyed for per capita hostel dwellers' waste generation. Senior students of Visva-Bharati stay in 16 girls' hostels and 22 boys' hostels in Visva-Bharati, with a total accommodation of 966 girls and 918 boys (Visva-Bharati University, n.d.). A sample of two kitchens, serving a total of 1950 meals per day and a canteen serving 170–225 meals per day were surveyed for assessing a total of 15 kitchens and 5 canteens at the university.

20.3 RESULTS AND DISCUSSION

Kitchen waste mixed with cow dung can provide a generous amount of biogas. If the kitchen waste is alkali pretreated and has a volatile solid (VS) amount about 80%, in 0.5:1 ratio of cow dung and kitchen waste, a total amount of 250.06 L/kg-VS biogas can be produced after 30 days of digestion. This ratio was obtained by experimenting with different cow dung to kitchen waste ratios, where 0.5:1 cow dung to kitchen waste produced the best results (Figure 20.3). With this configuration, inoculum cost is optimised with regard to biogas production.

The calorific value of 1 m^3 biogas is around 22 MJ, making the electricity generation potential 6.1 kWh. Considering the efficiency of conversion of 35%, 1 m^3 of biogas will produce 2.14 kWh of electricity (Mathew et al., 2015).

According to this study, the total waste produced by the university's 44 hostels (38 seniors' hostel and 6 school hostels), 15 kitchens, and 5 canteens is around 1.29 tonnes/day; within this only 0.58 tonnes (45% of total waste) are available daily for biogas generation. From this amount of waste, 248.19 kWh of daily energy can be produced after the first 30 days of biogas plant operation.

The 2014 census indicated that an electrified Indian household consumed about 90 units (kWh) of electricity per month (WEC, 2014). From the generated energy, 83 household can have a permanent electric supply from available food waste alone, which is enough to run four tube-lights, four ceiling fans, a television, a small refrigerator, and small kitchen appliances with typical usage hours and efficiency levels in India.

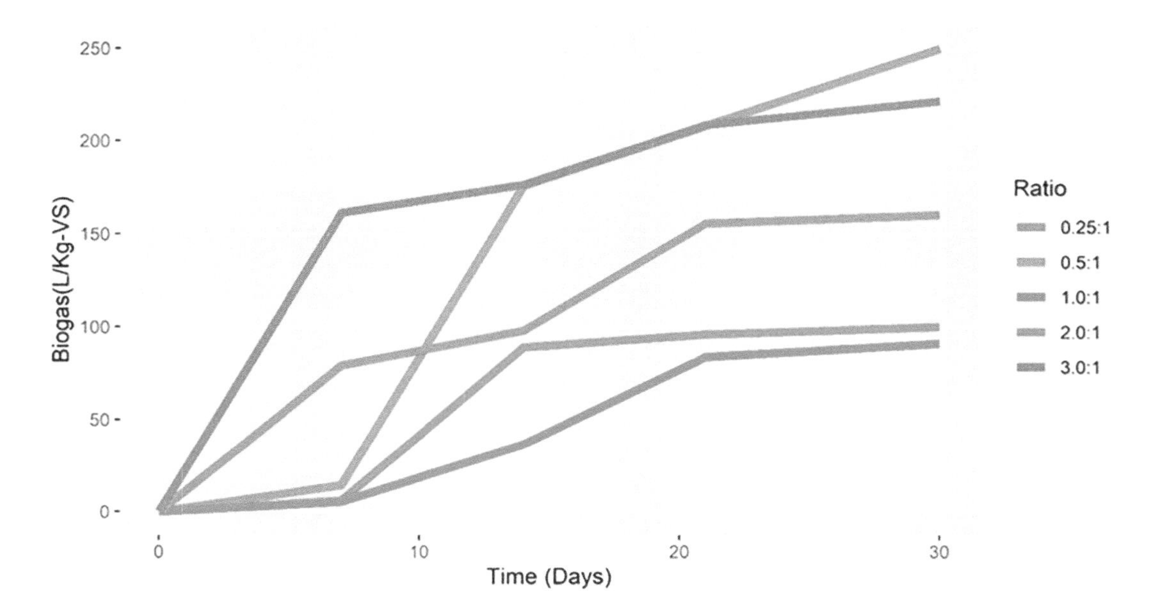

FIGURE 20.3 Biogas production in different inoculum ratios (cow dung to kitchen waste).

This amount of energy can be produced without getting into conflict with the ecosystem nutrient flow, as well as the economic systems of existing kitchens. Additionally, since 0.97 tonnes of carbon dioxide are emitted from northeastern regional grids of India for each megawatt hour of electricity (Government of India Ministry of Power, 2014), 0.24 tonnes of carbon dioxide emission can be reduced daily.

20.4 CONCLUSION

From this study, it can be observed that adopting biogas as an energy source can provide substantial benefits to society and the ecosystem. The 0.58 tonnes of food waste available daily from university hostels and kitchens can be utilised to produce 248.19 kWh of energy, enough to provide electricity to 83 typical Indian households. It can also provide a solution to food waste problems, while curbing 0.24 tonnes of daily carbon dioxide emissions. The challenges lie in the construction and sustenance of the food waste supply chain, which requires a lot of coordination among the different kitchens of the university and their management.

ACKNOWLEDGEMENT

Aman Basu is thankful for the financial support received from Biotechnology and Biological Sciences Research Council (BBSRC), United Kingdom, funded project BEFWAM (BB/S011439/1) for financial support and research

REFERENCES

Government of India Ministry of Power (2014). CO_2 Baseline Database for the Indian Power Sector (User Guide Ver.10). Retrieved from https://cea.nic.in/wp-content/uploads/baseline/2020/07/user_guide_ver10.pdf

Electricity use per household | Electricity Consumption Efficiency| WEC. (2014.). Retrieved August 27, 2019, from https://wec-indicators.enerdata.net/household-electricity-use.html

Food and Agriculture Organization of the United Nations. (2017). Key facts on food loss and waste you should know! | SAVE FOOD: Global Initiative on Food Loss and Waste Reduction | Food and Agriculture Organization of the United Nations. https://doi.org/10.1177/027243168200200405

CSR Journal (2018). Food Wastage in India, And What You Can Do about It. Retrieved August 27, 2019, from https://thecsrjournal.in/food-wastage-in-india-a-serious-concern/

Mathew, A. K., Bhui, I., Banerjee, S. N., Goswami, R., Shome, A., Chakraborty, A. K., Balachandran, S., Chaudhury, S. (2015). Biogas production from locally available aquatic weeds of Santiniketan through anaerobic digestion. *Clean Technologies and Environmental Policy*, 1681–1688. https://doi.org/10.1007/s10098-014-0877-6

Milner-Gulland, E. J. (2012). Interactions between human behaviour and ecological systems. *Philosophical Transactions of the Royal Society of London. Series B, Biological Sciences*, *367*(1586), 270–278. https://doi.org/10.1098/rstb.2011.0175

Visva-Bharati University. (n.d.). Student Welfare. Retrieved July 8, 2019, from http://www.visvabharati.ac.in/StudentWelfare.html

Index

P

R

S

T

U

V

W

X

Z